AF332949

Probabilistic safety assessment for optimum nuclear power plant life management (PLiM)

Related titles:

Nuclear corrosion science and engineering
(ISBN 978-1-84569-765-5)
Understanding corrosion mechanisms, the systems and materials they affect, and the methods necessary for accurately measuring their incidence is of critical importance to the nuclear industry for the safe, economic and competitive running of its plants. *Nuclear corrosion science and engineering* reviews the fundamentals of nuclear corrosion, looking at the different types of both aqueous and non-aqueous corrosion mechanisms and the nuclear materials susceptible to attack from them, along with applicable monitoring and control methodologies and modelling and lifetime prediction tools. The book also explores corrosion issues across the range of current and next-generation nuclear reactors, as well as nuclear fuel and radioactive waste facilities.

Nuclear decommissioning: Planning, execution and international experience
(ISBN 978-0-85709-115-4)
Over the course of their operational life, a variety of components in both nuclear power plants and other civilian nuclear facilities will become contaminated by irradiation from the radioactive sources present. Once these power plants or facilities have come to the end of their operational lifetime, the need to decommission and decontaminate them arises. *Nuclear decommissioning* critically reviews the decommissioning and decontamination processes and technologies available for rehabilitating sites used for nuclear power generation and civilian nuclear facilities, from fundamental issues and best practices, to procedures and technology, and onto decommissioning and decontamination case studies.

Understanding and mitigating ageing in nuclear power plants: Materials and operational aspects of plant life management (PLiM)
(ISBN 978-1-84569-511-8)
Plant life management (PLiM) is a safety-based methodology for the management of nuclear power plants over their entire lifetime. It is used by plant operators and regulators to assess the condition of nuclear power plants, and to establish the technical and economic requirements for safe, long-term operation. This book discusses the fundamental ageing-degradation mechanisms that affect nuclear power plant structures, systems and components (SSC), along with relevant analysis modelling methods and mitigation paths. Coverage of plant maintenance and replacement routes is extended through chapters on the development of advanced materials and components, as well as through reactor-type specific PLiM practices.

Details of these and other Woodhead Publishing materials books can be obtained by:

- visiting our web site at www.woodheadpublishing.com
- contacting Customer Services (e-mail: sales@woodheadpublishing.com; fax: +44 (0) 1223 832819; tel.: +44 (0) 1223 499140 ext. 130; address: Woodhead Publishing Limited, 80 High Street, Sawston, Cambridge CB22 3HJ, UK)

If you would like e-versions of our content, please visit our online platform: www.woodheadpublishingonline.com. Please recommend it to your librarian so that everyone in your institution can benefit from the wealth of content on the site.

Woodhead Publishing Series in Energy: Number 49

Probabilistic safety assessment for optimum nuclear power plant life management (PLiM)

Theory and application of reliability analysis methods for major power plant components

Gennadij V. Arkadov, Alexander F. Getman and Andrei N. Rodionov

Woodhead Publishing Limited
in association with
Cambridge International Science Publishing Limited

Oxford Cambridge Philadelphia New Delhi

Published by Woodhead Publishing Limited in association with Cambridge International Science Publishing Limited

Woodhead Publishing Limited, 80 High Street, Sawston, Cambridge CB22 3HJ, UK
www.woodheadpublishing.com; www.woodheadpublishingonline.com

Woodhead Publishing, 1518 Walnut Street, Suite 1100, Philadelphia, PA 19102-3406, USA

Woodhead Publishing India Private Limited, G-2, Vardaan House, 7/28 Ansari Road, Daryaganj, New Delhi – 110002, India

Cambridge International Science Publishing Limited, 7 Meadow Walk, Great Abington, Cambridge CB21 6AZ, UK
www.cisp-publishing.com

First published 2012, Woodhead Publishing Limited and Cambridge International Science Publishing Limited. This work is a revised and translated version of an original Russian-language version published in 2010 by Energoatomizdat, Russia.

British Library Cataloguing in Publication Data
A catalogue record for this book is available from the British Library.

Library of Congress Control Number: 2011935983

ISBN 978-0-85709-398-1 (print)
ISBN 978-0-85709-399-8 (online)
ISSN 2044-9364 Woodhead Publishing Series in Energy (print)
ISSN 2044-9372 Woodhead Publishing Series in Energy (online)

The publisher's policy is to use permanent paper from mills that operate a sustainable forestry policy, and which has been manufactured from pulp which is processed using acid-free and elemental chlorine-free practices. Furthermore, the publisher ensures that the text paper and cover board used have met acceptable environmental accreditation standards.

Typeset by Butterfly Info Services, India
Printed by TJ International Ltd, Padstow, Cornwall, UK

Contents

Part Two: Practical application of probabilistic methods for strength reliability

Woodhead Publishing Series in Energy

This monograph has been prepared within the framework of the European international program 'Ageing and PSA' by experts of the Russian Scientific and Research Institte of Operation of Nuclear Power Plants (VNIIAES), Moscow, and IRSN, Paris.

The monograph provides an overview of probabilistic methods for calculating the strength and operating life of equipment and pipelines of nuclear power plants (NPPs), including using the criteria of resistance to full or partial destruction with the formation of breaks, leaks or defects in the metal. These methods are discussed, taking into account the ageing of equipment and pipelines in service, which is important in predicting the operating life and safety of nuclear power plants.

Particular attention is paid to the practical application of the results of calculations of the strength and operating life of equipment and pipelines in the probabilistic safety analysis of NPP (PSA), the optimisation of in-service inspection (ISI) and maintenance and repair.

In discussing the practical application of probabilistic methods it is described how to optimise the volume and frequency of ISI, optimise the frequency and conditions of hydraulic tests, optimise the frequency and technology of technical inspection of equipment and pipelines, optimisation of preventive maintenance of the plant.

The aim of this book is to:

1. Describe briefly the main methods used in nuclear power engineering for the determination of the quantitative characteristics of the reliability of equipment and piping of NPP;

2. Clarify the mechanisms of ageing of equipment and piping in service, discuss the impact of these mechanisms on reliability and determine the quantitative methods of taking into account ageing in predicting the reliability and residual life of equipment and piping of the working nuclear power stations,

3. Describe the methods of practical application of the reliability characteristics for both the evaluation of the actual level of reliability and safety of the NPP which on its own is evidently of considerable

importance, and also for solving the current problems of operation. These problems include the optimisation of non-destructive testing, the development and justification of measures taken to ensure the acceptable level of reliability and safety of the individual components of equipment and piping, optimisation of technical certification, and optimisation of repair and technical maintenance of equipment and piping;

4. Evaluate the possibility of extensive practical application of the methods for determining the reliability characteristics and the use of these methods for solving the current problems in operation, including from the viewpoint of providing initial data for these methods, including statistical data;

5. Illustrate on a number of examples the applicability of the most promising methods.

The monograph does not pretend to be an exhaustive coverage of the problems, but gives an idea of their present state and the possibility of practical application of technologies to optimise measures for the maintenance of the safety and operating life of equipment and piping during plant operation.

1

Terminology, concepts and definitions

1.1   Terminology, abbreviations, symbols

Terminology

Equipment and pipelines (EP) – mechanical nuclear power plant (NPP) equipment, including pressure vessel shells, pump and valve housings, main and auxiliary pipelines.

Strength – the property of the materials parts and structures to resist mechanical and thermomechanical loads in a particular environment, characterised by temperature, chemical composition and condition, gravity, radiation, electromagnetic and other physical fields, without destruction, while preserving the shape and integrity within the limits sufficient to fulfil their functions.

Useful life of EP NPP – the total operating time of the object from the beginning of its operation or resume after repairs to the transition to the limiting state (GOST 27.002-89). For the power plant EP operating under thermomechanical loading, the limiting state occurs with the loss of strength properties, or obsolescence, or with a reduction of reliability below the acceptable level established in technical conditions.

γ-percent yield – the total operating time during which the object does not reach the limiting state with probability γ, expressed in percentage (GOST 27.002-89).

Useful life of the first kind – the useful life of EP or their elements according to the criterion of resistance to complete failure (formation of a critcial size crack).

Useful life of the second kind – the useful life of EP or their elements according to the criterion of resistance to partial destruction with the formation of a leak (a continuous stable defect).

Useful life of the third kind – the useful life of EP or their elements according to the criterion of formation of dead-end defects in metal with the size larger than permissible size.

Useful life of the fourth kind – the useful life of EP or their elements according to the criterion of fatigue strength, creep strength, corrsion resistance, etc. loss of stability, shape changes.

Fitness for purpose – a set of technical characteristics of EP determining the possibility of their exploitation.

Note. Fitness for purpose is the characteristic of only EP, while the service life of EP depends on fitness for purpose of EP and the service conditions.

Reliability – the property of the object to maintain in time and within the established ranges the values of all parameters characterising its ability to perform the required functions in specified terms and conditions for use, maintenance, storage and transportation (GOST 27.002-89).

For EP of NPP working in conditions of thermomechanical loading, the ability to perform the required functions is the property of strength.

Note

1. The concept of reliability includes the concept of failure-free operation, durability, maintainability, and storability.

2. Reliability of the EP of NPP can be evaluated using different criteria. The most important criteria are:

– Criterion of resistance to complete destruction of the structure or element (*reliability of the first kind*);

– Criterion of resistance to partial destruction of the structural element with formation of leaks, loss of integrity, etc. (*reliability of the second kind*);

– Criterion of the resistance of the structure or its elements of to the formation of defects in metal (*reliability of the third kind*).

– Criterion determined by the resistance to fatigue crack initiation, creep, stress corrosion cracking, or unacceptable shape variations (*reliability of the fourth kind*).

Safety of EP – reliability of EP or their components in respect of to human life, health and ecology of the environment.

Note:

1. Resistance to total destruction of the elements of EP defines nuclear, radiation and industrial safety of NPP, as well as the efficiency of their operation (in terms of costs to eliminate accidents and to reduce load factor) is measured by indicators of the reliability of the first kind;

2. Resistance to partial destruction of the elements of EP with the formation of a leak determines radiation and partially nuclear and industrial safety and effectiveness of plant operation in part of the costs for eliminating the consequences of damage and a reduced load factor; it is measured as the reliability of the second PLiM;

3. Resistance to the formation in the elements of EP of discontinuity defects identifies only the efficiency of nuclear plant operation in terms of the costs to eliminate defects and the reduction of the load factor; it is measured by the reliabilities of the third and fourth kind;

Endurance – the property of EP or their elements to retain strength, service life, reliability and safety in the presence of cracks or crack-like defects or the property of EP to maintain for the limited period of time the performance under conditions not covered by the design.

Metal of EP – the base metal, weld metal, heat-affected zone and cladding (if no special explanations are provided).

Discontinuity of metal – generic name of cracks, peeling, burns, holes, pores, lack of fusion and inclusions (PNAEG-7-010-89), as well as other defects of the operational nature.

Continuity defect (defect) – a discontinuity whose size exceeds the size permitted by the standard documents (SD). These defects are characterised by type, location, orientation.

Note:

This document distinguishes the defects of the first, second and third kind.

Defect of the first kind – a defect which causes fast (almost instantaneous) destruction of the structure or element.

Defect of the second kind – a continuous stable defect through which the coolant leaks out;

Defect of the third kind – a defect whose size is greater than the permissible size but smaller than that of the defects of the first and second kind.

Defectiveness of EP – the total number of continuity defects of metal found in EP or their elements.

Residual defects – defectiveness of EP or their elements after non-destructive testing and removal (repair) of detected defects.

Non-destructive testing (NDT) – non-destructive inspection of the state of structural elements for the detection of discontinuities in metal; non-destructive testing of defects; defectoscopy.

In-service inspection (ISI) – includes the input, pre-operation and periodic inspection during service, and inspection after repair or reconstruction.

Detectability of NDT – the degree of conformity of the NDT results with the actual characteristics of discontinuities of the structure.

Note:

1. Quantitatively, the accuracy of NDT can be defined by the probability of error-free decision-making when assessing the quality of the object or a batch of components. The quantitative characteristic of the reliability of inspection, determined by this procedure, is associated with the norms of the defects.

2. Measure of detectability of defects – is the probability of detecting a discontinuity on the basis of the specified parameter.

3. An important indicator of the detectability of inspection is the reproducibility of its results. This characteristic can be defined as the frequency of coincidence of inspection results in different conditions. Reproducibility may be a partial characteristic of reliability.

Crack – a flat discontinuity in metal, usually with sharp edges.

Critical crack size – a crack having the dimensions at which it becomes unstable. The crack grows in a stable manner to the critical size; after reaching the critical size the rate of crack growth rapidly increases and growth becomes uncontrolled.

Permissible crack size – a crack whose propagation over the remainder of the operating live or to repair does not lead to destruction of the structure.

Note:

In the manufacture of products the permissible dimensions of discontinuities are established based on the achieved level of technology and the possibilities of non-destructive testing. Typically, the size of discontinuities that are allowed at the manufacturing stage and during operation differs significantly.

Destruction – The process taking place in the material resulting in the loss of its strength.With single loading, destruction can occur by brittle, ductile or quasi-brittle mechanism.

Fatigue – Accumulation of material damage under cyclic loading, leading eventually to the formation of a fatigue macrocrack, its propagation and destruction of the structure.

Creep – Plastic deformation under long-term mechanical stress and temperature, causing thermally activated processes leading ultimately to destruction of the structure. The property of the structure (material) to resist destruction by creep is the rupture strength.

Object – the technical product for a specific purpose, considered during design, manufacture, testing and service.

Objects can be different systems and their elements.

Element – the simplest part of the product, fo the purpose of reliability may consist of many parts.

System – a set of co-operating elements, designed for independent fulfillment of prescribed functions.

The concepts element and system are transformed, depending on the task at hand.For example, a machine tool, in establishing its own reliability, is considered as a system consisting of individual elements – mechanisms, components, etc., and to study the reliability of a production line – as an element.

Reliability of the object has the following basic *conditions* and *events*:

Good state the state of the object at which it meets all the requirements established by regulatory and technical documentation (RTD).

Up state – the state of the object in which it is able to perform specified functions, while maintaining the basic values of the parameters specified by RTD.

The main parameters characterise the operation of the object in carrying out the assigned tasks.

The concept of the *good state* is broader than the concept of the *up state*. The object capable of operation must meet only those RTD requirements which ensure that the normal use of the object according to its purpose. Thus, if the object is not capable of operation, this indicates a malfunction. On the other hand, if the object is defective, it does not mean that it cannot be used in service.

Limiting state is the state of the object in which its intended use is unacceptable or inappropriate.

Application (use) of the object is terminated in the following cases:

• unavoidable breach of safety;

• unavoidable deviations of the given parameters;
• unacceptable increase of operating costs.

For some subjects, the limiting state is the final stage in its operation, i.e. object is removed from service, and for others it is a certain phase in the operating schedule requiring repair work.

In this context, objects can be:

• *non-recoverable*, for which performance in the event of failure cannot be restored;

• *recoverable*, their performance can be restored, including by replacement.

The non-recoverable objects include, for example: anti-friction bearings, semiconductor products, gears, etc. Objects consisting of many elements, such as metal-working machines, automobiles, electronic equipment, are restorable since their failure is associated with damage to one or a few elements that can be replaced.

In some cases the same object depending on the features and stages of operation or purpose may be recoverable or non-recoverable.

Failure – an event consisting of the breach of the operational state of the object.

Failure criterion – the distinguishing feature or a combination of features used for establishing the moment of failure.

Types of failure are sub-divided into:

• functioning failure (the objects stops fulfilling the core functions, e.g. broken gear teeth);

• parametric failures (some object parameters are changed in unacceptable limits, such as loss of precision machine tools).

Reliability measure quantifies the extent to which this object has certain properties ensuring reliability. Some reliability measures (e.g. technical resources, operating life) may have a dimension, a number of others (for example, the probability of failure-free operation, availability factor) are dimensionless.

Assigned operating lifetime – the total operating time of the object after which operation should be terminated regardless of its condition.

Useful lifetime – calendar duration of service (including storage, maintenance, etc.) from its beginning to the limiting state.

Figue 1.1 shows the graphical interpretation of these indicators, in particular:

$t_0 = 0$ – the beginning of operation; t_1, t_5 – points at which operatin is interrupted for technical reasons; t_2, t_4, t_6, t_8 – the moment when service of the object is restarted; t_3, t_7 – the moment when the objected is to be repaired; t_9 – the moment of cessation of operation; t_{10} – the time of reaching thec limiting state.

Technical operating life

$$T_t = t_1 + (t_3 - t_2) + (t_5 - t_4) + (t_7 - t_6) + (t_{10} - t_8).$$

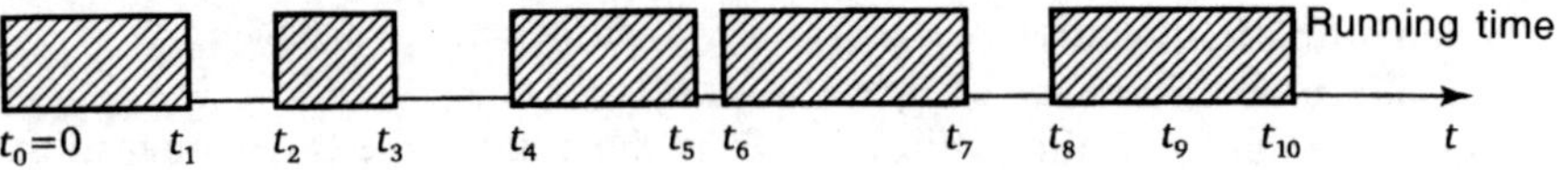

1.1 Graphical interpretation of the running time to failure.

The assigned operating lifetime

$$T_a = t_1 + (t_3 - t_2) + (t_5 - t_4) + (t_7 - t_6) + (t_9 - t_8).$$

The operating lifetime of the object

$$T_c = t_{10}.$$

Abbreviations

AS – accidential situation
DBA – maximum design based accident
DNOC – deviation from normal operation conditions
EC – eddy current inspection
EP – equipment and pipelines
HT – hydraulic tests
ICC – intercrystalline corrosion
ISI – in-service inspection
MCP – main circulation pipeline
MDE – maximum design earthquake
NDT – non-destructive testing
NOC – normal operation conditions
NPP – nuclear power plant;
PC – pitting corrosion
PDF – probability density function
PSA – probabilistic safety assessment
PWR – pressure water reactor
PZ – pressurizer
RPF – rated power factor
RPV – reactor pressure vessel
SCC – stress corrosion cracking
SG – steam generator
SS – safety systems
STD – standard-technical documentation
TND – truncated normal distribution
UT – ultrasonic testing
VSG – vertical steam generator
VVER – pressurised water power reactor;

Symbols

σ – stress

σ_m – membrane stress

σ_b – the total bending stress

σ_a – stress amplitude

σ_{QF} – local stress amplitude, taking stress concentration into account

$(\sigma_F)_{max}$ – maximum reduced conventional elastic stress cycle, taking into account the concentration of conventional elastic stresses

$[\sigma]$ – nominal permissible stress

R_m – tensile strength

$R_{p0.2}$ – yield stress

R_{-1} – endurance limit for symmetric tension–compression cycles

Z – the relative reduction in area of the cross-section of the sample for tensile tests to failure

E – modulus of elasticity (Young' modulus)

N – number of loading cycles of the structural element in service

N_p – number of cycles to fatigue crack initiation

a – cumulate fatigue damage

F – neutron fluence with energy > 0.5 MeV

A_p – coefficient of radiation embrittlement

K_1 – stress intensity factor

K_{1C} – critical stress intensity factor;

$n_{0.2}$ – safety factor for yield strength;

n_m – the safety factor for ultimate strength;

n_σ – safety factor for stress when calculating fatigue resistance

n_N – safety factor for the number of loading cycles when calculating fatigue resistance

n_{1K} – the safety factor for K_{1C} in the calculation of brittle fracture resistance.

1.2 Basic terms and formulas of reliability theory, probability theory and mathematical statistics

1.2.1 Reliability measures and information from probability theory

The most important measures of the reliability of non-repairable items are *reliability parameters* which include:

- reliability function;
- probability density function;
- failure rate;
- mean time to failure.

Reliability measures are presented in two forms (definitions):

- statistical (sample estimate);

* probabilistic.

Statistical definitions (*sample estimates*) are derived from the results of reliability tests.

Suppose that the testing of a number of similar objects yields a finite number of observations of random variable parameters of interest – operating time to failure. These values represent a sample of a certain volume of the general which has an unlimited amount of data on the time to failure of the object.

Quantitative parameters defined for the general population are *true* (*probabilistic*) indicators as they objectively characterise a random variable – time to failure.

The parameters identified for the sampling and used to draw conclusions about a random variable are *selective* (*statistical*) *estimates*. It is obvious that at a sufficiently large number of tests (large sample) the estimates approach the true probabilistic parameters.

The probabilistic indicators are useful in analytical calculations, and statistical – in experimental studies of reliability.

The statistical estimates are denoted by the ^ sign.

The following *scheme of testing* is adopted to assess reliability.

Let N identical mass-produced objects are sent for testing. Test conditions are identical, and each of these objects is tested up to failure.

The following notation is introduced:

$T = \{0, t_1, ... t_N\} = \{t\}$ – a random variable of the operating time to failure;

$N(t)$ – number of objects functioning at the operating time t;

$n(t)$ – number of objects that failed at the operating time t;

$\Delta n\ (t,\ t + \Delta t)$ – number of objects that fail in the operatine time range $[t,\ t + \Delta t]$;

Δt – operating time interval.

Since the further definition of the sample estimates is based on mathematical models of probability theory and mathematical statistics, the following are the basic (minimum required) concepts from the theory of probability.

Basics of mathematical models for calculations in the probability theory

Probability theory is a mathematical science that studies the patterns in random phenomena.

One of the basic concepts of probability theory is the random event.

The event is any fact (the outcome) which may or may not occur as a result of experiments (trials).

Each of these events can be associated with the number called its probability and is a measure of the possible completion of this event.

Probability theory is based on the axiomatic approach and builds on the concepts of set theory.

The set is any set of objects of arbitrary nature, each of which is called an element of the set.

Suppose that some experiments (trial) are carried out and the result is not known beforehand. Then the set Ω of all possible outcomes of the experiment is a space of elementary events, and each of its elements $\alpha \in \Omega$ (individual outcome of the experiment) is an *elementary event*. Any set of elementary events (any combination) is considered a *subset* (part) of the set Ω and is a *random event*, i.e. any event A is a subset of the set Ω: $A \subset \Omega$.

In general, if the set Ω contains n elements, then it is possible to specify 2^n subsets (events).

Several definitions will be introduced.

Joint (incompatible) events are such events where the occurrence of one of them does not preclude (exclude) the possibility of occurrence of another.

Dependent (independent) events – events where the occurrence of one them affects (does not affect) the occurrence of another event.

The opposite event relative to a selected event A is an event which does not lead to the occurrence of the selected event (indicated by $\bar{A}$).

The complete group of events is a collection of events at which at least one of the events of this set should occur as a result of experiments.

Axioms of probability theory. The probability of event A is denoted by $P(A)$ or $P\{A\}$. The probability is selected so that it meets the following conditions or axioms:

$$P(\Omega) = 1; \; P(\varnothing) = 0; \qquad\qquad [1.1]$$
$$P(\varnothing) \le P(A) \le P(\Omega) = 0. \qquad\qquad [1.2]$$

If A_i and A_j are mutually exclusive events, i.e. $A_i \wedge A_j = \varnothing$, then

$$P(A_i \vee A_j) = P(A_i) + P(A_j), \qquad\qquad [1.3]$$

where $\vee$ is the sign of logical addition of the events; $\varnothing$ is the empty (no events).

Axiom (3) can be generalised to any number of mutually exclusive events $\{A_i\}_i^n = 1$:

$$P\left\{\bigcup_{i=1}^{n} A_i\right\} = \sum_{i=1}^{n} P(A_i). \qquad\qquad [1.4]$$

The frequency definition of probability of any event A:

$$P(A) = m_A/n, \qquad\qquad [1.5]$$

represents the ratio of the number of cases (m_A), enabling the occurrence of an event A, to the total number of cases (the number of possible outcomes of an experiment) n.

Unlimited increase of n is associated with statistically ordering when the variation of *the frequency of event A* (*sample estimate*) becomes less and less marked and approaches a constant value – *the probability of event A*.

Basic rules of probability theory. Probability addition theorem. If A_1, A_2, ..., A_n are incompatible events, and A is the sum of these events, the probability of event A equals the sum of the probabilities of events A_1, A_2, ..., A_n:

$$P(A) = P\left\{\bigcup_{i=1}^{n} A_i\right\} = \sum_{i=1}^{n} P(A_i).$$

[1.6]

Because the opposite events A and $\bar{A}$ are incompatible and form a complete group, the sum of their probabilities

$$P(A) + P(\bar{A}) = 1.$$

[1.7]

Probability multiplication theorem. The probability of the product of two events A_1 and A_2 is the probability of one of them, multiplied by the conditional probability of another, assuming that the first event occurred:

$$P(A_1 \wedge A_2) = P(A_1)P(A_2 \mid A_1) = P(A_1)P(A_1 \mid A_2),$$

[1.8]

where the conditional probability of an event A_1 at the beginning of event A_2 is the probability of event A_1, calculated on the assumption that the event occurred A_2:

$$P(A_1 \mid A_2) = P(A_1 \cdot A_2)P(A_2).$$

[1.9]

For any finite number of events multiplication theorem takes the form

$$P\left\{\bigcap_{i=1}^{n} A_i\right\} = P(A_1 \mid A_2...A_n)P(A_2 \mid A_3...A_n)...P(A_{n-1} \mid A_n)P(A_n).$$

[1.10]

If the events A_1 and A_2 are independent, then the corresponding conditional probability is

$$P(A_1 \mid A_2) = P(A_1); P(A_2 \mid A_1) = P(A_2),$$

so the multiplication theorem of probability [1.8] takes the form

$$P(A_1 \wedge A_2) = P(A_1)P(A_2),$$

[1.11]

and for a finite number of n independent events

$$P\left\{\bigcap_{i=1}^{n} A_i\right\} = \prod_{i=1}^{n} P\{A_i\}.$$

[1.12]

The consequence of the main theorems – the formula of total probability (FTP) and the Bayes formula – are widely used in solving a large number of tasks.

The formula of total probability. If the results of experiments can be used to propose n mutually exclusive hypotheses H_1, H_2,...H_n, representing the complete group of incompatible events (for which $\Sigma i = I^n P(i) = 1$), the probability of event A, which can only come from one of these hypotheses, is defined by:

$$P(A) = P(H_1)P(A \mid H_i),$$
[1.13]

where $P(H_i)$ is the probability of hypothesis H_i; $P(A|H_i)$ is the conditional probability of event A under hypothesis H_i.

Since event A can occur with one of the hypotheses H_1, H_2, ... H_n, then $A = AH_1 \vee AH_2 \vee ... \vee AH_n$, but H_1, H_2,...H_n are incompatible, so

$$P(A) = P(A \wedge H_i) + ... + P(A \wedge H_n) = \sum_{i=1}^{n} P(AH_i).$$

If event A depends on validity of hypothesis H_i $P(AH_i)=P(H_i)\cdot P(A|H_i)$, and this leads to expression [1.13].

The Bayes formula (*the formula of probability of hypotheses*). If the probabilities of hypotheses H_1, H_2, ... H_n prior to the experiment were equal to $P(H_1)$, $P(H_2)$, ..., $P(H_n)$, and event A took place as a result of the experiment, then the new (conditional) probabilities of the hypotheses are evaluated:

$$P(A \mid H_i) = \frac{P(H_i)P(A \mid H_i)}{\sum_{i=1}^{n} P(H_i)P(A \mid H_i)} - \frac{P(H_i)P(A \mid H_i)}{P(A)}.$$
[1.14]

The probabilities of the hypothesis before the start of the experiment (initial) $P(H_1)$, $P(H_2)$, ..., $P(H_n)$ are called *apriori*, and tjose after the experiment $P(H_1|A)$, ... $P(H_n|A)$ *aposteriori*.

The Bayes formula is used to reconsider the possibility of hypotheses in the light of the experimental result.

The proof of the Bayes formula follows from the material discussed previously. Since $P(H_i \wedge A) = P(H_i)\cdot P(A|H_i) = P(H_i)\cdot P(H_i|A)$:

$$P(H_i \mid A) = \frac{P(H_i \wedge A)}{P(A)} = \frac{P(H_i)P(A \mid H_i)}{P(A)}.$$
[1.15]

If another experiment is caried out after the experiment which gave the event A, and this second experiment can be carried whether or not the event A_1 took place, then the conditional probability of the second event is computed from [1.13], which does not include the former hypotheses $P(H_i)$ and instead includes the new ones – $P(H_i|A)$:

$$P(A_1 \mid A) = \sum_{i=1}^{n} P(H_i \mid A)P(A_1 \mid H_i A).$$

[1.16]

Expression [1.16] is called the formula for the probabilities of future events.

1.2.2 The cumulative distribution function, probability density function, failure rate

Statistical evaluation of the cumulative distribution function (c.d.f.) – empirical reliability function – is defined by the ratio of the number $N\,(t)$ of objects which worked flawlessly up to operating time t, to the number of objects repaired up to the beginning of the tests ($t = 0$) and the total number of objects N:

$$\hat{P}(t) = \frac{N(t)}{N}.$$

[1.17]

Assessment of c.d.f. can be regarded as an indicator of the proportion of good state terms at the operating time t.

Since $N\,(t) = N{-}n(t)$, then c.d.f. from [1.17] is

$$\hat{P}(t) = 1 - \frac{n(t)}{N} = 1 - \hat{Q}(t),$$

[1.18]

where $\hat{Q}\,(t) = n\,(t)\,/\,N$ is the estimate of failure probability (FP).

In statistical evaluation the FP estimate is the empirical distribution function of failures.

Since the events consisting in the occurrence or non-occurrence of failure at operating time t, are opposite, then

$$\hat{P}(t) = \hat{Q}(t) = 1.$$

[1.19]

It is easy to verify that c.d.f. is decreasing and FP increasing function of operating time. In fact:

- at the beginning of trial $t = 0$, the number of working objects is equal to their total number $N(t) = N(0) = N$, and the number of failed objects is $n\,(t) = n\,(0) = 0$, so $\hat{P}(t) = \hat{P}(0) = 1$ and $\hat{Q}(t) = \hat{Q}(0) = 0$;
- at the service life $t \to \infty$ all the objects put to the test fail, i.e. $N\,(\infty) = 0$ and $n\,(\infty) = N$, so $\hat{P}(t) = \hat{P}(\infty) = 0$ and $\hat{Q}(t) = \hat{Q}(\infty) = 1$.
- Probabilistic determination of c.d.f.

$$P(t) = P\{T \le t\}.$$

[1.20]

Thus, the c.d.f. is the probability that a random value of the operating time to failure T will not be less than some specified operating time t.

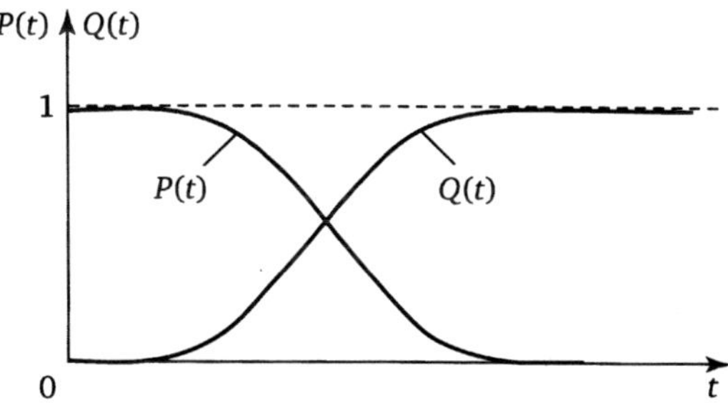

1.2 Graph of c.d.f. and FP.

It is obvious that the FP is a function of the distribution of T and represents the probability that the time to failure is less than some specified operating time t:

$$Q(t) = P\{T < t\}. \qquad [1.21]$$

The graphs of c.d.f. and FP are shown in Fig. 1.2.

In the limit as the number N (an increase of the sample) of test objects increases, $\hat{P}(t)$ and $\hat{Q}(t)$ converge in probability (their values become similar) to $P(t)$ and $Q(t)$.

Convergence in probability is as follows:

$$P\left\{\lim_{N\to\infty} | \hat{P}(t) - P(t)| = 0\right\} = 1. \qquad [1.22]$$

The determination of c.d.f. in the operating time range $[t, t + \Delta t]$ is of interest for practice, provided that the object had worked flawlessly y to the beginning of the interval t. This probability is determined using the multiplication theorem of probabilities and highlighting the following events:

A = {reliable operation of the object until the moment t};

B = {reliable operation of the object in the range Δt};

$C = A \cdot B$ = {reliable operation of the object until the moment $t + \Delta t$}.

Obviously $P(C) = P(A \cdot B) = P(A) \cdot P(B|A)$, since the events A and B are dependent.

The conditional probability $P(B|A)$ is c.d.f. $P(t, t + \Delta t)$ in the interval $[t, t + \Delta t]$, so

$$P(B|A) = P(t, t + \Delta t) = P(C)/P(A) = P(t + \Delta t)/P(t). \qquad [1.23]$$

Failure probability in the operating time period $[t, t + \Delta t]$, taking into account [1.23], is:

$$Q(t, t + \Delta t) = 1 - P(t, t + \Delta t) = [P(t) - P(t + \Delta t)]/P(t). \qquad [1.24]$$

Statistical evaluation of the failure probability density function (p.d.f.) is determined by the ratio of the number of objects $\Delta n (t, t + \Delta t)$, failed in the operating time range $[t, t + \Delta t]$ to the product of the total number of objects N and the operating time range Δt.

$$\hat{f}(t) = \frac{\Delta n(t, t + \Delta t)}{N \Delta t}. \qquad [1.25]$$

Since

$$n\,(t,\, t + t) = n\,(t + \Delta t) - n\,(t),$$

where $n\,(t + \Delta t)$ is the number of objects that failed during the operating time $t + \Delta t$, then the estimate of the p.d.f. is:

$$\hat{f}(t) = \frac{\Delta n(t + \Delta t) - n(t)}{N \Delta t} = \frac{1}{\Delta t}\left[\hat{Q}(t + \Delta t) - \hat{Q}(t)\right] = \frac{\hat{Q}(t, t + \Delta t)}{\Delta t}, \qquad [1.26]$$

where $\hat{Q}(t, t + \Delta t)$ is the estimate of the FP in the operating time range, i.e. the increment of FP in Δt.

The estimate of the p.d.f. is the frequency of failures, i.e. the number of failures per operating time related to the initial number of objects.

Probabilistic definition of p.d.f. follows from [1.26] as the operating time interval $t \rightarrow t_0$ and increase of the sample size $N \rightarrow \infty$

$$f(t) = \lim_{\Delta t \to 0} \frac{\hat{Q}(t, t + \Delta t) - n(t)}{\Delta t} = \frac{dQ(t)}{dt} = \frac{d[1 - P(t)]}{d(t)} = -\frac{dP(t)}{dt}. \qquad [1.27]$$

The failure distribution density is essentially the distribution density (probability density) of the random variable T of the operating time of the object to failure.

Since $Q(t)$ is a non-decreasing function of its argument, then $f(t) \geq 0$.

One of the possible types of graph $f(t)$ is shown in Fig. 1.3.

As seen from Fig. 1.3, p.d.f. $f(t)$ characterises the failure rate (or reduced FP) with which the specific values of the operating time of all N objects $(t_1, ..., t_N)$, forming the random value of operating time to failure T pf the given object, are distributed. Let us say the tests show that the value of operating time t_i inherent to the greatest number of objects as indicated by the maximum value of $f(t_i)$. On the other hand, longer operating time t_j was recorded only for a few objects and, therefore, the frequency $f(t_j)$ of suc long operating tim being recorded on the general background is small.

Some operating time t and the infinitesimally interval of operating time of width dt, adjacent to t, are plotted on the abscissa. Then the probability that the random value of operating time T fits in the elementary section of width dt is:

$$P\{T \in (t, t + dt)\} = P\{t < T < t + dt\} \approx f(t)dt, \qquad [1.28]$$

where $f(t)dt$ is the element of the FP of the object in the interval $[t,\ t +$

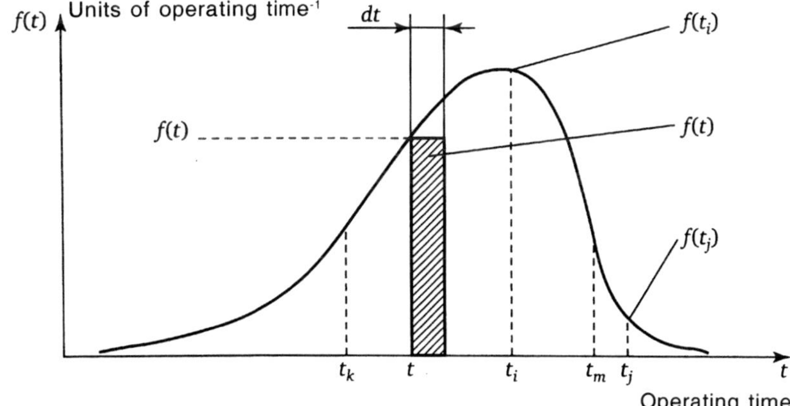

1.3 One of the possible types of graph *f*(t).

dt] (geometrically this is the area of the shaded rectangle 'resting' on the interval *dt*).

Similarly, the probability of operating time T fitting in the interval $[t_k, t_m]$ is:

$$P\{T \in (t_k, t_m)\} \approx \sum_{t_i \in (t_k t_m)} f(t_i)dt_i \approx \int_{t_k}^{t_m} f(t)dt, \qquad [1.29]$$

which is interpreted geometrically by the area under the curve $f(t)$ on the plot $[t_k, t_m]$.

Failure probability and c.d.f. can be expressed as a function of p.d.f.. Since $Q(t) = P\{T < t\}$, then using the expression [1.29]

$$Q(t) = P\{0 < T < t\} = P\{T \in (0, t)\} = \int_0^t f(t)dt. \qquad [1.30]$$

The extension of the interval to the left to zero is due to the fact that T cannot be negative.

Because $P(t) = P\{T \geq t\}$, then

$$P(t) = P\{t \leq T < \infty\} = \int_t^\infty f(t)dt. \qquad [1.31]$$

It is obvious that $Q(t)$ is the area under the curve $f(t)$ to the left of t, and $P(t)$ is the area under $f(t)$ the right of t. Since all values of the operating time obtained by testing lie under the curve $f(t)$, then

$$\int_0^\infty f(t)dt = \int_0^t f(t)dt + \int_t^\infty f(t)dt = Q(t) + P(t) = 1. \qquad [1.32]$$

Statistical estimation of failure rate (FR), expressed in units of inverse operating time, is defined by the ratio of the number of objects $\Delta n(t, t +$

Δt), failed in the operating time period $[t, t + \Delta t]$, to the product of the number N of efficiently working object at time t by the duration of the operating time period Δt:

$$\hat{\lambda}(t) = \frac{\Delta n(t, t + \Delta t)}{N(t)\Delta t}.$$

[1.33]

Comparing [1.25] and [1.33] it may be noted that the failure rate characterises slightly better the reliability of the object at the operating time t, since it shows the failure rate related to the acutal number of working objects at the operating time t.

Probabilistic definition of the failure rate is obtained by multiplying and dividing the right-hand side of expression [1.33] by N

$$\hat{\lambda}(t) = \frac{\Delta n(\overline{t}, t + \Delta t)}{N(t)\Delta t} \frac{N}{N} = \frac{\Delta n(t, t + \Delta t)}{N\Delta t} \frac{N}{N(t)}.$$

The estimate of the failure rate $T_{0|t \leq t_1}(t)$ is

$$\hat{\lambda}(t) = \frac{\hat{Q}(t, t + \Delta t)}{\Delta t} \frac{1}{\hat{P}(t)},$$

where at $\Delta t \to 0$ and $N \to \infty$

$$\lambda(t) = \lim_{\Delta t \to 0} \frac{\hat{Q}(t, t + \Delta t)}{\Delta t} \frac{1}{\hat{P}(t)} = \frac{dQ(t)}{dt} \frac{1}{P(t)} = \frac{f(t)}{P(t)}.$$

[1.34]

Possible changes of the failure rate $\lambda(t)$ are shown in Fig. 1.4.

1.2.3 Relationship of reliability indicators

Since the failure rate $\lambda(t)$ is a more complete characteristic of reliability, it is interesting to express c.d.f. $P(t)$ through FR.

Using the expression for the failure rate

$$\lambda(t) = f(t) / P(t),$$

we write

$$dP(t)/dt = -\lambda(t)P(t).$$

Separating the variables (multiplying both sides by $dt/P(t)$), we obtain

$$dP(t)/P(t) = -\lambda(t)\, dt.$$

Integrating from 0 to t and taking into account that at $t = 0$ c.d.f. of the object is $P(0) = 1$ leads to

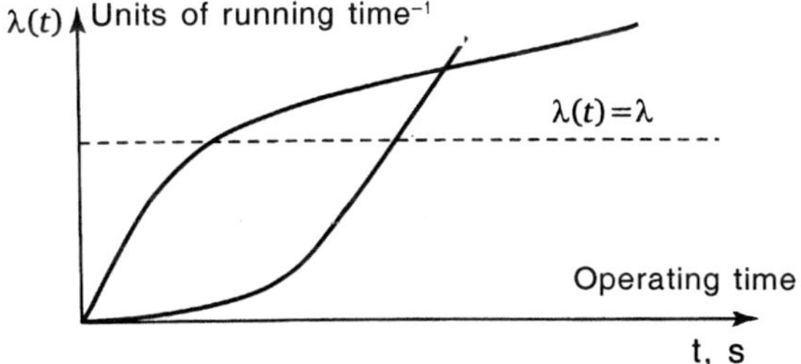

1.4 Possible changes of FP $\lambda(t)$.

$$\int_0^t \frac{dP(t)}{P(t)} = \ln P(t)\Big|_0^t = \ln P(t) = -\int_0^t \lambda(t)dt,$$

from the basic equation of the relationship of the main reliability indicators is as follows:

$$P(t) = \exp\left\{-\int_0^t \lambda(t)dt\right\}. \qquad [1.35]$$

The value of $\lambda(t)\,dt$ is the likelihood that the element worked flawlessly in the operating time range $[0, t]$ and fails in the interval $[t, t + dt]$.

The equation [1.35] shows that all the reliability indicators $P(t)$, $Q(t)$, $f(t)$ and $\lambda(t)$ are equal in the sense that knowing one of them, we can define others.

Numerical characteristics of the reliability of non-renewable items.
Mean operating time to failure

The previously discussed functional reliability indicators $P(t)$, $Q(t)$, $f(t)$ and $\lambda(t)$ completely describe the random value of operating time $T = \{t\}$. At the same time, to solve some practical problems of reliability it is enough to know some numerical characteristics of this random variable and, above all, *the mean operating time to failure*.

The statistical estimate of the mean operating time to failure

$$\hat{T}_0 = \frac{1}{N}\sum_1^N t_i, \qquad [1.36]$$

where t_i is the operating time to failure of i-th object.

The *probabilistic definition* of the mean operating time to failure is the *expected value* (EV) of the random value T defined by:

$$T_0 = M\{T\} = \int_0^\infty tf(t)dt. \qquad [1.37]$$

Using the expression for the density distribution of failures

$$f(t) = -dP(t)/dt,$$

and integration by parts, equation [1.37] can be transformed to the form

$$T_0 = \int_0^{\infty} P(t)dt, \qquad\qquad [1.38]$$

taking into account the fact that $P(0) = 1$, $P(\infty) = 0$.

From [1.38] it follows that the mean operating time to failure is geometrically interpreted as the area under the curve $P(t)$ (Fig. 1.5).

It is obvious that with increasing test sample $N \to \infty$ the arithmetic mean operating time (estimate $\hat{T}_0$) converges in probability to the EV of the operating time to failure.

The mathematical expectation of the operating time T_0 is the mathematically the expected operating time to failure of similar elements, i.e. the average operating time to first failure.

In practice, the **conditional mean operating time** is of interest:

1) *the mean useful operating time* $(T_{0|t \le t_1})$ determined under the condition that when the operating time t_1 is reached all remaining objects are removed from service;

2) *the mean duration of impending work* $(T_{0|t > t_1})$, provided that the item was working smoothly in the interval $(0, t_1)$.

Reasons for using these indicators:

1. Highly reliable objects, usually operated over a shorter period than T_0 $(t_{ser} < T_0)$, i.e. replaced due to obsolescence before the end of the operating time T_0.

2. Frequently, the test period for these facilities is reduced (carried out for the operating time corresponding to their obsolescence), so T_0 in this case is understood as the mean time which would occur in reality, if FR remained the same as in the initial period of testing.

The mean net operating time $T_{0|t \le t_1}$ (by analogy with T_0):

$$T_{0|t \le t_1} = \int_0^{t_1} P(t)dt.$$

The mean duration of impending work $T_{0|t > t_1}$

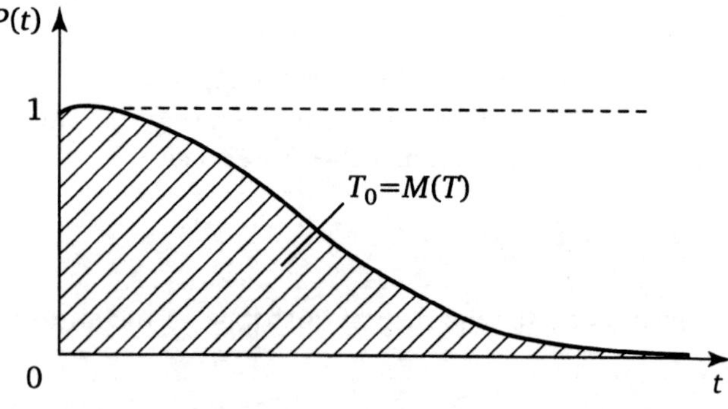

1.5 Mean operating time to failure.

$$T_{0|t\leq t_1} = M\{T - t_1\} = \frac{1}{P(t)}\int_{t_1}^{\infty} P(t)dt.$$

The relation between $T_{0|t\leq t_1}$, $T_{0|t>t_1}$ and T_0:

$$T_{0|t\leq t_1} + T_{0|t>t_1}\cdot P(t_1).$$

The graphic concepts $T_{0|t\leq t_1}$ and $T_{0|t>t_1}$ are illustrated in Figure 1.6.

At the same time, the mean operating time can not fully characterise the malfunction-free service of the object.

So, for the equal mean operating times to failure T_0 the reliability of the objects 1 and 2 can be quite significantly different (Fig. 1.7). Clearly, in view of the large dispersion of operating time to failure (curve p.d.f. $f_2(t)$ below and wider), the object 2 is less reliable than the object 1.

Therefore, to assess the reliability of an object by value $\hat{T}_0$ it is necessary to know and measure the dispersion of the random variable $T=\{t\}$, near the mean operating time T_0.

The dispersion indices include the *dispersion* and the *standard deviation* (SD) *of the operating time to failure.*

The dispersion of the random operating time:
– statistical evaluation

$$\hat{D} = \frac{1}{N-1}\sum_{1}^{N}(t_1 - \hat{T}_0)^2; \tag{1.39}$$

– probabilistic definition

$$D = D\{T\} = M\{(T - T_0)^2\} = \int_{0}^{\infty}(t - T_0)^2 f(t)dt. \tag{1.40}$$

The standard deviation of the random value of operating time:

$$\hat{S}^2 = \hat{D} \text{ or } \hat{S}^2 = \hat{S}^2\{T\} = D\{T\}. \tag{1.41}$$

The mean operating time to failure T_0 and the standard deviation of

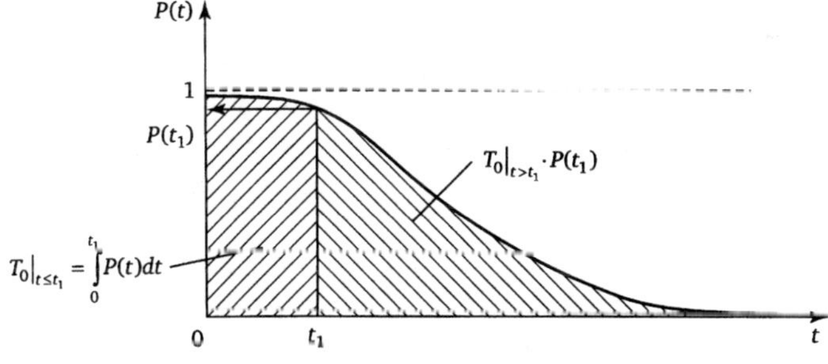

1.6 Graphic concepts $T_{0|t\leq t_1}$ and $T_{0|t>t_1}$.

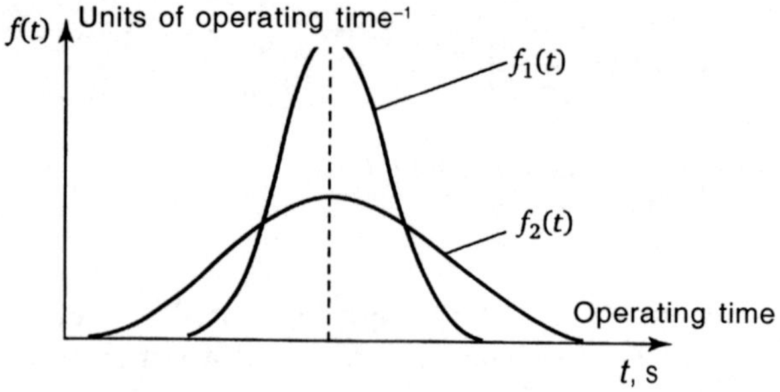

1.7 Reliability of objects at different operating times to failure.

operating time S have the dimension [units of operating time], and the dispersion D – [units of square operating time].

1.2.4 Mathematical models of reliability theory. Statistical processing of test results

To solve the problems of estimating the reliability and its prediction, it is necessary to develop a mathematical model that is represented by analytical expressions of one of the parameters $P(t)$ or $f(t)$ or $\lambda(t)$. The main procedure to construct such a model is based on testing, calculation of the statistical estimates and approximations of these estimates by analytic functions.

The models used in reliability theory will be investigated.

Let us explain how the failure-free operation of objects changes during service so that the models can be classified and modalities for their application can be identified.

Experience has shown that changes of FP $\lambda(t)$ of the vast majority of the objects can be described by the U-shaped curve (Fig. 1.8).

The curve can be divided into three characteristic regions:

first – burn-in period;

second – normal operating period;

third – ageing period.

The burn-in period of the object has a higher failure rate caused by burn-in failures due to defects in manufacturing, assembly and adjustment. Sometimes the end of this period is also the end of the warranty period for the object during which failuresare put right by the manufacturer.

During the normal operating period the failure rate decreases and remains almost constant, while failures are random and appear suddenly, primarily due to violation of the service conditions, random load changes, adverse external factors, etc. This period is consistent with the main operating time of the object.

Failure rate increases mainly in the ageing period of the object and is caused by an increase in the number of failures from wear, ageing and other factors associated with prolonged operation.

The type of analytic function describing the change of reliability

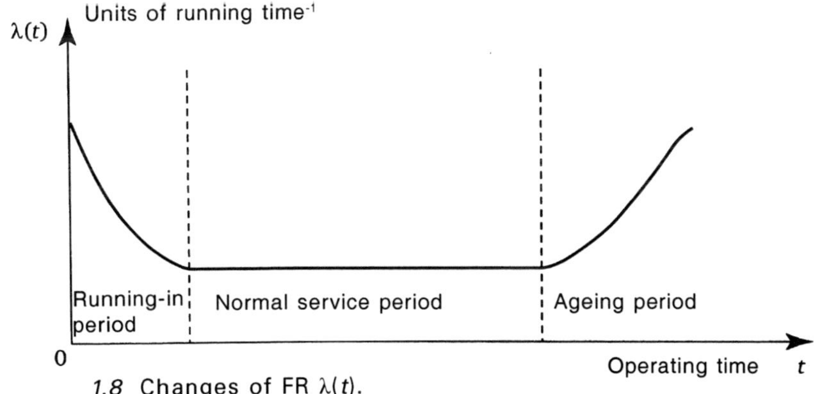

1.8 Changes of FR $\lambda(t)$.

measures $P(t)$, $f(t)$ or $\lambda(t)$, determines the law of the random variable which is chosen depending on the properties of the object, its working conditions and the nature of failures.

Formulation of the problem of statistical processing of test results and determination of reliability indcators. The results of tests of non-renewable N identical objects yield the statistical sample – a set of operating times (in any units) to failure of each of N test objects. Sampling characterises the random value of the operating time to failure of the object $T = \{t\}$.

The law of distribution of T must be selected and the correctness of the choice of appropriate criteria verified.

Selection of the distribution is based on approximations (smoothing) of the experimental data on operating time to failure which must be presented in the most compact graphical form. The approximating function is selected on the basis of the hypothesis put forward by the researcher. Experimental data can confirm or not confirm the validity of a hypothesis with different probability. Therefore, it is necessary to answer the question: are the results of an experiment in agreement with the hypothesis that the random value of operating time is governed by the distribution law selected on the basis of the experiment? The answer to this question given by result of calculations using specific criteria.

The algorithm for data processing and calculation of reliability indicators. *Formation of a statistical series*

With a large number of test objects the set of the operating time values $\{..., t_i, ...\}$ is a cumbersome and insufficiently visual form of expressing the random variable T. Therefore, for brevity and clarity, the sample is represented in the graphical representation of a statistical series – the histogram of the operating time to failure. For this purpose it is necessary to:

– set the operating time interval $[t_{min}, t_{max}]$ and its length $\xi t = t_{max} - t_{min}$, where

$$t_{min} \le \min_{i} \left\{ ...,t_i,... \right\}, t_{max} \le \max_{i} \left\{ ...,t_i,... \right\};$$

– split the operating time interval $[t_{min}, t_{max}]$ into k intervals of equal width Δt (the step of the histogram)

$$\Delta t = \frac{\xi t}{k}, \Delta t = t_{i+1} - t_i = t_i - t_{i-1};$$

– to calculate the frequency of occurrence of failures in all k intervals

$$\hat{P}_i = \frac{\Delta n(t_i, t_i + \Delta t)}{N} = \frac{\Delta n(t_i, t_{i+1})}{N},$$

where $\Delta n(t_i, t_i + \Delta t)$ is the number of objects that fail in the interval $\left[t_i, t_i + \Delta t \right]$. It is obvious that

$$\sum_{1}^{k} \hat{P}_i = 1.$$

The resultant statistical series is represented as a histogram which is constructed as follows. The intervals Δt are plotted on the abscissa and each interval is used as a base for constructing a rectangle whose height is proportional to (in the chosen scale) the corresponding frequency. The possible form of a histogram is shown in Fig. 1.9.

Calculation of empirical functions. The data of the generated statistical series are used to determine statistical estimates of reliability indicators, i.e. the empirical functions:

– the distribution function of failure (estimate of FP)

$$\hat{Q}(t_{min}) = n(t_{min}) / N = 0;$$

$$\hat{Q}(t_1) = n(t_1) / N = \Delta n(t_{min}, t_1) / N = \hat{P}_1;$$

$$\hat{Q}(t_2) = n(t_2) / N = \Delta n(t_{min}, t_1) + \Delta n(t_1, t_2) / N = \hat{P}_1 + \hat{P}_2;$$

$$...$$

$$\hat{Q}(t_{max}) = n(t_{max}) / N = \sum_{1}^{k} \hat{P}_i = 1;$$

– the reliability function (estimate of c.d.f.) (Fig. 1.10)

$$\hat{P}(t_{min}) = 1 - \hat{Q}(t_{min}) = 1;$$

$$...$$

$$\hat{P}(t_{max}) = 1 - \hat{Q}(t_{max}) = 0;$$

– the density of distribution of failures (estimate of p.d.f.) (Fig. 1.11)

$$\hat{f}(t_i) = \Delta n(t_i, t_{i+1}) / N\Delta t = \hat{P}_i / \Delta t;$$

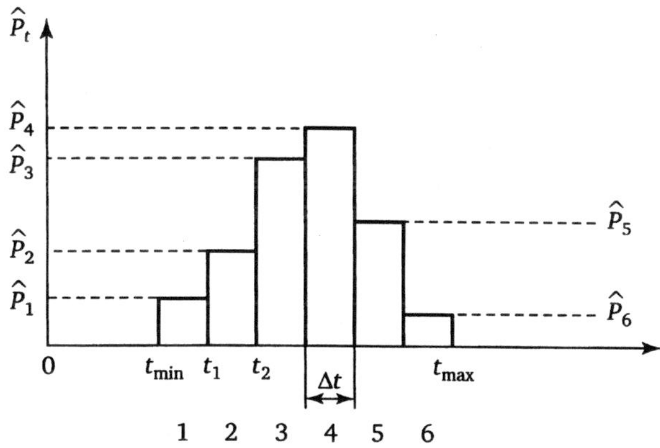

1.9 Possible type of histogram of operating time to failure, $k = 1 - 6$.

– failure rate (estimate of FR)

$$\hat{\lambda}(t_i) = \frac{\Delta n(t_i, t_{i+1})}{N(t_i)\Delta t} = \frac{\Delta n(t_i, t_{i+1})}{[N - n(t_i)]\Delta t}.$$

Figures 1.10–1.12 are the plots of statistical estimates $\hat{Q}(t)$.

The rules for constructing graphs are clear from the above calculation formulas. Each of the graphs has its scale.

Calculation of statistical estimates of numerical characteristics.

The statistical estimates of numerical characteristics can be calculated using the data of the generated statistical series.

Estimates of the characteristics are defined as follows:

– *estimate of mean operating time to failure* (*statistical mean operating time*):

$$\hat{T}_0 = \sum_1^K \hat{t}_i \hat{P}_i;$$

– *estimate of the dispersion of operating time to failure* (*empirical dispersion of operating time*):

$$\hat{D} = \sum_1^K (\hat{t}_i - \hat{T}_0)^2 \hat{P}_i;$$

where $\hat{t}_i = (t_i + \Delta t)/2 = (t_{i+1} - \Delta t)/2$ is the middle of the i-th interval of operating time, i.e. the mean operating time in the interval.

The estimate of dispersion $\hat{D} = \hat{S}^2$.

It is advisable to calculate the estimates and some subsidiary characteristics of dispersion of the random variable T:

– sampling coefficient of asymmetry of operating time to failure

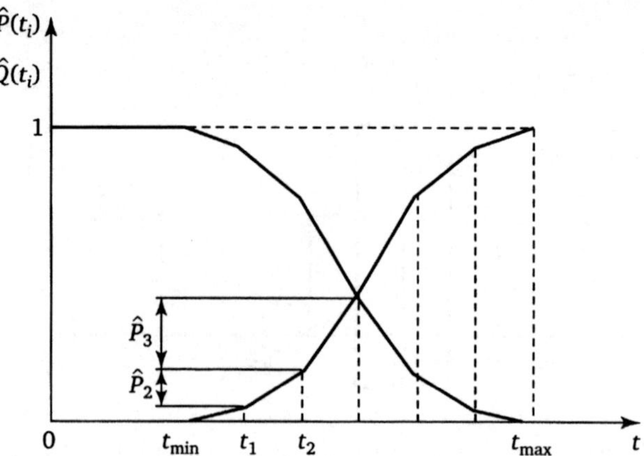

1.10 Evaluation of c.d.f..

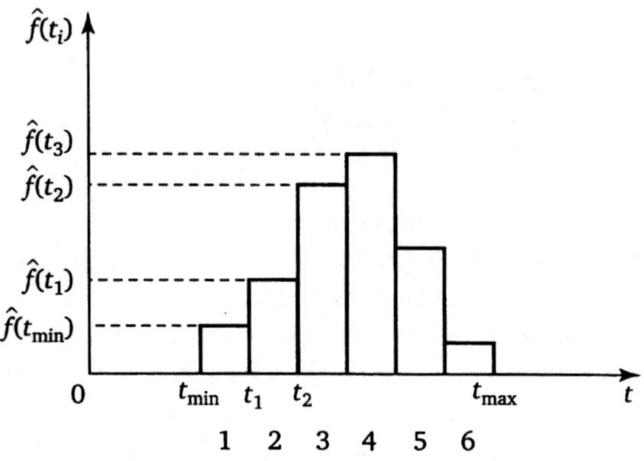

1.11 Evaluation of p.d.f..

$$A = \sum_{1}^{K} (\hat{t}_i - \hat{T}_0)^3 \frac{\hat{P}_i}{\hat{S}^3};$$

– sampling excess operating time to failure

$$E = \left[\sum_{1}^{K} (\hat{t}_i - \hat{T}_0)^4 \frac{\hat{P}_i}{\hat{S}^4} \right] - 3.$$

These characteristics are used to select the approximating function.

So, the asymmetry coefficient is a characteristic of 'skewness' of the distribution, for example, if the distribution is symmetric with respect to the EV, then $A = 0$.

In Fig. 1.13a, the distribution $f_2(t)$ has a positive asymmetry $A > 0$ and $f_3(t)$ negative $A < 0$.

The excess characterises the 'steepness' (sharp or flat-tipped) of the distribution. For the normal distribution $E = 0$.

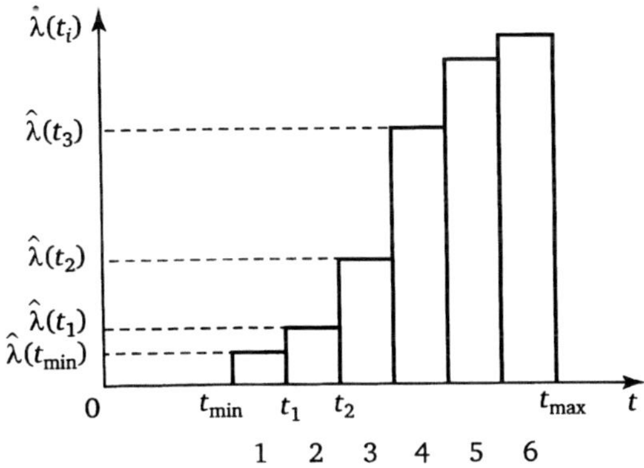

1.12   Estimate of failure rate.

The curves $f(t)$ which have sharper tips than the normal curve have $E > 0$, and vice versa – flat-tipped curves $E < 0$ (Fig. 1.13b).

The selection of the distribution law consists of the selection of an analytical function which best approximates the empirical reliability function.

Selection is to a large extent an uncertain and largely subjective procedure and much depends on apriori knowledge about the object and its properties, operating conditions, as well as analysis of graphs $\hat{P}(t)$, $\hat{f}(t)$, $\hat{\lambda}(t)$.

It is obvious that the choice of the distribution depends primarily on the type of empirical p.d.f. $\hat{f}(t)$, as well as on the type of $\hat{\lambda}(t)$. , the choice of the distribution law has the nature of the process of adoption of a hypothesis.

Suppose that for one reason or another, a hypothetical distribution law given by theoretical p.d.f.is selected

$$f(t) = \Psi(t, a, b, c...).$$

where $a, b, c, ...$ are unknown distribution parameters.

It is required to choose these parameters so that the function $f(t)$ smoothes out most efficient the stepped graph $\hat{f}(t)$. The following method is used here: the parameters $a, b, c, ...$ are selected so that several important numerical characteristics of the theoretical distributions are equal to the corresponding statistical estimates.

On the graph, the theoretical p.d.f. $f(t)$ is plotted together with $\hat{f}(t)$ so that the results of approximation (differences between $\hat{f}(t)$ and $f(t)$) can be visually assessed. Because these differences are inevitable, the question arises: are they explained by random circumstances associated with the fact that the wrong theoretical distribution was chosen? The answer to this question is the calculation of the goodness of fit criterion.

Calculation of the goodness of fit criterion. The goodness of fit criterion

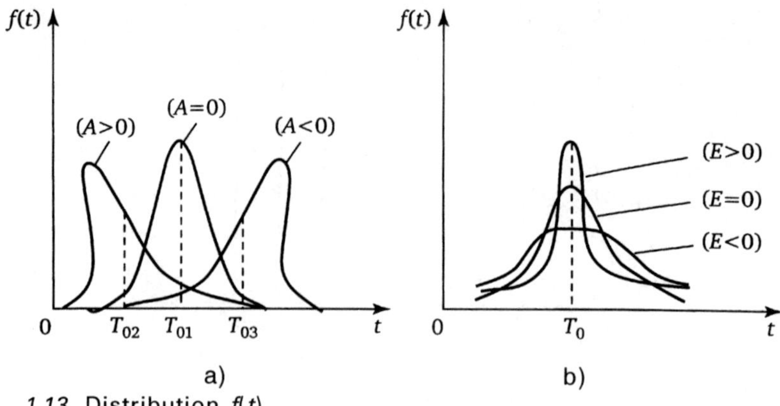

1.13 Distribution *f*(*t*).

is a criterion for testing the hypothesis of the random variable T, represented by its sample, has the distribution of the expected type.

The following procedure is used for verification. The criterion is calculated as a measure of the discrepancy between the theoretical and empirical distributions, and this measure is a random variable. The higher the measure of discrepancy, the larger the difference between the empirical and theoretical distributions, i.e. the hypothesis for the choice of the distribution should be rejected as highly unlikely. Otherwise the experimental data do not contradict the accepted distribution.

Of the known criteria, the Pearson criterion χ^2 (chi-square) is used most widely. The consistency of distributions using the χ^2 criterion is verified as:

– criterion χ^2 is calculated (a measure of divergence)

$$\chi^2 = N \sum_1^K \frac{(\hat{P}_i - P_i)^2}{P_i},$$

where $P_i = \hat{f}(t_i)\Delta t$ is the theoretical frequency (probability) of getting a random variable in the interval $[t_i, t_i + \Delta t]$;

– the number of degrees of freedom is determined $R = k - L$, where L is the number of independent conditions imposed on frequency $\hat{P}_i$, for example:

a) condition; $\Sigma \hat{P}_i = 1$;

b) the condition of coincidence; $\Sigma \hat{t}_i \hat{P}_i = T_0$;

c) the condition of coincidence $\Sigma\left(\hat{t}_i = \hat{T}_0\right)^2 \hat{P}_i = D$, etc.

In most cases, $L = 3$. The greater the number of degrees of freedom, the greater the random variable χ^2 that obeys the Pearson distribution;

– the calculated χ^2 and R are used to determine the probability P that the value having the Pearson distribution with R degrees of freedom exceeds the calculated value of χ^2.

The answer to the question: how small must probability P be to reject the hypothesis that the choice of a distribution law is largely undefined.

In practice, if $P < 0, 1$, it is recommended to find another distribution law.

In general, using the criterion of goodness of fit, it is possible to refute the selected hypothesis, and if P is large enough, then it cannot serve as proof of the correctness of the hypothesis, but merely indicates that the hypothesis does not contradict the experimental data.

1.2.5 The normal distribution law

The normal distribution, or Gaussian distribution, is the most versatile, convenient and widely used.

It is assumed that the operating time is subject to the normal distribution (normally distributed), if the density distribution of failures is given by:

$$f(t) = \frac{1}{d\sqrt{l\pi}} \exp\left\{-\frac{(t-a)}{2b^2}\right\},$$ [1.42]

where a and b are the parameters of the distribution, respectively, mathematical expectation and standard deviation which are expressed on the basis of the test results as follows:

$$a \approx \hat{T}_0; \quad b^2 \approx \hat{D},$$

where $\hat{T}_0, \hat{D}$ are the estimates of mean operating time and variance.

The graph of the reliability indices for the normal distribution is shown in Fig. 1.14.

The meaning of the parameters T_0 and S of the normal distribution will be clarified. The graph $f(t)$ shows that T_0 is the centre of symmetry of the distribution, since the sign of the difference $(t - T_0)$ does not change the expression. At $t = T_0$ the p.d.f. reaches its maximum

$$f(t)_{\max|_{t-T_0}} = \frac{1}{S\sqrt{2\pi}}.$$

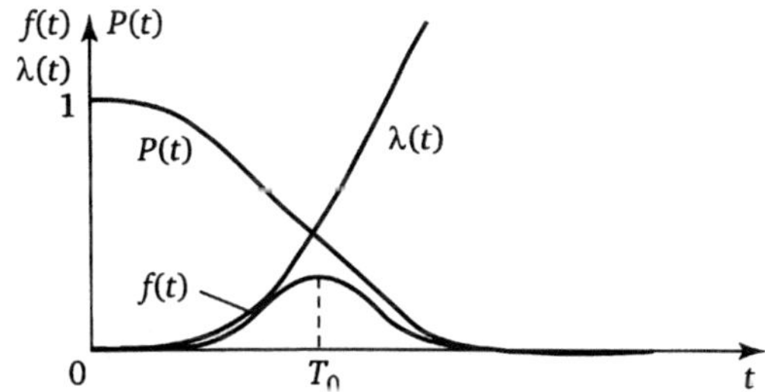

1.14 Graphs of changes in the reliability indices for the normal distribution.

When shifting T_0 to the left/right on the horizontal axis, the curve $f(t)$ moves in the same direction without changing its shape. Thus, T_0 is the centre of dispersion of the random variable T, i.e. mathematical expectation.

The parameter S characterises the shape of the curve $f(t)$, i.e. the dispersion of the random variable T. As S decreases the p.d.f. curve $f(t)$ moves upwards and becomes sharper.

Changes of the graphs of $P(t)$ and $\lambda(t)$ at different standard deviations of operating time $(S_1 < S_2 < S_3)$ and $T_0 = \text{const}$ are shown in Fig. 1.15.

Using the previously obtained relations between the reliability indicators, the expressions for $P(t)$; $Q(t)$ and $\lambda(t)$ can be derived from the well known expression [1.1] for $f(t)$. It is clear that these integral equations are very cumbersome and, therefore, the calculation of integrals for in practice is replaced by tables.

To this end, we transfer from the random variable T to a certain random variable

$$x = (t - T_0)/S,\tag{1.43}$$

distributed normally with parameters, respectively, $M\{X\} = 0$ and $S = \{X\} = 1$ and the distribution density

$$f(x) = \frac{1}{\sqrt{2\pi}}\exp\frac{-x^2}{2}.\tag{1.44}$$

Expression [1.44] describes the density of the so-called normalised normal distribution (Fig. 1.16).

The distribution function of random variable X is written in the form

$$F(x) = \int_{-\infty}^{x} f(x)dx,\tag{1.45}$$

and the symmetry of the curve $f(x)$ with respect to the EV $M\{X\} = 0$ shows that $f(-x) = f(x)$, from which $F(-x) = 1 - F(x)$.

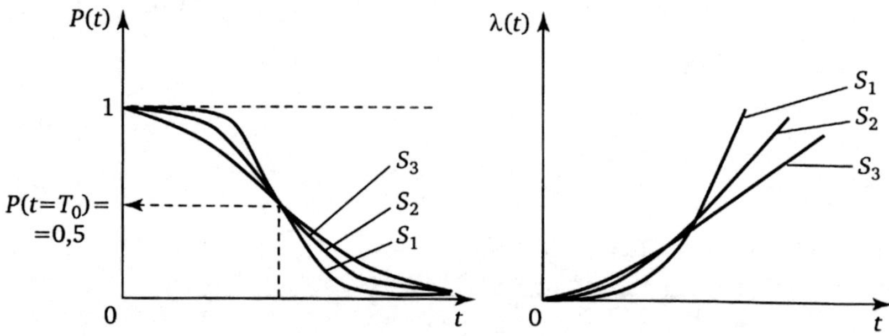

1.15 Changes in graphs $P(t)$ and $\lambda(t)$ at different standard deviations of operating time $(S_1 < S_2 < S_3)$ and $T_0 = \text{const}$.

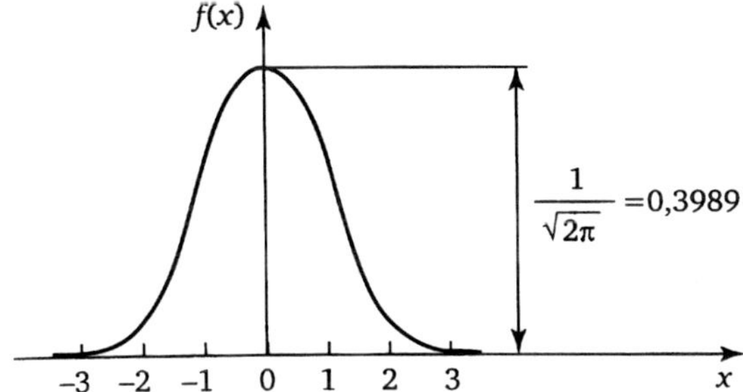

1.16 Density of the normalised normal distribution.

Reference literature shows the calculated values of the functions $f(x)$ and $F(x)$ for different $x = (t - T_0)/S$.

The reliability indices of the object expressed by the tabulated values of $f(x)$ and $F(x)$ are defined by the expressions:

$$f(t) = f(x) / S;$$ [1.46]

$$Q(t) = F(x);$$ [1.47]

$$P(t) = 1 - F(x);$$ [1.48]

$$\lambda(T) = f(x) / S[1 - F(x)].$$ [1.49]

In practical calculations, the function $F(x)$ is often replaced by the Laplace function, representing the distribution of positive values of the random variable X as:

$$\Phi(x) = \int_0^x f(x)dx = \frac{1}{\sqrt{2\pi}} = \int_0^x \exp\frac{-x^2}{2}dx.$$ [1.50]

It is obvious that $F(x)$ is related to $\Phi(x)$ as follows:

$$F(x) = \int_{-\infty}^x f(x)dx = \int_{-\infty}^0 f(x)dx + \int_0^x f(x)dx = 0.5 + \Phi(x).$$ [1.51]

Like any distribution function, the function $\Phi(x)$ has the properties.

$$\Phi(x)(-\infty) = -0.5; \quad \Phi(x)(\infty) = 0.5; \quad \Phi(x)(-x) = -\Phi(x).$$

In the literature there are also other expressions for $\Phi(x)$

The reliability indices of the object can be determined through $\Phi(x)$, using expressions [1.46]–[1.49] and [1.51]:

$$Q(t) = 0.5 + \Phi(x); \qquad\qquad [1.52]$$

$$P(t) = 0.5 - \Phi(x); \qquad\qquad [1.53]$$

$$\lambda(t) = f(x)/S\,[0.5 - \Phi(x)]. \qquad\qquad [1.54]$$

Most often, when assessing the reliability of an object it is necessary to solve the *direct problem* – at the given parameters T_0 and S of the normally distributed operating time to failure to determine a reliability indicator (for example, c.d.f.) for the given operating time t. However, in the course of design work it is also necessary to solve the *inverse problem* – determination of operating time required by the technical task for c.d.f. of the object.

These problems are solved using the quantiles of the normalised normal distribution.

Quantile is the value of the random variable corresponding to a given probability.

Denote:

t_p – the operating time corresponding to c.d.f. P;
x_p – the value of a random variable X corresponding to probability P.
Then from the constraint equation of x and t:

$$x_p = (t_p - T_0)/S.$$

At $x = x_p;\ t = t_p$:

$$t_p = T_0 + x_p\,S.$$

t_p, x_p is the non-normalised and normalised quantiles of the normal distribution, corresponding to probability P.

Values of the quantiles x_p values are given in literature for $P \geq 0.5$. For a given probability $P < 0.5$

$$x_p = -x_{1-p}.$$

For example, when $P = 0.3$

$$x_{0.3} = -x_{1-0.3} = -x_{0.7}$$

The probability of random value of operating time T fitting in a given operating time interval $[t_1, t_2]$ is determined by:

$$P\{T \in (t_1, t_2)\} = F(x_2) - F(x_1) = \Phi(x_2) - \Phi(x_1), \qquad\qquad [1.55]$$

where $x_1 = (t_1 - T_0)/S,\ x_2 = (t_2 - T_0)/S$.
Note that the time to failure is always positive, and the curve of c.d.f.

$f(t)$, in general, starts from $t - -\infty$ and extends to $t = \infty$.

This is not a significant disadvantage if $T_0 \gg S$, since [1.55] shows clearly that the probability that a random variable T fits in the interval $P\{T_0 - 3S < T < T_0 + 3S\} \approx 1.0$ with the accuracy up to 1%. This means that all possible values (with an error not exceeding 1%) of the normally distributed random variable with the ratio of the characteristics $T_0 > 3S$ are located in the section $T_0 \pm 3S$.

When the scatter of the values of the random variable T is large, the range of possible values is limited to the left $(0,\infty)$ and a truncated normal distribution is used.

Truncated normal distribution. It is well known that the classic normal distribution of operating time is used efficiently at $T_0 \geq 3S$.

For small values of T_0 and high S, there may be cases in which the c.d.f. $f(t)$ 'covers' by its left branch the region of negative operating time values (Fig. 1.17).

Thus, the normal distribution is a general case of distribution of the random variable in the range $(-\infty; \infty)$ and can be used for reliability models only in some cases (under certain conditions).

The truncated normal distribution is the distribution derived from the classic normal distribution with the limited range of possible values of operating time to failure.

In general, the truncation can be:
– left $(0; \infty)$;
– bilateral (t_1, t_2).

The meaning of a truncated normal distribution (TND) was considered for the case of restricting the random value of operating time to interval (t_1, t_2).

The density of the TND

$$\overline{f}(t) = cf(t),$$

where $f(t) = \dfrac{1}{S\sqrt{2\pi}} \exp\left\{-\dfrac{(t - T_0)}{2S^2}\right\};$

c is a normalising factor determined from the condition that the area under the curve $\overline{f}(t)$ equals 1, i.e.

$$\int_{t_1}^{t_2} \overline{f}(t)dt = \int_{t_1}^{t_2} cf(t)dt = c\int_{t_1}^{t_2} f(t)dt = 1.$$

Therefore

$$c = \dfrac{1}{\displaystyle\int_{t_1}^{t_2} f(t)dt},$$

where

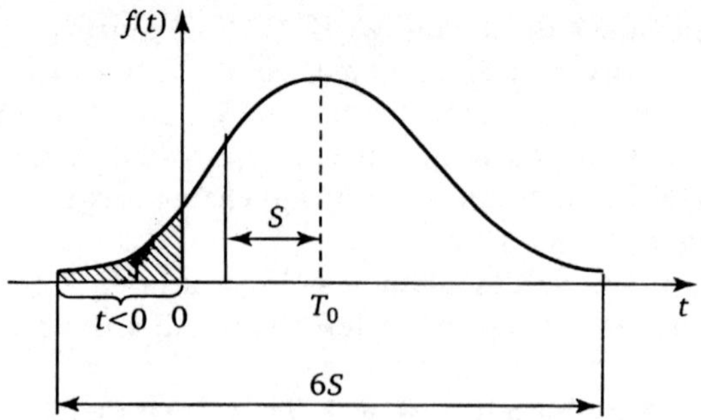

1.17 Range of negative operating time values.

$$\int_{t_1}^{t_2} f(t)dt = P(t_1 < T < t_2) = F(t_2) - F(t_1) = Q(t_2) - Q(t_1).$$

After transition from the random variable $T = \{t\}$ to the value $X = \{x\}$:

$$x_2 = (t_2 - T_0) / S; \; x_1 = (t_1 - T_0) / S,$$

this leads to

$$\int_{t_1}^{t_2} f(t)dt = Q(t_2) - Q(t_1) = 0.5 + \Phi(x_2) - 0.5 - \Phi(x_1) = \Phi(x_2) - \Phi(x_1).$$

and, therefore, the normalisation factor c is equal to:

$$c = \frac{1}{\Phi(x_2) - \Phi(x_1)}.$$

Since $[\Phi(x)(x^2) - \Phi(x)(x^1)] < 1$, then $c > 1$, so that $\hat{f}(t) > f(t)$. Curve $\hat{f}(t)$ is higher than $f(t)$, since the areas under the curves $\hat{f}(t)$ and $f(t)$ are the same and equal to 1 (Fig. 1.18):

$$\int_{T_0 - 3S}^{T_0 + 3S} f(t)dt = \int_{t_1}^{t_2} \overline{f}(t)dt$$

Reliability indices for the TND in the range (t_1, t_2):

$$\overline{f}(t) = cf(t) = cf(x) / S;$$

$$\overline{P}(t) = \int_{t}^{\infty} cf(t)dt = c \int_{t}^{\infty} f(t)dt = c[0.5 - \Phi(x)];$$

$$\overline{Q}(t) = 1 - c[0.5 + \Phi(x)];$$

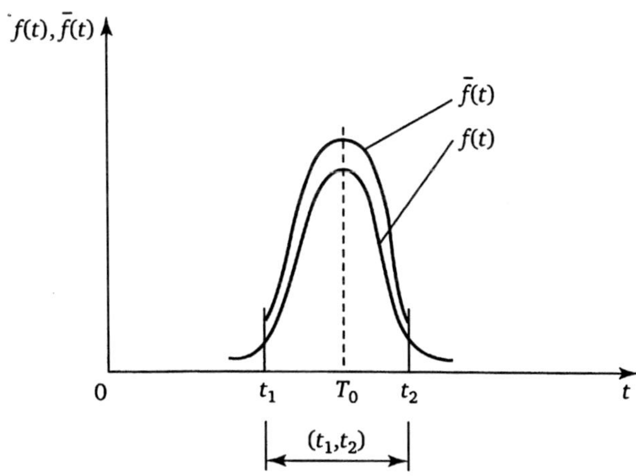

1.18 Curves $\bar{f}(t)$ and $f(t)$.

$$\bar{\lambda}(t) = \bar{f}(t)/\bar{P}(t) = f(x)/S\left[0.5 - \Phi(x)\right] = \lambda(t).$$

The TND for the positive operating time to failure – the range of $(0; \infty)$ has the c.d.f.

$$\bar{f}(t) = c_0 f(t),$$

where c_0 is the normalising factor determined by the condition:

$$c_0 \int_{t}^{\infty} f(t)dt = 1,$$

and is equal to (as above):

$$c_0 = \frac{1}{\int_{t}^{\infty} f(t)dt} = \frac{1}{Q(\infty) - Q(0)} = \frac{1}{\Phi(\infty) - \Phi(-T_0/S)} = \frac{1}{0.5 + \Phi(T_0/S)}.$$

Reliability indices of the UNR $(0;\infty)$

$$\bar{f}(t) = c_0 f(x)/S;$$

$$\bar{P}(t) = c_0\left[0.5 - \Phi(x)\right];$$

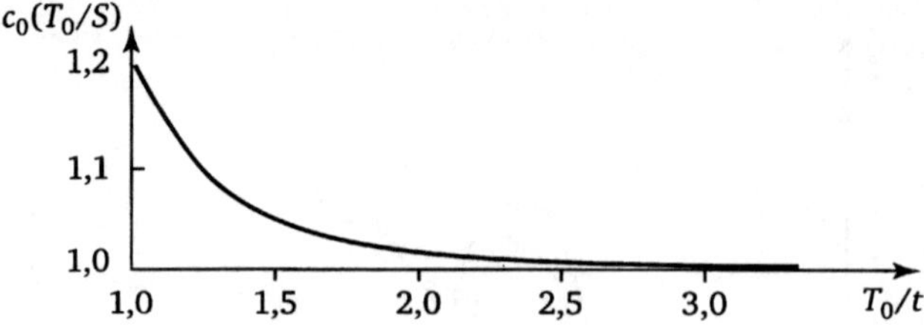

1.19 Variation of the normalising factor c_0.

$$\bar{Q}(t) = 1 - c_0\left[0.5 - \Phi(x)\right];$$

$$\bar{\lambda}(t) = \lambda(t) = f(t)/S\left[0.5 - \Phi(x)\right] = \lambda(t),\ x = (t - T_0)/S.$$

Changes of the normalising factor c_0 depending on the ratio T_0/S are shown in Fig. 1.19.

At $T_0 = S$, $T_0/S = 1$ $c_0 = $ max (≈ 1.2), at $T_0/S \geq 2.5$ $c_0 = 1.0$, i.e. $\bar{f}(t)(t) = f(t)$.

1.2.6   Binomial distribution (Bernoulli distribution)

To conclude this section, another distribution that arises in cases where the question: how many times some event occurs in a series of a certain number of independent observations (experiments) performed under identical conditions, will be described. For convenience and clarity, it is assumed that we know the value of p – the probability that the component (part) taken out of production is defective, and $(1 - p) = q$ is the probability that the part will not be rejected.

If X is the number of rejected parts from the toal number n of items, then probability that among n parts k parts will be defective is equal to:

$$P(X = k) = \frac{n!}{k!(n-k)!}\,p^k q^{n-k} = C_n^k p^k q^{n-k},\ \text{where } k = 0,1,\ldots n \quad [1.56]$$

Equation [1.56] is called the Bernoulli formula. With a large number of tests the binomial distribution tends to normal.

1.3   Safety of nuclear power stations. Active and passive safety features

The nuclear power plant consists of a large number of elements. All the elements of nuclear power plants can be classified in several groups: the

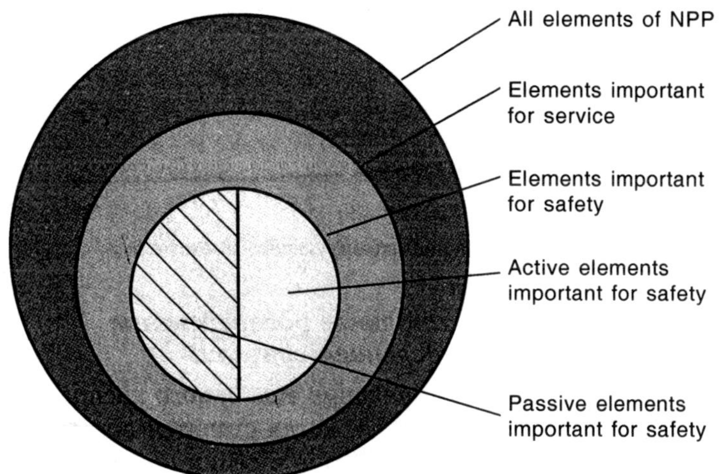

1.20 Ranking of equipment, pipelines and nuclear plant structures.

elements that are important to safety; elements which degrade with age in service and which can limit the useful life of NPP; the elements that are important to safety and could limit the useful life.

Ranking the elements and systems of NPP with VVER-440 is shown in Fig. 1.20. All the structural elements listed in Fig. 1.20, can be divided into four groups:

– elements of the core;
– pressure vessels and piping;
– mechanisms and internals;
– building structures.

Vessels and pipelines play the important role in the safety and lifetime assessment due to:

– the greatest impact on nuclear safety of nuclear power plants;
– they are the most numerous group of structures and components on the power unit:

they are in many cases high energy, representing a major threat in terms of not only nuclear, radiation, but also industrial safety;

– their replacement (or repair) is connected, usually with high material costs, and some structural elements, such as reactor vessels, are virtually impossible to replace.

In the design stage, materials, vessels and pipelines are selected so that their strength is ensured throughout the entire life cycle. In this case, nucleation of any cracks and also the transition of the section of the wall of the vessel or pipeline to the plastic state[3,4,etc] are not allowed. The margin of yield strength for the membrane (i.e. the average over the wall cross section) stress is $n_{0.2} = 1.5$. This means that the level of allowable membrane stress $[\sigma]$ is lower than the yield stress and considerably lower than the tensile strength of the material from which the structure is produced.

As indicated in Fig. 1.20, nuclear power plant safety is ensured by active and passive elements that are important for safety.

The passive safety elements include, first of all, all the elements of the safety barries 1,2,3, and 4 of the NPP as well as the mechanical elements of systems important to safety (Figs. 1.21 and 1.22).

The most important passive elements which determin in many case the reliability and safety of nuclear power plants, as well as their economic characteristics during operation, are equipment and piping.

Suffice to say that the maximum design accident starts with the rupture

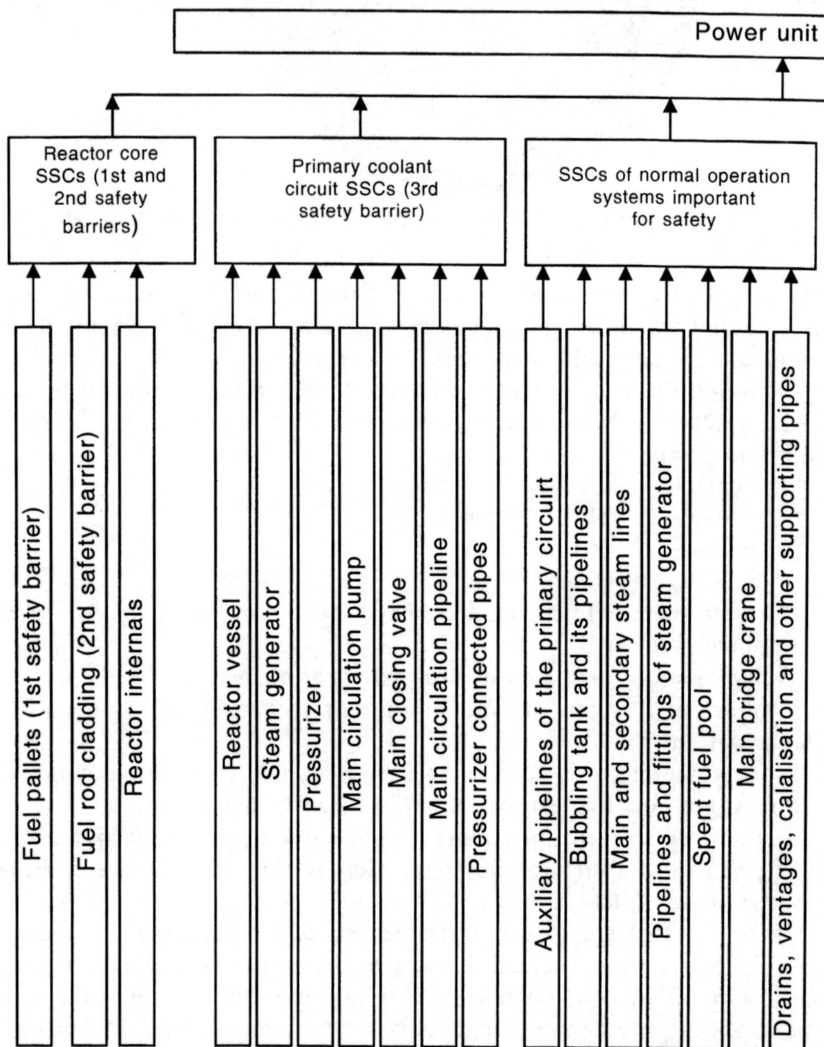

1.21 Systems, structures and components (SSC) of a power unit of a cooling system; FSS – frontline safety systems; LSS – localising safety

of the main circulation pipeline; mitigation of this accident requires the solicitation of many safety systems.

Destruction of the of the pressure vessel of the VVER- or PWR-type reactors is the beginning of the failure not foressen in design. The probability of such failure is not greater than 10^{-7} 1/ (reactor · year).

1.4 Strength reliability and its connection with nuclear safety and service life of NPP

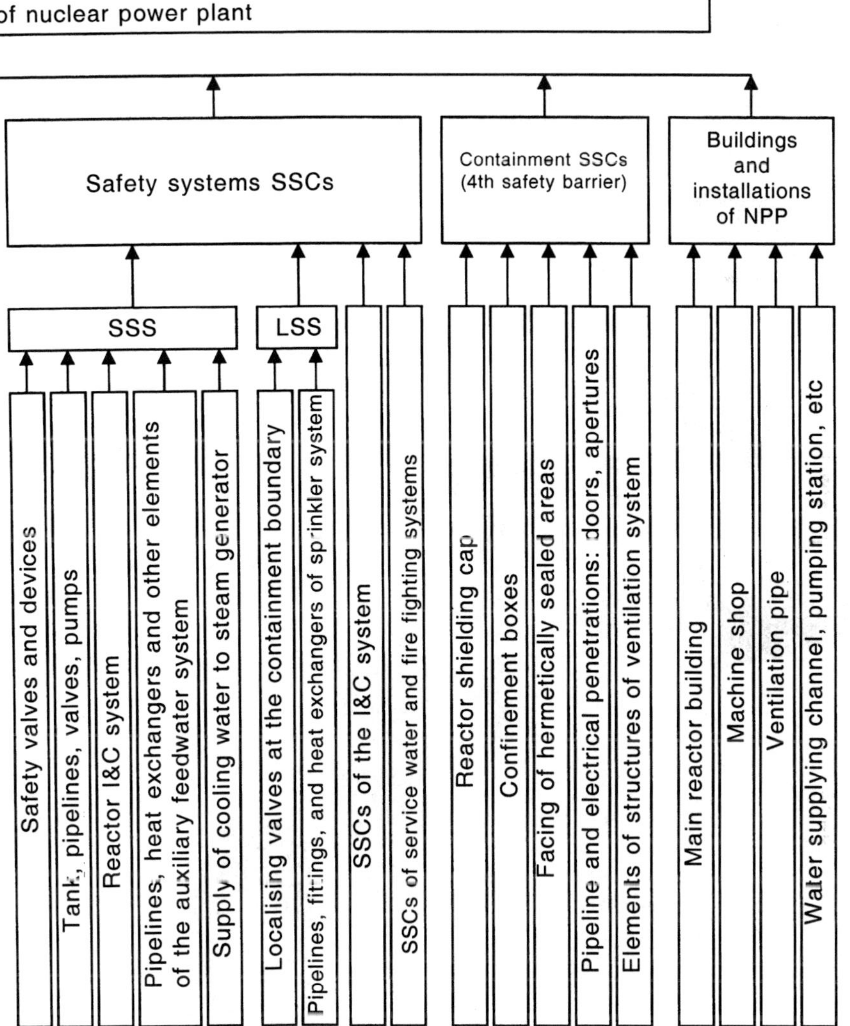

nuclear power plant with VVER 440 [6] reactor: ECCS – system for emergency systems; CSS – controlling safety systems; SS – safety systems;

Every element of the construction, equipment or pipeline of NPP can operate as long as it preserves its integrity and shape (i.e. strength). Integrity is the most important property of structural elements of nuclear power plants. The relationships of the integrity and propagation of fracture (i.e. violation of integrity) of structural elements are studied by the science of strength.

Strength is the property of materials and products produced from them to resist mechanical loads over a specific period of time and in a particular environment, characterised by temperature, chemical composition and physical fields, without destruction, while preserving the shape and integrity to the extent sufficient to perform their functions.

Strength is a fundamental property. For most of the products of modern technology, including nuclear power plants, strength is the most important property after the functionality property and determines the properties of products such as reliability, service life, safety.

The strength properties of the elements of nuclear power plants such as steam pipes, turbines, bearing elements of structures, lifting mechanisms,

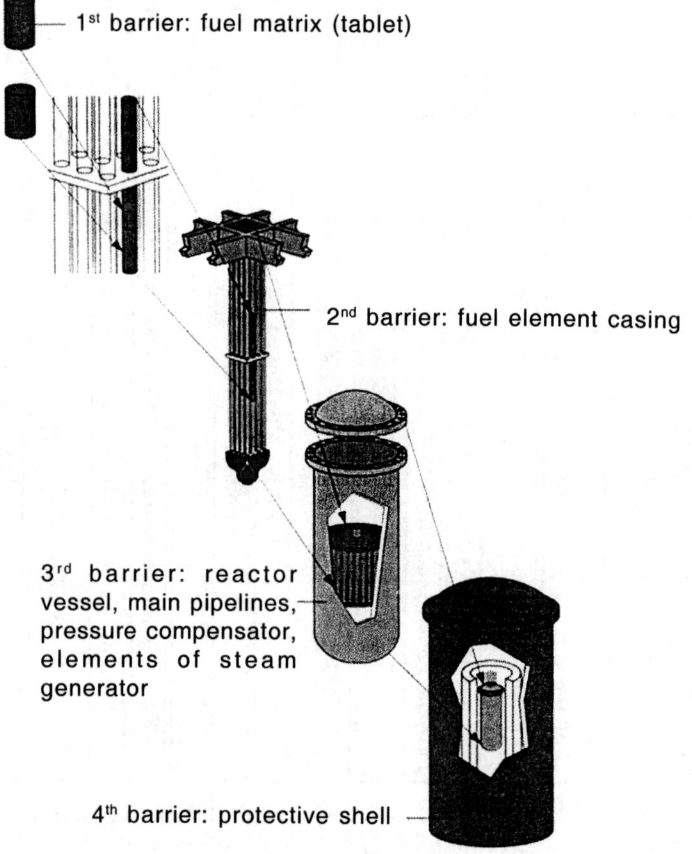

1.22 Physical security barriers and their location.

and others, subjected to mechanical loading determine their technical safety (it is enough to remember the accident with fatalities at the Mihama NPP in Japan in 2004).

Radiation and nuclear safety of the nuclear power plants are based on the concept of deeply layered protection in the path of propagation of ionising radiation and radioactive substances[3–6, etc]. Physical barriers to the spread of ionising radiation and radioactive substances are called safety fence. The structural elements of the safety barriers belong in the group of passive safety elements of NPP (Fig. 1.22) and their strength is obviously associated with radiation and nuclear safety.

The processes leading to damage which are taken into account in substantiating the strength at the design stage of structures, pipelines and equipment of NPP, include:
– corrosion;
– radiation damage;
– plastic deformation;
– failure under static loading;
– fatigue;
– creep;
– wear, abrasion, erosion.

In accordance with these processes, strength and service life are determined by the following criteria[4]:

1. resistance to plastic deformation over the ensire section, including the appearance of residual deformation which makes further operation impossible;

2. fracture resistance under static loading (by the viscous and brittle mechanisms);

3. fatigue strength;

4. creep resistance;

5. resistance to buckling.

The strength and service life of a structure with corrosion and radiation damage taken into account are ensured on the one hand by the choice of material and, on the other side, by the safety factor and allowances for the wall thickness of the construction.

Equipment, pipelines and nuclear power plant construction elements must not fail during operation. Such a requirement is contained in the Norms of trength of nuclear power plants[4], and other regulatory documents[2–4,5–8] etc., which also identify technologies and tools that ensure the above norm.

At the same time, nuclear power plants all over the world show every year damage in structural elements, including cracks, irrespective of their types, design features and operating conditions. Some of these cracks lead to the formation of leaks or even the complete destruction of the structural element, including fatalities, as the already mentioned Mihama nuclear power plant in Japan in 2004. Some summary data on the destruction of pipelines at nuclear power plants produced in the western countries, collected by GRS (Germany), are given in Figs. 1.24–1.26[9].

It should be noted that the justification of the strength and service life is conducted in deterministic formulation using a safety factor. For example, a 10-fold margin of time (number of cycles of repetition of various modes of operation) to the appearance of fatigue cracks is used, i.e. crack should form no sooner than after 300 years of operation at the assigned operating time of 30 years.

In fact, as mentioned above, cracks and failure occur even in the design life. This is due to the *probabilistic nature of strength*. The probabilistic nature of strength and lifeability of structural elements served as one of the reasons for creating Farmer charts (Fig. 1.23), reflecting the relationship between the probability of occurrence of accidents and their radiological consequences.

The section of the science of strength which studies the probabilistic laws of resistance of structural elements to destruction can be called strength reliability or the statistical theory of structural strength. To describe the probabilistic laws of strength reliability we can use conceptual and mathematical tools of mathematical reliability theory, formulated within the framework of the so-called system reliability theory for electronic and radio systems. This approach can be effective for studying the strength of mass production of items, such as automobiles and in the manufacture of agricultural machinery. The formal mathematical approach cannot be used in nuclear power engineering to understand the probabilistic laws of strength because of the lack of sufficient statistical data on the damage of similar structural elements. There are almost no data on large-scale destruction.

1.5 Ageing of equipment and pipelines. Ageing considered and not considered in design

Ageing is a term that has long been used in metals science where it refers to changes in metal under the influence of strain, temperature, other effects, or without them. Typically, ageing changes the mechanical properties of the metal.

In nuclear industry the term ageing has a broader meaning. Here ageing includes any changes to the metal or structure that occur during operation. In this case, the term 'ageing' is often replaced the term 'degradation'.

According to the documents[3,4,6–8,etc.] all ageing processes must be considered in the design. Equipment and piping should operate without damages and destructions during the whole lifetim. As a rule, all ageing degradations considered in the design processes and their limits are specified in designn strength calculations.

In fact, during the operation some elements could be damaged to the extent of design limits, and even destruction of individual elements and piping may take place. Typical ageing mechanisms excluded by the design process, but appeared during the operation of NPP are:

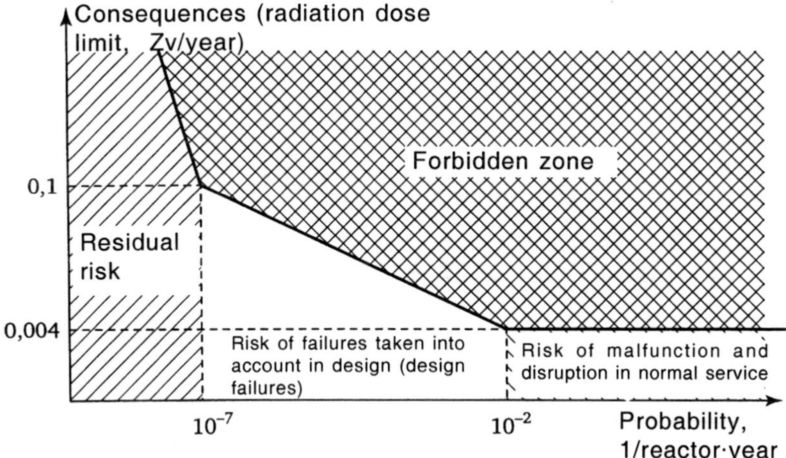

1.23 The relationship between consequences of failure the probability of its occurrence (residual risk is the risk that exists despite all the measures taken).

– stress corrosion cracking;
– intercrystalline corrosion;
– corrosion–erosion wear;
– fretting corrosion;
– accelerated radiation metal embrittlement of the reactor vessel;
– propagation of latent defects of continuity of metal.

Tables 1.1 and 1.2 list the main types of damage (ageing) for VVER (PWR) and BWR reactors[10,11].

The most dangerous types of the damage not considered in the design are associated with the appearance of cracks which may give rise to leaks and to very large leaks at destruction. Examples of such cases are shown in Figs. 1.24–1.26.

There are many causes of cracks and damage in service. In the chronological order they can be divided into three groups[5,etc.]:

– design flaws;
– shortcomings of manufacture;
– shortcomings of operation.

It was shown in Ref. 12 that the greatest contribution to the total amount of damage detected during operation comes from deficiencies of manufacturing technology (so-called technological defects) (Fig. 1.27).

Technological defects grow during service and can lead to small leaks (continuous stable defects) or large leaks due to destruction.

Defects associated with deficiencies in production or operation are removed usually by repair (Fig. 1.27). Upgrading or modernisation is usually required to address the design shortcomings.

Ageing of equipment and pipelines with defects is associated with the kinetics of their propagation under operational loads. From a safety perspective, the residual life assessment and evaluation of probabilistic reliability characteristics of ageing processed not considered in design are of greatest interest.

1.6 Quantitative characteristics of reliability and their implications for safety analysis and optimisation of operating costs

Table 1.1 Major components of nuclear power plants

Components		Mechanisms of		
	Radioactive	Creep embrittle-ment	Hydrogen embrittlement	Corrosion cracking under stress
Nuclear reactor pressure vessel	X			X
Reactor containment and base plate		X	X	X
The pipe reactor coolant				
Steam generator tubes				X
Circulation pump				
Pressurizer				X
Control rod				X
Cables	X			
Emergency diesel generator				
Reactor internals		X		X
Support reactor	X			X
Feed-pipe, nozzle and housing the steam generator				X

Using the safety factor in the justification of strength and service life of equipment and piping makes it impossible to assess the reliability of their elements. From a formal point of view, the probability of partial or complete destruction in this case should be taken equal to zero.

In fact, the practice of operation of NPP (and other branches of engineering) indicates that this probability is greater than zero. Therefore, since it is difficult to evaluate the actual probability of failure (or accuracy of such estimates is not sufficient high) the concept of maximum design-

with PWR and their degradation mechanisms

degradation

Low-frequency thermal fatigue	High-frequency mechanical and thermal fatigue	Corrosion fatigue	Thermal embrittle-ment	Mechanical wear, decay and fatigue	Corrosion and FAC
					X
					X
X	X		X	X	
	X	X		X	X
	X		X		X
X					
			X	X	X
			X		X
				X	X
	X			X	
X	X				X

Table 1.2 Major components of NPP with BWR

Components	Radioactive	Creep embrittlement	Hydrogen embrittlement	Corrosion cracking under stress
				Mechanisms of
Nuclear reactor pressure vessel	X			X
Reactor containment and base plate		X	X	X
The pipe reactor coolant				
Steam generator tubes				X
Circulation pump				
Volume compensator				X
Control rod				X
Safety and communications csbles	X			
Emergency diesel generator				
Reactor internals		X		X
Support reactor	X			X
Feed-pipe, nozzle and housing of steam generator				X

and their degradation mechanisms

| degradation | | | | | |
Low-frequency thermal fatigue	High-frequency mechanical and thermal fatigue	Corrosion fatigue	Thermal embrittlement	Mechanical wear, decay and fatigue	Corrosion and FAC
					X
					X
X	X		X	X	
	X	X		X	X
	X		X		X
X					
			X	X	X
			X		X
				X	X
	X			X	
X	X				X

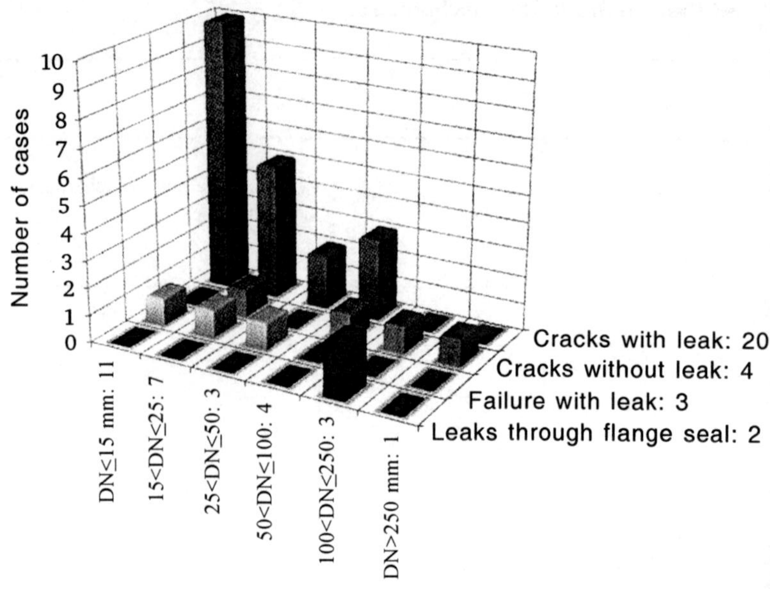

1.24 Types of damage to pipelines of the first circuit of NPP with PWR reactors.

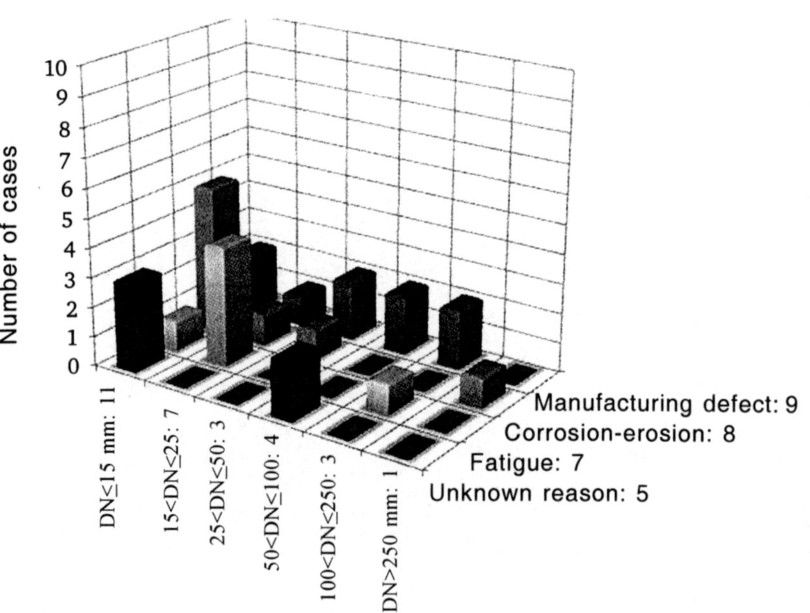

1.25 Causes of damage to the primary coolant pipe at NPP with a PWR reactor.

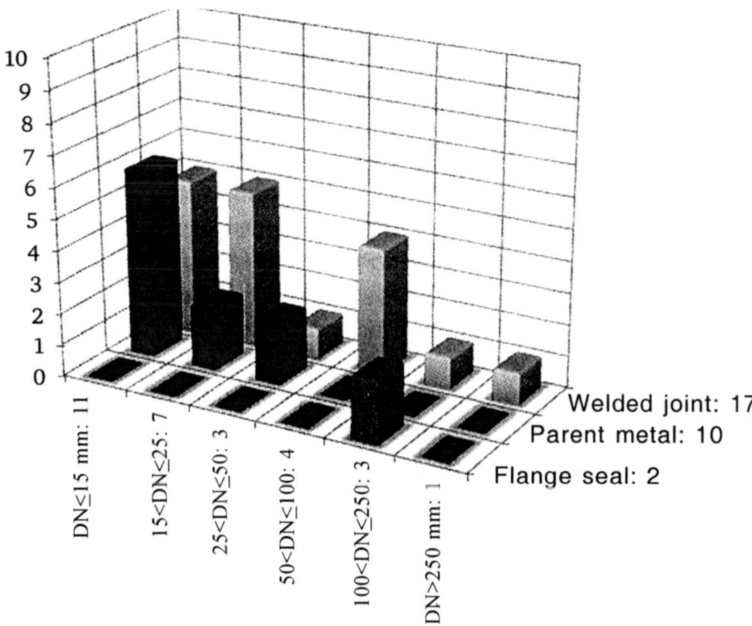

1.26 Location of damage to the pipelines of the primary circuit of NPP with PWR reactors.

based accident was applied in the design of NPP which is based on the assumption that the main pipeline of the NPP fails extensively during service plant, i.e. it is assumed that the probability of failure of the pipeline is equation to $1^{3,\ etc.}$.

It is obvious that underestimation of the unreliability of passive elements may result in lower estimates of the risk indicators NPP, and if the data on the reliability of passive components are overestimated the PSA will lead to overly conservative safety requirements.

Quantitative data on the probability of failure of equipment and pipelines are needed not only for PSA. These data are of interest also for other tasks.

The limiting states of equipment and pipelines of the NPP are determined by the criteria of strength, criteria of destruction, operating conditions, and the state of the structure. Calculation of the maximum permitted stress makes it possible to solve only one part of the problem of rational selection of sizes of installations, namely: to determine the conditions under which the installation should fail. The second, equally important part of the problem is the question how to ensure that these conditions do not occur, i.e. provide sufficient safety of equipment for the entire period of its operation. This issue was solved, until recently, by introducing more or less arbitrary factors – safety factors. A number of studies was published in recent decades in which it was attempted to theoretically develop more or less accurate methods for determining the safety factors

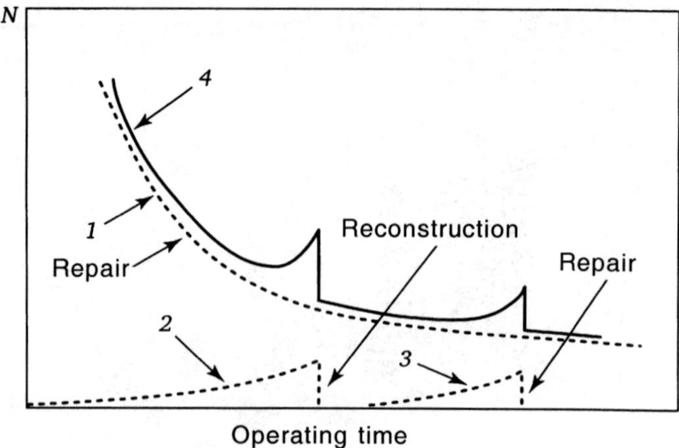

1.27 The defects detected during operation: 1) flaws inherent in the manufacturing stage and missed in inspection; repair is required; 2) increase in the number of defects associated with deficiencies of design, require reconstruction; 3) increase of the number of defects due to deficiencies of operation; require repair; 4) the total number of defects.

and allowable stresses. Procedures and methods of solutions used in these studies somewhat differed from those used than in the traditional tasks of ensuring the strength, namely, they were based on probability theory and mathematical statistics. Here we must bear in mind that most of the calculated values with which we have to operate in the calculations of structural elements are not strictly defined and may have different random deviations from their mean values. Therefore, each calculated value should be expressed not by the number but by a distribution function that characterises the probability of occurrence of all possible numerical values of this quantity. It is also assumed that the given calculation situation will be repeated many times and that the law of large numbers can be used here to judge the frequency of occurrence of certain events on the basis of the theoretically calculated probability of their appearance.

Another circumstance that requires estimates of the probability of destruction of equipment and pipelines is the need to optimise the costs of operating equipment and pipelines which, to ensure their safety, are subjected to periodic non-destructive testing, hydraulic testing, technical inspection, maintenance and repair work, upgrades. Obviously, in the case of excessive optimistic estimates of strength reliability (failure probability is zero) the above work seemed superfluous, and if exaggerated pessimistic estimates are obtained (the probability of failure is 1), these measures are insufficient. Obviously, the optimum organisation of operational technologies of non-destructive testing, maintenance and repairs is only possible on the basis of revised estimates of reliability and lifetime of equipment and pipelines. As shown below, only the quantitative safety assessment (probability of failure) determines the optimum frequency of non-destructive testing in nuclear power plants, and is used to prepare in a timely manner the schedule of repair and supply of spare parts, and decide on the appropriate test of strength by hydraulic tests.

1.7 Formal–statistical and physico–statistical approaches to predicting the reliability of technical systems

Reliability issues were first raised in connection with the statistical interpretation of the strength margin (safety factor) in the 30's of last century. However, in its present form the reliability theory developed in connection with the rapid development of electronics and computing. Electronic systems are composed of many elements and their reliability was known from the bench tests. The main task of the theory of reliability was to determine the reliability of the system using the known characteristics of the reliability of its elements. The mechanism of failure of individual elements are not considered. Such an approach in reliability theory is called the *system reliability theory*. It can also be called the formal-mathematical or formal–statistical or (due to the fact that the mechanisms of damage and failure are not considered in this theory) *phenomenological.*

Mechanical elements such as plant equipment and piping, are usually small scale products. Their failure in service is a rare event and, therefore, in most cases it is not possible to obtain their characteristics of reliability from operating experience. Bench tests of the reliability with adequate simulation of operating conditions cannot usually be carried out. Therefore, to obtain estimates of the reliability characteristics of mechanical equipment of NPP, it is necessary to study the processes of damage, ageing and degradation. The results of these studies as well as construction of a physical model and introduction to this model of the statistical characteristics of these variables which have a significant impact on the reliability characteristics of the element, provide reliable estimates of reliability. This approachis called the *structural approach*[17,etc.], in contrast to the phenomenological approach. The structural approach requires an understanding of the processes of damage (ageing, degradation) occurring in service and construction of the corresponding physical models. These models can also be called *physical–statistical.*

2

Formal–statistical methods

2.1 The simplest model

In Ref. 17–26 it was attempted to assess the reliability of individual items of equipment and pipelines through the analysis of their operation. This approach can be called simple, because it is based on a simple analysis and statistical processing of results of operation.

In Ref. 17 it was pointed out that nuclear power plants as large technical systems use a large number of different materials which operate under different conditions. Many factors could cause degradation of the functional ability of their elements. According to operating experience, failure of mechanical elements has been due to degradation processes, such as general and local corrosion, erosion, radiation and thermal effects, which caused embrittlement, fatigue and wear of materials[17,18].

Examples of some of the failures associated with ageing are as follows:
• Rupture of carbon steel feed water pipelines caused by erosion-corrosion wear;
• Reduction of the wall thickness of pressure vessel walls (loss of metal 1–9 mm) caused by erosion–corrosion wear;
• Fracture of thermocouple pressure pipes caused by blistering of the hydride;
• Degradation of the elements of engine valves due to cavitation-induced erosion;
• Corrosion of the exhaust valve of high-pressure reactor cooling systems, caused by boric acid;
• Failure of steam generator tubes due to intercrystalline stress corrosion, pitting;
• Degradation of plant cells due to erosion of the plates to the communication bus;
• Failures of stop valves;
• Failures of safety injection pumps due to erosion and vibration;
• Damage to pipes due to vibration.
The authors draw attention to the fact that the above failures, associated

with ageing, may significantly reduce plant safety, as they may damage one or more levels of protection. Ageing can lead to large-scale degradation of the physical safety barriers and their components and increase the probability of common cause failure. It may also lead to a reduction in safety margins provided by the design and statutory requirements and, ultimately, reduce safety.

Mechanical elements also pose a safety threat associated with ageing because they can be damaged by unnoticed, almost catastrophic failure. Mechanical items do not have enough failure statistics for analysis. Acceptable data for statistical analysis could be obtained from the monitoring data. Since the collection of data for mechanical components is very costly and uninformative, they are sometimes studied by simulation.

Taking into account the fact that each nuclear power plant is composed of thousands of items, reliability and ageing cannot be assessed for each of them, so that the elements for analysis and management should be very carefully selected. This also applies to probabilistic safety analysis (PSA), which deals with the components important for safety. In this case, the choice must be made so as to take into account the effects of ageing to assess the level of risk. This can ensure greater efficiency of analysis and reduce the costs associated with safety.

There are three qualitatively different approaches to data collection and analysis:

(1) frequent periodic data collection over a long period of time. The purpose of the analysis is to identify the existence and magnitude of the trend.

(2) A single data collection under current conditions. The obtained measurement data are used as input data for reliability and ageing analysis. The purpose of the analysis is to estimate the probability of failure in the future.

(3) data collection, such as the results of non-destructive testing of materials. Analysis may have two objectives: to assess the current state of the material or, as in (1), the detection and quantitative assessment of current trends.

Approach (1) is not new and includes developed methods of analysis. Approach (2) requires the development of mechanical models. Approach (3) requires special consideration since the gradual development of defects such as cracks can be detected by periodic in-service inspection (ISI). ISI data are extremely complex and difficult to interpret. A probabilistic approach to evaluate a random error must be used. Probabilistic models and methods of solutions considered are Refs. 16, 20–27.

It is evident that the data that should be collected and stored are determined by the tasks of controlling the reliability, ageing and operating life of nuclear power plant[19]:

• Prediction of residual life;
• Preventative maintenance of equipment and pipelines;

• Identification and evaluation of degradation and failures and failures of components and systems caused by the effects of ageing;

• Optimisation of operational modes and technologies aimed at reducing ageing and degradation, increase of reliability;

• Identification of new effects of ageing;

• Evaluation of possible long-term plant operation, including preparation of applications for renewal of a license;

The existing databases for nuclear power plants are usually insufficient to assess the reliability and ageing of components of equipment and pipelines, but they could be used in conjunction with international databases on failures, for example, those described in NUREG/CR-5750 (Poloski et al., 1999). This might solve the problem of lack of data for specific components and small intervals of observation time.

Typically, databases used in the ageing management programs, reliability and service life, consists of three categories of data:

• *Basic data* – population, the expected degradation mechanisms, installation data, qualification data, modification of the project, etc;

• *Service data* – system and component operating conditions, verifiable data, failure components, etc;

• *Maintenance data* – conditions and results of monitoring, service data (type, date and duration, job description, etc.), cost of repairs, replacements, etc.

What data are actually needed depends on the capabilities and types of problems, but mainly on models and approaches used to account for the effects of ageing in the PSA model.

Studies that were performed in Ref. 20–26 to demonstrate the impact of ageing effects on the reliability and safety of components and systems of nuclear power plant included qualitative assessments of periodic tests and service data, as well as data on failures. In this case, the phenomenon of ageing could be considered in different ways, depending on the rate of degradation of the functional components, availability and quality of data on failures and control conditions.

The general theory of reliability and ageing of components distinguishes three stages:

•   Failures in the early stages of operation;

•   A relatively steady stream of failures;

•   Increasing the flow of failures after long-term operation.

Identification and evaluation of reliability and ageing in the first (early) stage are based mainly on operating experience. Quantitative reliability evaluation is based on the simulation of physical degradation process or statistical data about the failure. Models of early ageing processes include the resumption of operation and the non-homogeneous Poisson process.

There are a lot of references describing and discussing in detail ageing and failures of various types of different components and their mechanisms, e.g. Refs. 20–26. Below is an example of such approaches to a specific

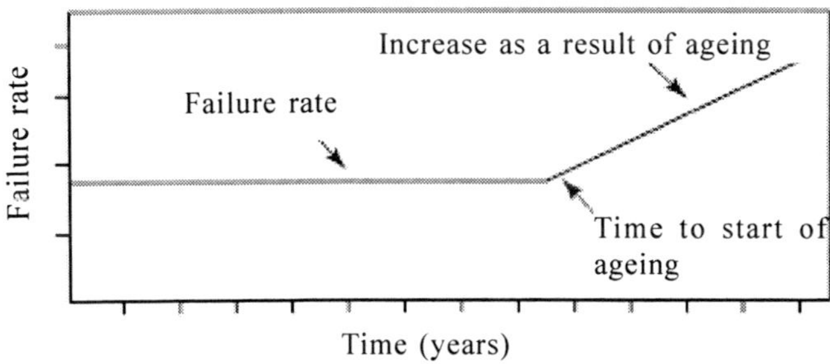

2.1 Scheme of evaluation of the probability of failure during operation.

component of equipment of nuclear power plant[26]. A linear model of ageing was used in this case. As shown in Ref. 25, it is necessary to determine two parameters from the operational statistics:

- failures associated with ageing;
- constant failure rate, when ageing is absent (Fig. 2.1).

The approach described can also be used with a non-linear dependence of reduction of reliability due to ageing (or time of operation). It simplifies the method for using the existing performance data. However, as shown in Refs. 16, 28, etc, such an approach has proved fruitful for the active elements of nuclear power plants. For mechanical elements (equipment, piping, etc.) it is complicated to use the above-mentioned simple models. This is due to:

1. Lack of sufficient information which can be subjected to statistical analysis.

2. Low quality of available information, which often lacks detailed description of failure and cause of the defect, the reasons for failure, etc.

3. Scatter of data for various nuclear power plants and the lack of an uniform data collection system. In this regard, a task of developing a data collection system and use it to study ageing and the prediction of reliability and safety with ageing taken into account was formulated within the framework of OSCE. 14 countries take part in the programme.

4. The disadvantage of the above models is also that the conditions of the transition of the defects from the monotonic, quasi-static process to the process of rapid, almost instantaneous destruction are not taken into account.

2.2   Markov processes

It is very convenient to describe the appearance of random events in the form of transition probabilities from one state of the system to another, since it is assumed that by going to one of the states the system should not continue to consider the circumstances of how it came to this state.

A random process called a Markov process (or *process without aftereffect*), if for every moment *t* the probability of any state system in

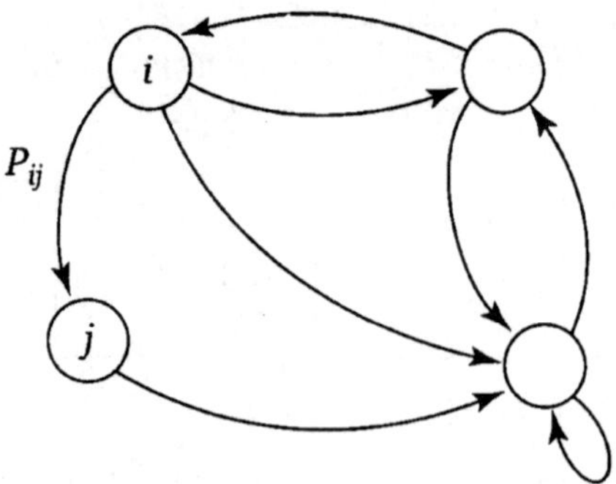

2.2 Example of a transition graph.

the future depends only on its state at present and does not depend on how the system has come to this state[28].

Thus, the Markov process is described conveniently by the graph of transitions from state to state. Consider two options for describing Markov processes – *with discrete and continuous time.*

In the first case, the transition from one state to another occurs in pre-known points in time – cycles (1, 2, 3, 4, ...). Transition occurs at each cycle, that is, the researcher is only interested in the sequence of states which is a random process in its development, and is not interested in exactly when each of the transitions occurred.

In the second case, the researcher is also interested in the chain of states changing each other, and in the time at which such transitions occur.

If the transition probability does not depend on time, then the Markov chain is called *homogeneous.*

Thus, the model of a Markov process has the form of a graph in which the states (tips) are linked together by bonds (the transitions from the i-th state to the j-th state), see Fig. 2.2.

Each transition is characterised by transition probability P_{ij}. Probability P_{ij} indicates how often after being transition to the i-th state transition to the j-th state takes place. Of course, such transitions occur randomly but if the frequency of transitions is measured over a sufficiently long time it turns out that this frequency coincides with a given probability of the transition.

It is clear that for each state the sum of the probabilities of all transitions (outgoing arrows) from this to other states must always be equal to 1 (see Fig. 2.3).

For example, the complete graph might look like the one shown in Fig. 2.4.

Implementation of a Markov process (the process of its modelling) is a computation sequence (chain) of transitions from state to state. The circuit in Fig. 2.5 is a random sequence and may also have other options for implementation.

To determine the new state to which the process moves from the current

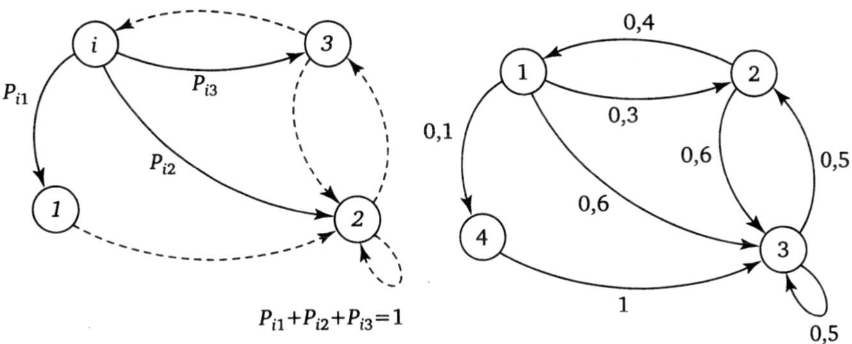

2.3 (left) Detail of the graph of transitions (transitions from the *i*-th state form a complete group of random events).
2.4 (right) Example of a Markov transition graph.

2.5 Example of a Markov chain, simulated on the Markov graph shown in Fig. 2.4.

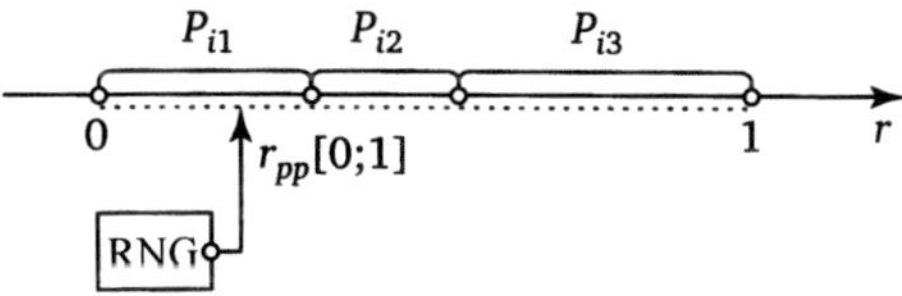

2.6 The process of modelling the transition from the *i*-th state of the Markov chain to the *i*-th state using a random number generator (RNG).

i-th state, it suffices to divide the interval $[0; 1]$ into subintervals value P_{i1}, P_{i2}, P_{i3}, ... ($P_{i1} + P_{i2} + P_{i3} + ... = 1$), see Fig. 2.6. Then, using a random number generator (RNG), it is necessary to determine the random number r_{pp} uniformly distributed in the interval $[0; 1]$ and determine which of the tanges it fits in.

After this the transition to the state defined by the RNG takes place and this procedure is repeated for the new state. The result of this model is a Markov chain (see Fig. 2.5).

Markov random processes with continuous time

So, again the model of a Markov process is described by a graph in which the states (tips) are linked together by bonds (the transitions from the *i*-th to *j*-th state), see Fig. 2.7.
Now, each transition is characterised by the probability density of transition

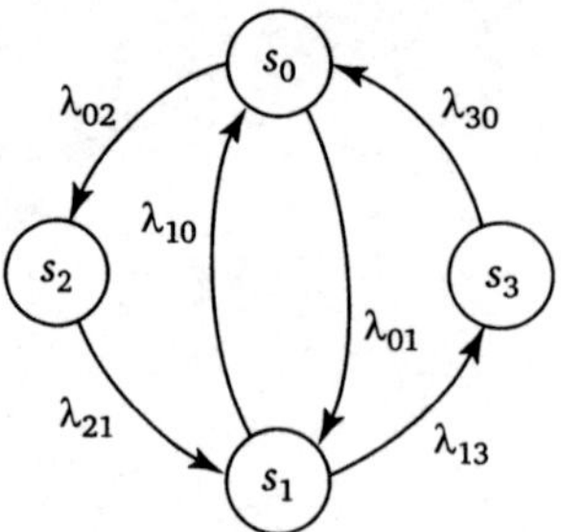

2.7 Example of a graph of the Markov process with continuous time.

λ_{ij}. By definition:

$$\lambda_{ij} = \lim_{\Delta t \to 0} \frac{P_{ij}(\Delta t)}{\Delta t}. \qquad [2.1]$$

The density is understood as a probability distribution over time.

The transition from the *i*-th to the *j*-th state occurs at random times, which are determined by the intensity of the transition λ_{ij}.

One passes on to the intensity of the transitions (here, the concept is identical in meaning with the probability density for time *t*) when the process is continuous, that is distributed over time.

Knowing the intensity λ_{ij} of occurrence of events generated by the flow, it is possible to simulate a random interval between two events in this flow:

$$\tau_{ij} = \frac{1}{\lambda_{ij}} \mathrm{Ln}(R), \qquad [2.2]$$

where τ_{ij} is the time between finding the system in the *i*-th or *j*-th state.

Furthermore, obviously, the system from any *i*-th state can go into one of several states $j, j + 1, j + 2,...$, related to transitions $\lambda_{ij}, \lambda_{ij} + 1, \lambda_{ij} + 2,...$.

In the *j*-th state it passes through τ_{ij}; in the $(j + 1)$-th state it passes through τ_{ij+1}; in $(j + 2)$-th state it passes through $\tau_{ij} + 2$, etc. Clearly, the system can move from the *i*-th state in only one of these states, and in fact the system can move to the state characterised by the first transition. Therefore, from a sequence of times: $\tau_{ij}, \tau_{ij+1}, \tau_{ij+2}$, it is necessary to choose the minimum time and to determine the index *j*, indicating in the state to which transition happens.

Markov-type models were used in Ref. 28, etc. The shortcomings of these studies in relation to the mechanical components is that they are not based on physical models of ageing (damage) so that they cannot adequately predict the onset of the most dangerous types of failures associated with the destruction of equipment components and pipes and the emergence of large-break coolant leaks.

2.3   The Monte Carlo method

2.3.1   The general characteristics of the Monte Carlo method

The Monte-Carlo metods is the common name of the group of numerical methods based on obtaining a large number of realisations of the stochastic (random) process, which is formed so that its probability characteristics coincide with similar values of the given task. The methods are used to solve problems in physics, mathematics, economics, optimisation, control theory, etc.[27,etc] In recent years the methods have been applied in several studies to ensure the reliability of NPP equipment and piping and maintenance optimisation[29, etc].

The principle of the Monte Carlo method is as follows: find the value a of some quantity to be investigated. To do this, choose a random variable X whose mathematical expectation is equal to a: $M(X) = a$.

In practice, the following procedure is used: carry out n tests which results in n possible values of X, compute their arithmetic mean and take $\bar{x} = (\Sigma x_i)/n$ as an estimate (approximate value) a^* of the unknown number a:

$$a \approx a^* = \bar{x}.$$

Since the Monte Carlo method requires a large number of tests, it is often called the method of statistical tests. The theory of this method indicates how to choose random variable X, how to find its possible values. In particular, procedures for reducing the variance of random variables are developed, resulting in reduced errors in replacing the unknown mathematical expectation a by its estimate a^*.

2.3.2   The procedure for estimating the inaccuracy of the Monte Carlo method

Suppose that n independent trials have been performed to estimating a^* expectation as a random variable X has been made (n possible values of X were used) and as a results the sample mean $\bar{x}$ was found which is accepted as the desired estimate $a^* = \bar{x}$. It is clear that if the experiments are repeated, other possible values of X are obtained and, therefore, another mean and hence a different estimate of a^*. This already implies that an exact estimate of the mathematical expectation is not impossible. The question naturally arises about the value of the permissible error. We restrict ourselves to finding only the upper limit δ of the permissible error with a given probability (reliability) γ:

$$P(|\bar{X} - a| \leq \delta) = \gamma.$$

The upper limit of error δ in which we are interested is in fact the accuracy estimate of the mathematical expectation of an average sample using confidence intervals. The following three cases will be considered.

Random variable X is normally distributed and its standard deviation δ is known.

In this case, with the reliability γ the upper limit of the error is

$$\delta = t\delta / \sqrt{n}, \qquad\qquad [2.3]$$

where n is the number of tests (played values of X); t is the argument of the Laplace function for which $\Phi(t) = \gamma / 2$, σ is the known standard deviation of X.

Random variable X is normally distributed, and its standard deviation σ is unknown.

In this case, with reliability γ the upper limit of the error is

$$\delta = t_\gamma S / \sqrt{n}, \qquad\qquad [2.4]$$

where n is the number of tests; s is the corrected standard deviation, t_γ is determined from tables.

Random variable X is distributed according to a law different from normal.

In this case, at a sufficiently large number of tests ($n > 30$) with the reliability which is approximately equal to γ, the upper limit of the error can be calculated from formula [2.3], if the standard deviation σ of the random variable X is known, but if σ is unknown, then we can substitute in the formula [2.3] its estimate s – the 'corrected' standard deviation, or use the formula [2/4]. Note that as n increases the difference between the results yielded by both formulas becomes smaller. This is explained by the fact that the Student distribution tends to normal at $n \to \infty$.

From the above it follows that the Monte Carlo method is closely connected with problems of probability theory, mathematical statistics and computational mathematics.In connection with the problem of simulation of random variables (in particular, uniformly distributed) the methods of number theoryplay a significant role.

Among other computational methods, the Monte Carlo method stands out for its simplicity and generality.

It has some obvious advantages:

a) The method does not require any assumptions about the regularity, except square integrability. This can be useful because very complex functions, whose regularity properties are difficult to establish, are used quite often;

b) It leads to a feasible procedure even in the multidimensional case, when numerical integration is not applicable, for example when the number of measurements is greater than 10;

c) It is easy to apply for small restrictions and without a preliminary analysis of the problem.

However, the methods have some disadvantages, namely:

a) The error bounds are not defined precisely and include some kind of randomness. However, this is a more psychological than real difficulty;

b) The static error decreases slowly;

c) The need to have a sufficient number of random numbers; this is difficult to achieve for the problem of forecasting the probability of destruction of equipment and piping of nuclear power plant.

Chapter 9 discusses an example of application of the Monte Carlo method to solve a local problem associated with the reliability of heat-exchange pipes NPP steam generators for the case when there are sufficient statistical initial data[29].

2.4   Risk theory

Specialists in various industries in their communications and reports constantly operate not only the definition of 'danger' but also such terms as 'risk'.

In the scientific literature there are different interpretations of the term 'risk' and they sometimes contain different contents. Several interpretations treat the risk as the probability of occurrence of an accident, danger, malfunction or disaster, under certain conditions (state) of production or the environment. These definitions emphasise the value of the activity of the subject and also the objective properties of the environment.

Common to all the above interpretations is that risk involves an uncertainty whether there will be an undesirable event and whether an adverse state will form. Note that in accordance with modern views the risk is usually interpreted as the probability measure of man-made or natural phenomena accompanied by the emergence, formation and action of the dangers and the resultant social, economic, ecological and other kinds of damage and injury.

Risk should be regarded as the expected frequency or probability of occurrence of hazards of a specific class, or the extent of possible damage (loss, degradation) caused by an undesirale event, or some combination of these variables.

Application of the concept of risk enables us to transfer the danger into the range of the measured categories. Risk, in fact, is a measure of danger. The term 'the level of risk' is used quite often and is not fundamentally different from the concept of risk, but only emphasises that we are talking about the measured quantity.

All these (or similar) interpretations of the term 'risk' are currently used in the analysis of hazards and safety management (risk) of technological processes and production in general.

An accurate understanding of using this term will become clear after further reading the contents of this chapter.

Formation of hazardous and emergency situations is the result of a specific set of risk factors posed by relevant sources.

Technical risk is a comprehensive reliability index of elements of the technosphere. It expresses the probability of an accident or disaster in the operation of machines, technological processes, construction and service of buildings and structures:

$$R_{\mathrm{T}} = \Delta T(t) / T(f), \qquad\qquad\qquad [2.5]$$

where R_{T} is technical risk; ΔT is the number of failures per unit time t of the identical technical systems and facilities; T is the number of identical technical systems and facilities, subject to general risk factor f.

Sources and technical risk factors are listed in Table 2.1.

The process of formation and development of risk is affected by a variety of factors and conditions specified on Fig. 2.8. The diagram show a range of underlying causes of risk: malfunction sites and equipment due to their structural deficiencies, poor manufacture or technical violations of maintenance, deviation from normal operating conditions, human error, external influences, etc. Because of these reasons, industrial facilities are constantly in an unstable state which in relation to production safety is especially critical for emergency situations at the facilities.

The risk occurs when the following necessary and sufficient conditions are fulfilled:

- The existence of risk factors (source of danger);

Table 2.1 Sources and technical risk factors

Source	Most common factors
Low level research	Wrong choice of directions of technological development and technologies for safety criteria
As above, experimental and design studies	Selection of potentially hazardous construction schemes and principles of technical systems. Errors in the determination of operating loads. Wrong choice of construction materials. Insufficient margin of safety. The absence of technical security measures in projects
Experimental production of new machinery	Poor finishing construction, technology, documentation for safety criteria
Series production of unsafe equipment	Variation of the chemical composition of constructional materials. Insufficient accuracy of structural materials. Violation of thermal and thermochemical treatment of parts. Violation of regulations of assembly and installation of structures and machines
Violation of the rules of safe operation of technical systems	Inappropriate use of technology. Violation of the passport (design) operating modes. Untimely prevention inspections and repairs. Violation of requirements for transportation and storage.
Human error	Poor skills in a difficult situation. Inability to assess the status of the process. Poor knowledge of the nature of the ongoing process. Lack of self-control under stress. Indiscipline.

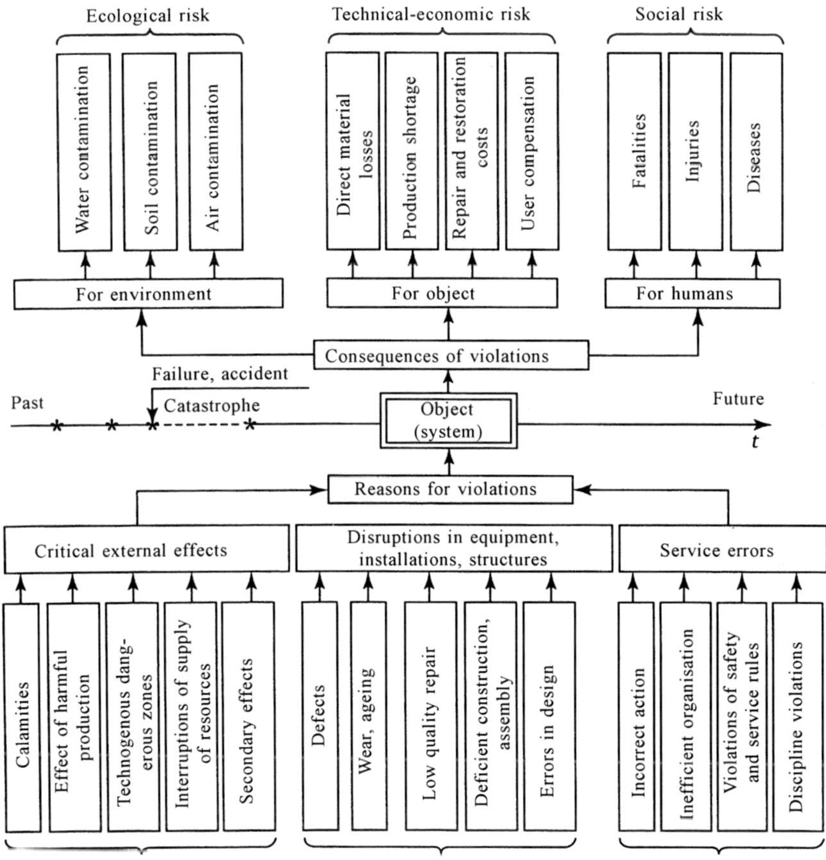

2.8 Functional model of risk of industrial systems.

- The presence of risk factors in a dose specific, dangerous (or harmful) for the objects;
- Exposure (sensitivity) of objects to the impact factors of the danger

There is a clear similarity between failures in different industries. Failures are usually preceded by the accumulation of defects in equipment or by deviations from the normal course of the processes. This phase can last for minutes, days or even years. By themselves, the defects or deviations do not cause failure but set the stage for it. Operators usually do not notice this phase because of inattention to the regulations or a lack of information on the object operation, so that they do not have a sense of danger.

The next phase is an unexpected or unusual event which significantly alters the situation. The operators try to restore the normal course of the process, but not having full information, often only aggravate the development of failure. Finally, in the last phase, another unexpected event – sometimes quite small plays the role of a shock, after which the technical system ceases to obey the operators and there is a disaster.

Risk is an unavoidable concomitant factor in industrial activities. The risk is objective, unexpected, occurs suddenly; this requires a prediction of risk, its analysis, evaluation and management – a number of actions to prevent risk factors or risk mitigation.

When working out the problems of risk and process safety special attention is given to a systematic approach to recording and studying the various factors affecting the risk factors, referred to as risk analysis.

Risk analysis is the process of hazard identification and risk assessment for individuals, groups, facilities, environmental and other objects of consideration.

Recall that the hazard means a source of potential harm or injury or a situation with the possibility of damage, and hazard identification is the process of identification and recognition of that hazard exists and of defining its characteristics.

There are many such formulations of this concept, but in general risk analysis is the process of hazard identification and assessment of possible negative consequences as a result of violations in specific technological systems and the representation of these effects in quantitative terms.

In the USA, 'process hazard analysis' is used instead of risk analysis and has almost the same value.

Risk analysis is a largely subjective process during which we take into account not only quantitative indicators but also indicators that are not amenable to formalisation, such as attitudes and opinions of different social groups, the possibility of compromise solutions, expert assessments, etc.

The variety of production activities, specific industrial facilities, and their affiliation to a variety of industries reflects the multifaceted problem of risk analysis.

A feature of technological risk analysis is that it considers potentially negative consequences that may result from the failure of technical systems, disruptions in production processes or errors on the part of staff. Of course, negative impacts on staff and the environment in accident-free operation of production (due to emission or leakage of hazardous or dangerous substances, raw sewage, etc.) can also be considered

The results of risk analysis are essential to making informed and rational decisions in determining the location and design of production facilities, during transportation and storage of hazardous substances and materials. Risk analysis uses widely formalised procedures and responses to diverse situations that can confront management staff in their work, especially in an emergency situation. The uncertainty under which management must make decisions in many cases affects the methodology, progress and outcomes of risk analysis. Methods used in the analysis should focus primarily on the identification and assessment of possible losses in the event of a failure, the cost of security and the benefits in implementing a project.

Risk analysis has a number of general provisions, irrespective of the methods of analysis and specific tasks. First, it is the general problem of determining an acceptable level of risk, safety standards for personnel, the public and environmental protection. Secondly, the acceptable level of risk is usually defined in areas with poor or unverified information, especially when it comes to new processes or technology.Thirdly, the task of analysis is to solve probabilistic problems that could lead to significant discrepancies in the results. Fourthly, risk analysis should be viewed as a process of solving multiobjective problems that may arise as a compromise between the parties interested in the outcome of the analysis.

Risk analysis should give answers to three basic questions:

1. Which harmful events can happen? (hazard identification),

2. How often may this happen? (frequency analysis),

3. What might the consequences be? (analysis of the consequences).

The main element of risk analysis is hazard identification (detection of potential violations) which can lead to negative consequences. The process of risk analysis expressed in the most general form can be presented as a series of sequential events:

1.   Planning and organisation of work.

2.   Hazard identification.

2.1. Hazard detection.

2.2. Preliminary assessment of the characteristics of hazards.

3.   Risk assessment.

3.1. Frequency analysis.

3.2. Analysis of the consequences.

3.3. Uncertainty analysis.

4.   Development of recommendations for risk management.

Firstly, the initial stage of any risk analysis is planning and organisation of work. Risk analysis is conducted in accordance with the requirements of regulations in order to provide input to the risk management process, but a more accurate selection of tasks, means and methods of risk analysis is usually not regulated. The documents emphasise that risk analysis should match the complexity of the processes, the availability of essential data and the qualification of specialists conducting the analysis. At the same time, more simple and understandable methods of analysis should be preferred for more complex methods that are not fully clear and methodically supported. Therefore, the first step is to:

- describe the reasons and problems that caused the need for risk analysis;
- identify and describe the analyzed system;
- choose the appropriate command for analysis;
- identify sources of safety information systems;
- specify the source data and restrictions governing the limits of risk analysis;
- clearly define goals of risk analysis and criteria for acceptable risk.

All the regulations require documenting this phase of risk analysis.

The next stage of risk analysis is the identification of hazards. The main task is to identify (based on information about the object, examination results and experience with similar systems) and describe accurately all the hazards inherent in the system. This is a crucial stage of analysis since the hazards undetected at this stage are not subject to further consideration and disappear from sight.

There are a number of formal methods of identifying hazards which are discussed below. Here is a preliminary assessment of hazards in order to select further activities:

- To stop further analysis because of low risk;
- A more detailed analysis of risk;
- Develop recommendations to reduce hazards.

Baseline data and the results of preliminary hazard assessment should also be properly documented. In principle, the process of risk analysis can already be terminated at the stage of hazard identification.

If necessary, after identifying the hazards we can transfer to the stage of risk assessment.

The final stage in risk analysis of technological systems is the formulation of recommendations for reducing the level of risk (risk management) if the level of risk is higher than acceptable.

In this type of work all the regulations prescribe the preparation of a report, with the requirements on the content of the report strictly defined and relating to the above questions.

The plurality of analysis results and the possibility of compromise solutions give reason to believe that risk analysis is not a strictly scientific process, verifiable by objective scientific methods.

Risk analysis is closely related to another process – risk assessment.

Risk assessment is the process used to determine the value (measures) of the analyzed risk of danger to human health, property, environment and other situations associated with the hazards. Risk assessment is an indispensable part of the analysis. Risk assessment includes an analysis of frequency, impact analysis and combinations thereof.

The English literature uses the terms 'risk estimation', 'risk assessment', 'risk evaluation', which often have different meaning but are translated as risk assessment.

Risk assessment is the stage at which the identified risk must be evaluated on the basis of the criteria of acceptable risk in order to identify hazards with an unacceptable level of risk, and this step will be the basis for the formulation of recommendations and measures to reduce hazards. In this case, both the criteria of acceptable risk and risk assessment can be expressed both qualitatively and quantitatively.

By definition, risk assessment includes an analysis of the frequency and impact analysis. However, when the effects are negligible and the frequency is very low, it suffices to estimate one parameter.

Several different approaches to risk assessment will be discussed:

Engineering – It relies on the statistics of breakdowns and accidents, on probabilistic safety assessment (PSA): construction and calculation of the so-called 'event trees' and 'fault trees' – a process based on oriented graphs. The former are used for predicting the state to which one or another technology failure may develop, and the fault trees, on the contrary, help to trace all the causes that can cause some adverse effects. When the trees are constructed, the probability of each scenario (each branch) is predicted, and the overall probability of a failure at the facility is then calculated.

Model – construction of the models of the impact of harmful factors on humans and the environment. These models can describe both the effects of normal operation of enterprises as well as damage from the accident to them.

The first two approaches are based on calculations; however, in most cases there are insufficient reliable baseline data for such calculations. In this case, the third approach – the *expert approach* should be used: the probabilities of various events, relationship between them and the consequences of a failure is determined not by calculation but by a survey of experienced experts.

There are many uncertainties associated with risk assessment. Uncertainty analysis is an essential part of risk assessment. Typically, the main sources of uncertainty are information on equipment reliability and human errors as well as the assumptions of the models of emergency processes. To correctly interpret the magnitude of risk, we should understand the uncertainty and their causes. Uncertainty analysis is the translation of uncertainty of initial parameters and suggestions for assessing risks, the uncertainty of the results.

Sources of uncertainty should be identified. The main parameters to which the analysis is sensitive must be included in the results.

It is important to emphasise that the complex and expensive calculations often give risk values with very low accuracy. The accuracy of calculations of individual risk for complex technical systems is not higher than one order, even in the presence of all necessary information. The full quantitative risk assessment is more useful for comparing different options (e.g., placement of equipment) than for conclusions about the safety of the facility. Overseas experience shows that the largest volume of safety recommendations is produced using high-quality (engineering) methods of risk analysis which enable us to achieve the main objectives of risk analysis using a smaller amount of information and labour costs. However, quantitative risk assessment techniques are always very helpful and in some situations they are the only acceptable techniques, in particular, when comparing risks of different nature or examining high-risk, complex and expensive systems.

The quantitative risk indicator is the numerical value of the likelihood of an undesirable event or (and) the results of undesirable consequences (damage).

The risk can be quantitatively defined as the frequency (dimension – the inverse time) of occurrence of hazard.

The study of statistical data reveals the frequency of occurrence of hazardous events.However, the seriousness of the events (even within a single class of failures) may vary considerably from event to event; consequently, it is necessary to introduce categories of events (for example, events with severe, moderate or minor effects) and to consider the frequency of each of these categories. The latter is achieved by allocating a risk indicator (number of events over a certain period of time divided by the duration of this period) to each class or subclass which has the dimension of inverse time. This figure is sometimes seen as a measure of 'probability' of the event.

For example, it is possible to characterise the phenomenon of random variable z by the incidence of events (phenomenon of implementation) for a certain period of time T, for example, for the year. It is well known that the mathematical expectation Mz of the random variable z is the average (expected) number of events in a year or the frequency of occurrence of the events. Then, in accordance with the accepted terminology in mathematical statistics, the number of events (which is taken from statistical data) is a sample, the ratio of the number of events to the duration of the observation period – statistics, which is obviously an unbiased and consistent estimator of the mathematical expectation Mz or the frequency of occurrence of events. For example, if we assume the Possion distribution of z, i.e. if we put

$$P(z = k) = e^{-rT}(r \cdot T)^k / k!, \qquad\qquad [2.6]$$

where r is a constant, it is possible to evaluate the condition when the input parameter can be considered as reliability. In fact, for the Poisson distribution $Mz = r \cdot T$. On the other hand, for the Poisson distribution the probability that at time T at least one event happens is $1 - e^{-rT}$. Therefore, the input parameter can be interpreted only at very low frequencies of an event as the probability of occurrence of at least one event during time T.

However, it must be noted that the parameter inputted in this way is not an indicator of the probability in the precise, mathematical sense of the word. The probability of occurrence (of an event in the final scheme in the classical sense) is the ratio of the power of the set of elementary outcomes that make up this event to the power of the entire set of elementary outcomes. The probability of an event is a real number lying in the range 0–1. Thus, in the case considered here the links between events A and B, where B occurs only if event A occurs, can be interpreted as the probability.

The risk can be defined quantitatively as the probability P of an event when the event A occurs (a dimensionless quantity that lies within 0–1).

Since hazard is an accidental phenomenon, the hazard risk (no matter how we define it – as frequency or probability) is a numerical characteristic of the corresponding random variable used to describe this risk. As a simple example of a possible formal approach consider the random variable s – duration of trouble-free operation of an industrial enterprise whose domain is the set of operating conditions for an arbitrary (possibly infinite) time. It is possible to explicitly calculate the distribution function of this quantity $F_s(t) = P(s \leq t)$, assuming that it is independent of the prior history of operation of the enterprise (this assumption is most optimistic in respect of the level of security). It is well known that there exists a unique solution that satisfies the formulated condition:

$F_s(t) = 1 - e^{-rT}$ for $t > 0$;

$F_s(t) = 0$ for $t < 0$, where $p > 0$ is a constant;

This is the so-called exponential distribution. Mathematical expectation Ms of the random variable s is $Ms = 1 / p$, which allows us to interpret the parameter p as the mean (expected) frequency of failures or the risk of failures in the sense of the discussed definitions.

The likelihood of failure p_T for a period of time not exceeding T is determined obviously as $P_T = P(s \leq T) = 1 - e^{-rT}$. Note that always $p_T < pT$, so the often made claim that the hazard risk is equal to $1/T$ would always result in a failure in period T is not correct (the probability of this event is $1 - e^{-1}$, i.e. approximately 0.632). Moreover, even in this simplest case of the exponential distribution it would be wrong to say that the probability of a failure p_T for a period of time less than or equal to T is defined as the product of failure frequency p for the period T. There is only an approximate equality in the case of small risks, i.e. rare failures. However, the functional relationship between the probability of failure and the frequency of its occurrence (for a fixed distribution) exists.

The quantitative measure of risk can be expressed not only by the probability value. Risk is sometimes interpreted as the expectation of damage arising from the given hazard.

In determining the mathematical expectation of damage it seems appropriate to take into account all possible types of dangerous occurrences for the given object and carry out risk assessment on the basis of the sum of products of the probabilities of these events in respect of the appropriate damage. In this case the following relationship:

$$R_{\text{me}} = \sum_{i=1}^{n} p_i U_i, \qquad [2.7]$$

where R_{me} is level of risk expressed in terms of mathematical expectation of damage; p_i is the probability of a hazardous event of the i-th class; Y_i is the damage caused by the i-th event.

Although the latter interpretation is used quite often, the probability measure of risk is more convenient and applicable in solving a wide range of problems of scientific and practical nature, especially problems related to industrial safety.

Risk as a measure of the mathematical expectation of the random damage M in the form of a discrete random variable has the form:

$$R = R_{\Sigma} = M[U]\sum_{i=1}^{n} P_i U_i,$$ [2.8]

where $U_1,..., U_n$ are possible values for accidental damage to a nuclear fuel cycle facility during its life cycle; $p_1,..., p_n$ is the likelihood that accidental damage from machinery during the life cycle takes the values of $U_1,...,U_n$, respectively.

In general, the number of possible values of the damage components of a complete group of events is infinite. Since the accidental damage caused by the object is a consequence of random adverse events with the object, then:

$$R_{\Sigma} = \sum_{i=1}^{n} p_i U_i,$$ [2.9]

where p_i is the probability of the i-th adverse event in the life cycle of a nuclear facility; U_i is the damage as a result of the i-th adverse event at the facility; n is the number of possible adverse events at the facility.

Obviously, the proposed generalised index is the sum of risks R_i of adverse events which can occur at the site of the nuclear cycle in the complex:

$$R_{\Sigma} = \sum_{i=1}^{n} R_i,$$ [2.10]

where $R_i = p_i U_i$ is the risk to the life support of a nuclear facility due to the possibility of i-th event (the risk of the i-th event).

Treating certain risks of a technical system in the broadest sense, the adverse event to the object of the nuclear cycle is any accidental event that results in damage. Any random deviation in the normal functioning: malfunction of an element, failure, disruption or destruction of an object – requires certain expenditure, i.e. damages and, strictly speaking, is the adverse event. Problems associated with risk, malfunction and failure of the nuclear cycle facility, differ from the problems of risk at the sites of the traditional technosphere. The danger from traditional technosphere objects is determined only by failure of the object itself, injuries of staff and operators and is not directly connected with the loss of human life and health and considerable environmental and material damage at a large distance from the object.

Hazardous events caused by the presence at a nuclear facility of fission materials and radioactive substances can be defined as nuclear and radiation hazardous events. According to this, the index containing the sum of risks expressed by [2.7]–[2.9] for only such events becomes an integral indicator of nuclear and radiation hazardous hazards of the nuclear cycle facility.

A list of the dangerous events and, accordingly, the content of the indicator of hazard of special machinery significantly differs for different types of objects – nuclear power plants, nuclear and conventional munition, objects in rocket and PSAce technology. The list is determined on the basis of objectives of the study and preliminary analysis of the structure and composition of the object, its purpose and process of operation. As the hazardous developments become more extensive and detailed, the corresponding integral index R becomes more representative. It is important to match the selected dangerous events with 'qualitative leaps' in changing the state of the object, the staff and operators, the nature and magnitude of the effect of damaging factors on the population and the environment. For example, for nuclear power plants dangerous events in the first approximation include an uncontrolled nuclear reaction (nuclear explosion), the destruction or explosion of a reactor core with the release of fissile material and radioactive substances, core meltdown, depressurisation of the nuclear installation circuits with emission of radioactive substances, etc.

Thus, the risk according to [2.8]–[2.10] quantifies the risk of the whole 'nuclear facility – individual – environment' complex and, strictly speaking, is an integral indicator of the dangers of this global complex. Under certain fixed characteristics of a nuclear facility, the environment, personnel and population, the integral indicator can be regarded as one of the most important indicators of the safety of the nuclear facility and is used for comparison and selection of preferred (as regards the level of hazard) options of the scheme-design version of the object, the organisation of its operation and protection from accidents and disasters.

The algorithm for risk assessment for nuclear facilities includes the following basic procedures.

1. Analysis of the structure and composition of the nuclear facility and determination on this basis of a limited list of catastrophic outcomes (hazardous events associated with a significant negative impact from a nuclear facility on the environment). Ranking of catastrophic outcomes from nuclear facilities on the basis of the reasons (mechanisms) for occurrence and the degree of severity. Allocation of payment options for catastrophic outcomes from a nuclear facility that are significantly different in severity and mechanisms.

2. Life cycle analysis of a nuclear facility. Allocation of stages (substage, fragments) of the life cycle which differ significantly in means and methods of operation, the placement of the nuclear facility, and according to other parameters determining the probability of catastrophic outcomes, and the magnitude of the damage resulting from them.

3. Estimating the probability p_i of occurrence of each calculated disastrous outcome from nuclear facilities at each of the selected stages (substage, fragments) of the life cycle of the nuclear facility by:

– Determining the list of possible ad-hoc situations in the nuclear facilities (including damage, failures, human error, natural anomalies, malicious and terrorist acts, etc.), their probabilities and the most important characteristics (levels, combinations, the duration of the primary anomalous effects on nuclear objects from the environment, the types and combinations of failures of elements and subsystems), highlighting the most characteristic and presumably the most dangerous of ad-hoc situations.

– Analysis of ad-hoc situations, constructing a 'tree' of events, determining the levels and probability distributions of impacts on critical elements of nuclear facilities, separation and quantitative description of the limited (necessary and sufficient) list of options for the anomalous effects on the critical elements of nuclear facilities;

– Analysis of the possible reactions of nuclear facilities (the critical elements of nuclear facilities) to the anomalous effects and failures, the definition of the probability of catastrophic outcomes from nuclear facilities in case of certain types of failures and abnormal influences.

4. Determination of types and levels of damaging factors from the nuclear facility in catastrophic outcomes. Evaluation of the aggravation or reduction of the damaging effects of the nuclear facility on the natural and social environment, technosphere, the actions of staff.

5. Estimation of probabilities p_i of the death of individuals in the event of catastrophic outcomes in each of the selected fragments of the life cycle of the nuclear facility

6. Description of the types of damage U to the public in the event of a catastrophic outcome of the nuclear facility. Definition of indicators and a single measure of damage U.

7. Estimation of damages U_i of different kind to the public in the event of catastrophic outcomes from nuclear facilities at each of the selected fragments of the life cycle of the nuclear facility.

8. Hazard assessment and risk R of the nuclear facility for social projects, including various types of risk at different stages of their life cycle, in accordance with the generalised and integral indexes of the probability p and damage U.

In the statistical and probabilistic risk analysis of nuclear power plant it is fundamentally important to estimate the probability of failures. The characteristic data obtained in at the end of 20th century include the data on the total number N of disruptions of operation of nuclear power plants and the coefficients of variation of the number of disruptions from year to year.

In this case, the recorded data include the total number of failures of systems (components), the number of failures in safety systems, the number of human errors, the number of deficiencies in the organisation of operations, the number of activation of emergency protection systems, and the number of independent and dependent 'overlapping' of failures.

The detected defects develop by the mechanism of intergranular stress corrosion cracking of metal. Most of the welds with unacceptable defects are repaired. Decisions on suitability for service are taken for welded joints with unacceptable defects and special measures are proposed for other welded joints with defects of the unacceptable size.

The formation of cracks in welded joints of pipelines increases the probability p of failures with depressurisation of the preliminary circulation circuit. The decision to allow welds with unacceptable defects increases the probability of such an accident, so it is necessary to investigate and take corrective measures to prevent pipeline rupture.

Unacceptable defects are also found in technological channels (longitudinal cracks in zirconium pipes, cracks in the welds between the clips and the upper tract).

The number of human errors tends to increase.

The relevant problems of nuclear safety are as follows:

- detection of cracks in pipelines and equipment of nuclear power plant;
- insufficient leaktightness of hermetic enclosure systems;
- problems of storing radioactive waste, spent fuel;
- lack of information about the residual life of equipment and pipelines;
- low reliability of certain types of equipment;
- the problem of depressurisation of fuel rods in fuel assemblies.

The scientific program associated with the development of the theory of risk can be illustrated by the scheme shown in Fig. 2.9. Solid arrows in the diagram show the most important logical relationships. In fact, after it becomes clear what we want to achieve using the risk criteria $R\,(t)$ and in which sort of future we expect to live, we can evaluate the 'corridor' of the capacity of the current scientific and technological sphere and the boundaries of acceptable global changes. In essence, it is necessary to understand how and what resources we have to bring the technosphere and society to a state in which it could safely exist with acceptable risks [$R(t)$], not only in the next few years but many decades.

After that it is necessary to define the priority scientific issues of risk management, based on our goals and opportunities, which must be solved. In other words, the systematic analysis of the structure of our ignorance of the

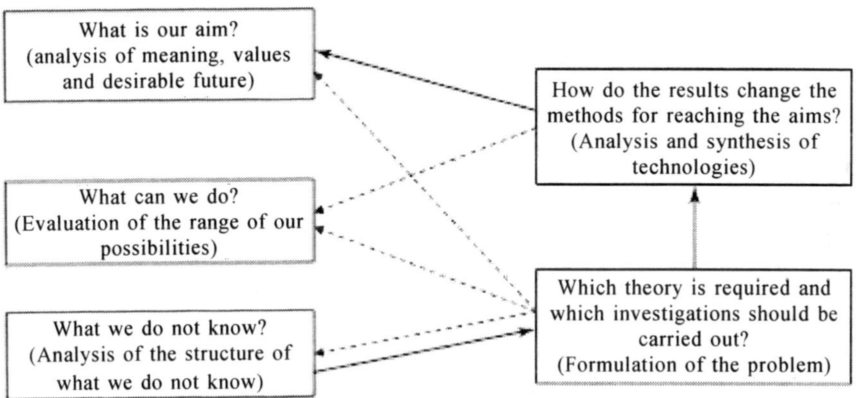

2.9 Scheme of the program associated with the construction of the theory of risk.

theory of risk and safety should be carried out. On this basis, one could rank the scientific problems in the order of their importance, set priorities and formulate a number of specific tasks to assess the necessary costs $Z(t)$ and their effective use. Their solution would figure out how the results change ways of achieving the goals. This part of the study could be called analysis and synthesis of technologies of protection against risks. On this basis we could adjust long-term goals, meanings, values, and the desired future. Also, we can return to the definition of expected parameters of the technosphere, predict future changes in the social sphere and the biosphere and redefine the corridor of our ability. The circle is now closed.

Unfortunately, this simplest and most logical way of constructing a theory is difficult to implement at the moment. First of all, while there are also problems with the definition of long-term strategic objectives, it is difficult to take into account the likelihood of dangerous states $P(t)$ and associated damage $U(t)$.

Many events that generate threats, dangers, risks $R(t)$, the probability $P(t)$ and losses $U(t)$, can be described by a large number of probability models. However, these probabilities are themselves often subject to very definite deterministic laws. Therefore, they can be assessed taking into account the prior history of the systems, the measures, a wide range of different factors, and can be manipulated. This gives a new probability of direct quantitative approaches to the prediction of emergencies, technological and sociogenic hazards, the new algorithm to improve security in contingency, emergency and catastrophic situations of many complex systems and provide a given level of human security, facilities and environment.

An example of applying the theory to address the operational risks of the problem at the Ignalinal nuclear power plant is given in sections 7.2 and 7.3.

2.5   Accounting for ageing in formal mathematical models

A detailed review of the ageing processes in models of reliability of components and systems of nuclear power plant safety was published in Ref. 31.

It is shown in Ref. 31 that most of the PSA carried out to date for existing, under construction and planned nuclear power units in the simulation of emergency consequences, unpreparedness of the safety functions (systems) and personnel actions, use the assumptions and results of the deterministic analysis based on a conservative approach. It is assumed that the strength and reliability safety margins for the components and the chosen strategy of maintenance are conservative enough to ensure reliable operation of the components during the entire period of service under any of design conditions (maximum design-basis accident is regarded as the most severe emergency mode). In this case, the majority of emergency situations taken into account in PSA are outside the range of the failure conditions considered in the project.

Analysis of the possible impact of ageing on modelling of failure consequences, the formulation of success criteria for safety functions, and erroneous actions of the personnel is given in Ref. 32.

Input parameters in the PSA model used at the frequencies of initial events averaged out over a specific period of time (e.g. 1 year) and factors of unavailability of equipment. In most cases, these average values are calculated assuming that the time to failure or probability of failure 'on demand' are random variables distributed according to the exponential and binomial laws, respectively. The parameters of reliability, which are required for the calculation of the average unavailability in the given period of time, failure rate and probability of failure 'on demand', do not depend on the age (operating time) of equipment and are constant. Similar assumptions are used for calculating the frequency of initial events.

The use of these assumptions is explained and justified by the following considerations:

• *simple models*: the use of exponential and binomial distribution laws greatly simplifies the process of calculating risk (unavailability of systems, the frequency of melting of the reactor core, etc.) and, in fact, of reliability indicators;

• *lack of representative statistical data*: given the high reliability of safety systems, the number of equipment failures is extremely small so that when developing the first PSA it is not possible to perform reliability analysis taking into account the time of operation;

• *adjustment of the negative impact of ageing on the reliability of equipment by organisational and technical measures*: from an engineering point of view, for the optimum control of ageing (the technical and operational measures and maintenance), equipment failure rate should remain constant for at least the design life of the plant.

In addition, PSA practice implies a periodic review of models, input data and results of calculations of risk, taking operating experience into account. Typically, such audit is conducted once in 10 years in the periodic reassessment of the security unit.

Nevertheless, possible errors in the management of ageing and/or extending the life of plants (i.e. beyond the limits originally set by the design) may cause deterioration of equipment due to ageing and as a consequence the above assumptions are baseless.

Thus, accounting for the effects of ageing in the PSA is primarily related to the revision of the basic assumptions made in modelling.

For nuclear power units approaching the design life and working outside of original design life, it is necessary at least to verify the assumptions made in modelling failure sequences, the formulation of criteria for the success of safety systems and constant failure rate (the probability of failure on demand) of the components over time.

If the effects of ageing of the components significantly affect the assumptions made in modelling failure sequences or safety features, this effect can be considered at the level of the structure of event trees (ET) and/or fault trees (FT) in the PSA model.

In an ageing trend is found in the reliability indicators, it is necessary to choose adequate models of reliability of equipment and to assess the impact of reliability parameters on the PSA results.

In Ref. 33, the authors identify two main approaches to understanding and modelling of the ageing phenomenon:

• ageing, as the concept of reliability associated with the behaviour of failure rate over time;

• ageing as the physical process of gradual degradation of equipment.

In the first case, reliability is modelled using applied statistical models with data from operating experience. This approach is mainly used for simulation of active elements.

In the second case, reliability modelling is based on understanding the physics and kinetics of the degradation process associated with a specific ageing mechanism. Such models correspond to the passive elements.

From the viewpoint of the sensitivity of risk assessment to the effects of ageing of the CCM, special attention should be given initially to failures of active elements of the safety systems. The active components are explicitly modelled in FT and ET, and changing the reliability indicators directly affects the results of PSA and the indicators themselves are calculated on the basis of statistical data from operating experience.

If sufficient statistical data are available, analysis of the effect of ageing of the active elements on the safety of a unit may be carried as follows:

• preliminary analysis of statistical data on reliability:

– graphical (visual) analysis of the trend;

– non-parametric methods for testing hypotheses about the presence or absence of a trend;

– parametric methods for testing hypotheses about the presence / absence of a trend;

• selection and validation of the reliability model, calculation of reliability indices;

• Prepare a set of source data for incorporation into models of PSA;

• Assess the risk with ageing taken into account.

Methods for graphical analysis of data on reliability, taking into account the specificity of the primary data from the operating experience of nuclear power plant, are presented in Ref. 34–38.

Atwood et al[35] suggests using the cumulative distribution of failures in time (cumulated failure plot) or interval estimates of failure rates over time.

Klyugel[36] and Prokachchia[37] give examples of using the Nelson–Aalan and Kaplan–Meier graphical analysis of indicators for the ageing trend.

Two factors that are important from the viewpoint of practical application of one or another approach are noted in Ref. 38:

• format of reliability data: operating time to failure or interval values of intensities of the flow of failures;

• an assumption about the type of reconstruction of equipment after failure and preventive maintenance: full recovery, partial recovery or no recovery.

The following non-parametric methods for checking the presence of a trend have been proposed and studied in Ref. 35:

• Laplace criterion (operating time to failure);
• test of two cells (operating time to failure);
• inversion method (the intensity of the flow of failures).

To test assumptions about the type of recovery equipment, the authors of Ref. 38 proposed to use the Lewis–Robinson test.

Graphical analysis and non-parametric methods for testing the trend are easy to implement and do not require special software, and the accuracy of the results is inferior to more powerful parametric methods of analysis. When the original data are accurate, these methods allow us to quickly evaluate the primary data and give information about important attributes such as the homogeneity of the sample, failure in running-in, the level of uncertainty and importance of estimates. These methods are well suited for preliminary data analysis and selection of components for further analysis.

Requirements for reliability models are formulated depending on the structure and volume of PSA and the available (or accessible) data on reliability[39].

The overall logic of the process shown in Fig. 2.10.

In Refs. 34, 35, 25, 40–42 the authors proposed and investigated cases of generalised linear models to account for the reliability of ageing safety equipment in PSA:

• linear model for the failure rate:

$$\lambda\,(t) = a + bt, \qquad\qquad [2.11]$$

• Weibull model:

$$\lambda\,(t) = at^{b}, \qquad\qquad [2.12]$$

• log-linear model:

$$\lambda\,(t) = a \exp\,(bt), \qquad\qquad [2.13]$$

Here $\lambda\,(t)$ is the failure rate, t is time (age) of equipment, a, b are parameters of the model, where b is a parameter characterising the presence or absence of ageing. If $b > 0$, the failure rate increases with time, $b = 0$ – corresponds to a constant failure rate and if $b < 0$ the failure rate decreases with time (running-in failure).

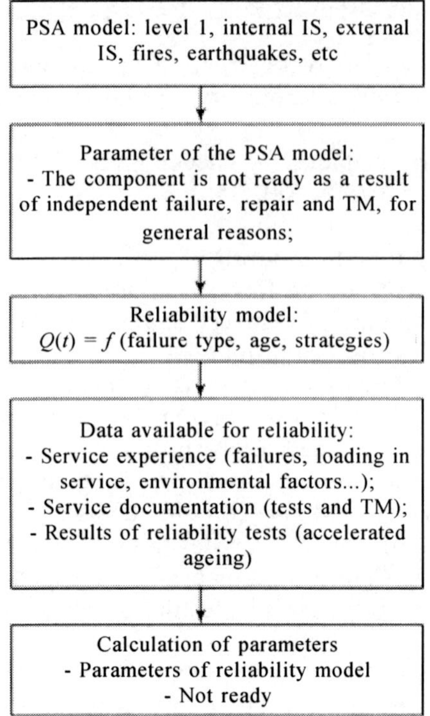

2.10 Selecting a model of reliability for use in PSA.

It should be noted that the linear model and the Weibull distribution are special cases of generalised linear models and can be used both for modelling failure in operation or standby mode (continuous distribution) and for failure on demand (discrete distribution).

Unavailability of the element at a specified time interval (an input parameter PSA models) can be expressed in the following form:

$$q(t) = q_0 + 1 - \exp\left[-\int \lambda(t')dt'\right], \qquad [2.14]$$

Here q_0 is the value of unavailability of equipment at time $t = 0$.

Wesley[25] proposes to calculate the unavailability of recoverable items following analytical relations obtained for the linear model:

$$q(t) = 1 - \exp\frac{1}{2a(t - t_N)^2} - \qquad [2.15]$$

for full recovery (as good as new) as a result of repair or technical maintenance (TM)

$$q(t) = 1 - \exp\frac{1}{2a(t^2 - t_N^2)} - \qquad [2.16]$$

for lack of recovery (as bad as old) as a result of repair or TM. Here t_N is the time the last repair or TM.

NUREG/CR-6508[40] gives more detailed formulae for calculating the unavailability of components in the standby mode. They take into account the frequency of testing and the frequency and type of preventive maintenance.

So, for the equipment in standby mode, the periodic tests and the type of recovery 'as good as new', the average unavailability over the interval between preventive maintenance/inspection will be:

$$q = 1 - \exp\left[-\left(\lambda_0\frac{T}{2} + \frac{a}{6}T^2 + a\frac{T}{2}\frac{L-T}{2}\right)\right] + \frac{d}{L+d}, \qquad [2.17]$$

λ_0 is the failure rate at the initial time ($\lambda_0 = a$ in [2.11]); L is the interval between preventive maintenance/inspections; T is the interval between periodic tests; d is the average time of unavailability during the preventive maintenance/inspection.

Accordingly, the average unavailability for equipment in standby mode without periodic testing and the type of recovery 'as good as new' is expressed by:

$$q = 1 - \exp\left(-\lambda_0\frac{T}{2} - \frac{a}{6}L^2\right) + \frac{d}{L+d}. \qquad [2.18]$$

For the elements of the safety system with the restoration 'as bad as old' (preventive maintenance/inspection control only the readiness of equipment and, from this point of view, are similar to periodic tests), the average unavailability over the life S is calculated as:

$$q = 1 - \exp\left[-\left(-\lambda_0\frac{T}{2} - \frac{a}{6}L^2 + a\frac{T}{2}\frac{S}{2}\right)\right] + \frac{d}{L+d}. \qquad [2.19]$$

In the absence of periodic testing, unavailability is defined as:

$$q = 1 - \exp\left[-\left(-\lambda_0\frac{T}{2} - \frac{a}{6}L^2 + a\frac{L}{2}\frac{S}{2}\right)\right] + \frac{d}{L+d}. \qquad [2.20]$$

The equations [2.17]–[2.20] are the first-order approximation for the exponential dependence (Taylor series). They give fairly accurate results for values less than 0.1 and a conservative estimate for higher values.

A more complicated case of partial recovery of equipment is also considered.

In addition, NUREG/CR-6508[40] includes formulas for calculating the unavailability in similar cases using the Weibull model. For example, for equipment in standby mode, the periodic tests and the type of recovery 'as good as new', the average unavailability over the interval between preventive maintenance/inspection will be:

$$q = 1 - \exp\left(-\lambda_0 \frac{T}{2} + \frac{a}{(b+1)(b+2)(b+3)} \frac{(L+T)^{b+3}}{LT} - \frac{a}{(b+1)(b+2)(b+3)} \right.$$
$$\left. \frac{a}{(b+1)(b+2)(b+3)} \frac{L^{b+2}}{T} - \frac{a}{(b+1)(b+2)} L^{b+1} \right) + \frac{d}{L+d}. \qquad [2.21]$$

It is clear that the use of such cumbersome dependence makes the analysis of initial data and calculations of PSA very difficult.

In some studies[42–44] the authors considered cases of expansion of the linear model or the Weibull model to simulate the mixed flow competing failures (risks concurrent model), i.e. random failures and failures caused by ageing.

So, for the linear model NUREG 5052[43] suggests using the following relationship:

$$\lambda(t) = \lambda_0 \text{ at } 0 < t < t_0$$

and

$$\lambda(t) = \lambda_0 [1 + \beta(t \, 2 \, t_0)] \text{ at } t > t_0 \qquad [2.22]$$

In this interpretation of the linear model λ_0 is the intensity of random failures and is constant in time, t_0 is the time at which the ageing process begin to affect the reliability; β is a coefficient

By analogy with the linear model, the Weibull model can be represented as follows:

$$\lambda(t) = \lambda_0 \text{ at } 0 < t < t_0$$

and

$$\lambda(t) = \lambda_0 (t/t_0)^\beta \text{ at } t > t_0. \qquad [2.23]$$

The unavailability values are calculated with the following assumptions:
• periodic testing does not affect the age of equipment ('bad as old');
• complete restoration after failure ('good as new');
• failure can be detected with probability $P = 1$ during periodic testing and with probability $P = 0$ at any other time;
• the duration of testing and restoration is negligible compared with the test interval T.

Unavailability of components is calculated at time t with $nT < t < (n+1)T$, $n = 0, 1,..., N$, $t/T < N$.

$$q(kT, t) = 1 - \exp[\lambda_0(t - nT)], \text{ if } (n-k)T < t;$$

$$q(kT, t) = 1 - \exp\left\{-\lambda_0(\tau - nT) - \frac{\lambda_0}{(b+1)\tau^b}\left[(t - kT)^{b+1} - \tau^b\right]\right\},$$

if $(n-k)T \geq \tau, t - kT < \tau;$

$$q(kT, t) = 1 - \exp\left\{-\frac{\lambda_0}{(b+1)\tau^b}\left[(t-kT)^{b+1} - (n-k)^{b+1b}T^{b+1}\right]\right\},$$

Here λ_0 is the intensity of random failures, T is the interval of periodic testing, τ is the beginning of ageing, B is the shape parameter for the Weibull distribution.

The uniform wear model, proposed in Ref. 38, belongs in the same class of models.

Random failures (not time-dependent) are characterised by a constant flow rate of failures λ_0, and it is assumed that in case of failure the element returns to the usable state without improving the reliability characteristics ('as bad as old'). Failures associated with ageing occur when the element reaches a limited number of states N, for example, start/stop, start, etc. It is assumed that each loading results in partial and uniform degradation of the element and when the N states are reached the element is completely degraded and changes to a failure state.

The unavailability of such an element can be expressed as

$$q(t) = 1 - \exp(-\lambda_0 t)\left[1 - H\chi^2(2w, 2N)\right], \tag{2.24}$$

where w is the frequency of loading of the element; N is the maximum number of loads, $H\chi^2$ is the distribution function χ^2 with $2N$ degrees of freedom.

The main factor limiting the use of all proposed models is the availability of representative and inexpensive initial data.

Comparative analysis of models in terms of information necessary for calculating the parameters of the model was carried out in Ref. 45. Some of the results of the comparison are presented in Table 2.2. In addition, certain difficulties are experienced in selecting algorithms for calculating the parameters depending on the structure and completeness of primary reliability data.

For example, in Ref. 44 and 46 it is reported that in most cases the data on the reliability of safety equipment in nuclear power plants are heavily censored. The examples considered in Ref. 35 show that the censorship on the right is a typical case where data are available for the entire period of operation of the unit. If the data were collected for a certain limited period of operation, we can speak of interval censoring, i.e. right and left. In practice, many PSA use the data collected for a limited period of operation.

An obligatory step in choosing a model is the validation stage, validation of the model. The use of parametric chi-square tests, Fisher, Kolmogorov-Smirnov tests, etc., is proposed in Refs. 34, 35, 37, 38. Since there are no requirements for testing the hypothesis for the constancy of the intensity of failures in the procedures to implement PSA and because of the specificity of data on the reliability of equipment in nuclear power plant the statistical

Table 2.2 Results of comparison

Model	Parameters	Data on failures (age, criticality)	Data on the nature of failure	Data on loading	Data on repairs and TM
Linear	τ, β	+	–	–	+
Linear for competing failures	τ, β, T_0	+	+	–	+
Weibull	β, η	+	–	–	+
Weibull for competing failures	β, η, η_0, T_0	+	+	–	+
Uniform wear	λ_0, ω, N	+	+	+	+

tests must be appropriately adapted together with appropriate preparation of the input data.

Given the above difficulties, Atwood *et al.*[34] recommend to choose the simplest two-parameter model of reliability and accurate approach to calculation of parameters, uncertainty analysis and extrapolation of the results.

Depending on the application of the PSA results, the initial data on the values of unavailability of elements can be represented as point or averaged over a certain range of time estimates. Average values of unavailability for a period of one year are used in most cases. It should be noted that for the elements in operating or standby mode unavailability is a function of time, even at a constant failure rate. For example, in RiskSpectrum[47] the unavailability for the elements in the standby mode is calculated as:

$$q\,(t) = 1 - \exp\,(-\lambda T) \approx \lambda T/2, \qquad [2.25]$$

where λ is the failure rate in the standby mode, T is the interval of periodic testing.

The unavailability function of an element failure rate governed by Weibull law [2.23] with different assessments is analysed in Ref. 42:

• the point estimate of the before test unavailability (BTU)) – the maximum value of unavailability in the interval $(n - 1)\,T < t < nT$;

• maximum before test unavailability (MBTU) – the maximum BTU per lifetime;

• point estimate of the forty-year value (FYV)) – BTU value in 40 years;

• average interval unavailability (AIU)) – the average estimate in the range;

• maximum average interval unavailability (MAIU)) – the maximum value of AIU for the entire term of service;

• year average unavailability (YAU)) – average estimate over a period of one year;

• maximum year average unavailability (MYAU)) – the maximum value YAU over the life of the unit.

These unavailability estimates are used for sensitivity analysis of models by varying the parameters b, T and t.

As mentioned in the previous chapter, the availability of qualitative and representative data on equipment reliability is a prerequisite for modelling the effects of ageing in PSA.

In general, there are three types of data sources:
• data from operating experience of nuclear power plants;
• data from operating experience of similar equipment in other industries,
• data from accelerated reliability tests.

Data from the experience of operating nuclear power plants give an idea about the history and current status of equipment reliability and, under certain assumptions, allow current trends to be approximated. But this is only an approximation based on current information.

Experience with operation of similar equipment in other industries can give an idea about the reliability of equipment over a longer period of operation, provided the operating time of equipment is longer than that in nuclear power plant. However, the direct use of data on the reliability of other industries always requires a careful analysis of the representativeness of data: comparing the characteristics of equipment, conditions and modes of operation, maintenance strategies, failure criteria, etc.

The data obtained in accelerated reliability tests can be used to judge the reliability of equipment for the period simulated in the tests. Nevertheless, the representativeness of the test results strongly depends on the reliability of theoretical models of accelerated ageing and completeness of the tests as regards the ageing mechanisms.

In this section, we consider, as a source of baseline data, only the data on the reliability of operating experience.

2.5.1 Data from operating experience of similar equipment in other industries

First of all, it should be noted that the vast majority of literature data on equipment reliability did not include the effects of ageing, i.e. reliability indices are calculated on the assumption that the failure rate or probability of failure on demand is constant (Poisson flow of failures). Therefore, to analyse reliability with ageing taken into account, it is necessary to consider treatment to the primary data to which access is usually limited . Furthermore, in reality there are not many facilities that have been in operation for a long period (more than 30 years). In this case, it is necessary to collect and analyse reliability data for these facilities.

As an example, data on the reliability of equipment of thermal power plants in North America and Western Europe can be analysed[48].

The integrated reliability indices of operating units and specific indicators of reliability for some types of equipment are analysed in Ref. 48. Analysis was performed for three groups of units:

- put into operation from 1960 to 1965;
- put into operation from 1965 to 1970;
- put into operation from 1970 to 1975.

This grouping of equipment allows to perform the analysis of developments in the operating time range from 10 to 40 years. Given the fact that the age of most nuclear power units is 10 to 30 years (for example, in France in 2008, just one block reaches the operating life of 30 years), trend analysis of reliability of similar equipment in the period from 30 to 40 years can be very useful to develop measures to manage the ageing nuclear power plant.

The main conclusion concerning the integral indices of reliability is that unplanned unavailability of blocks decreases or remains constant over time. In other words, further experience with the reliability of blocks is gained, i.e. there is an effective feedback of operating experience.

With regard to the specific reliability parameters, the increasing trend was identified for some types of equipment such as pumps, fans, valves, switches, heat exchangers, etc. The parameters of 'ageing' for the linear model are comparable with TIRGALEX data[49].

Another practical conclusion concerns application of the strategy of 'minimum maintenance' which can lead to an increase in failure rate.

In addition, we show that even with a constant or decreasing failure rate of equipment, the total unavailability may increase with time due to increased frequency and duration of repairs related to non-critical equipment failures, as shown in Fig. 2.11.

2.5.2 Data from operating experience of NPP equipment

The main sources of data on the reliability of nuclear power plant equipment are shown schematically in Fig. 2.12. In general, we can distinguish between generalised (at the national level, according to the type of reactor, of equipment, etc.) and specific data.

If we talk about specific data, in terms of analysis of ageing in the PSA, the most qualitative and representative data are the data on the frequency of initiating events, and reliability of components for use in PSA. These data can be the basis for further analysis. However, these data are insufficient for trend analysis of ageing and for developing time-dependent reliability models.

The required additional information can be obtained from the primary operation and maintenance of data and documents and/or summarised

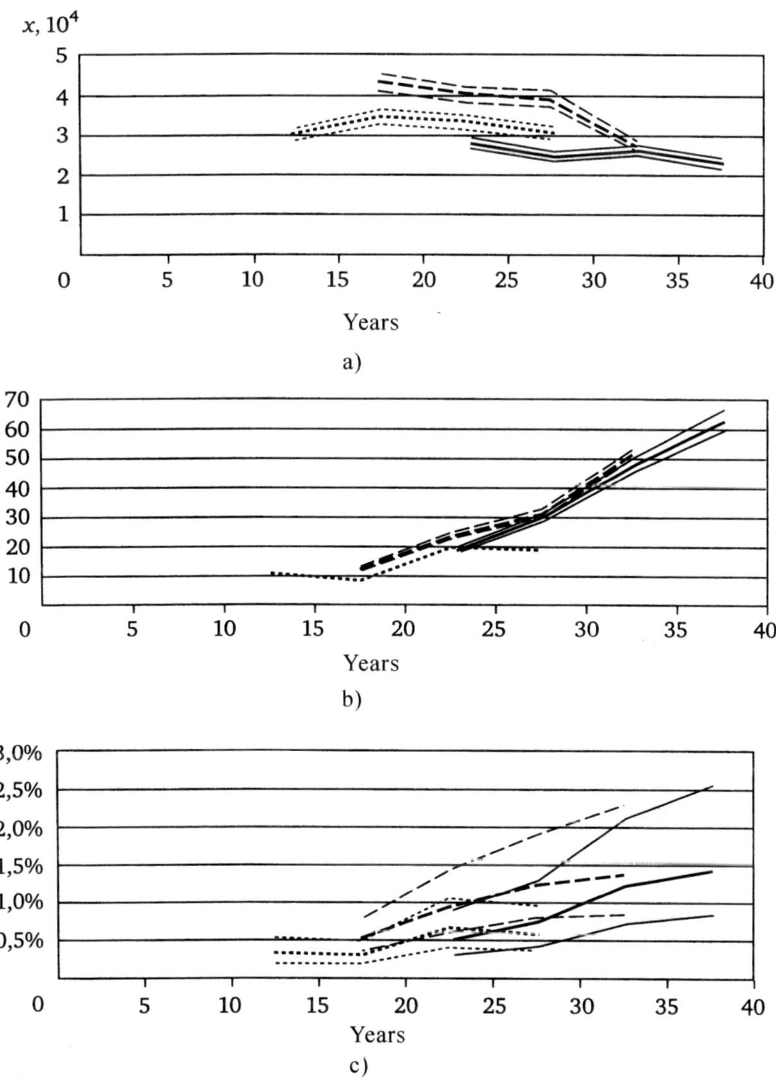

2.11 Dependence on the time failure in service (a), average recovery time (b) and total unavailability (c) of the systems for monitoring and control of the boiler: ———— 60; – – – – 65; --------- 70 (mean values and ranges)

data. In Ref. 25 it is proposed to determine the parameters of the linear model [2.11] using the following information from the data on equipment failures:

• fraction of failures associated with ageing in the general flow of failures;

• the rate of failures not associated with ageing.

This approach makes it easy to obtain the desired parameters from the

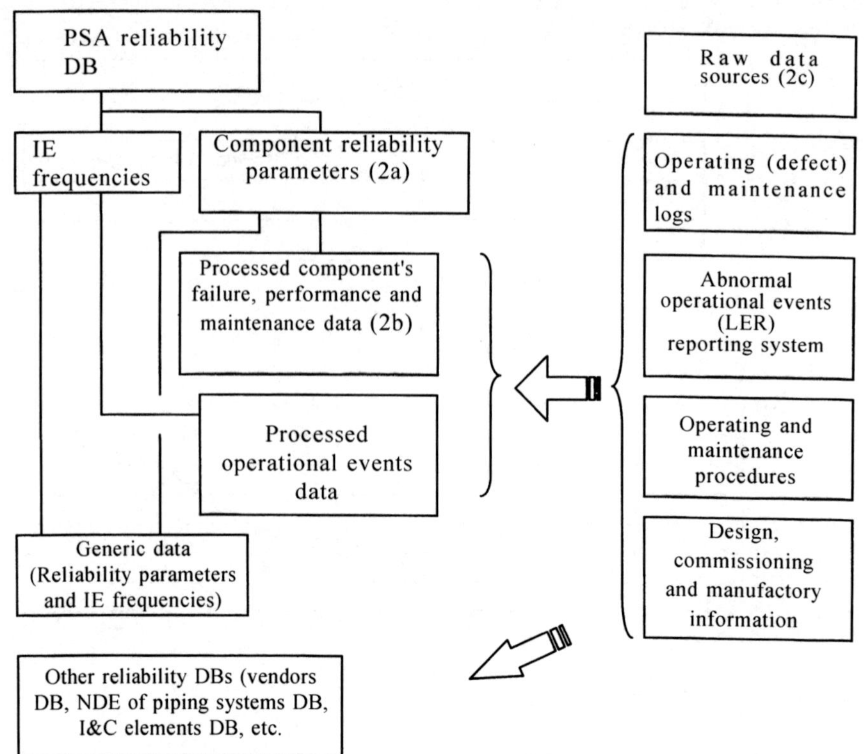

2.12   Data for the reliability of nuclear power plant equipment.

primary data sources considered in Refs. 25, 40, 41, 43:
  • licensing event reports (LER),
  • reliability data system for nuclear power plant INPO (NPRDS),
  • intrastation data system (IPDS).
Table 2.3 shows examples of the relative increase of the failure rate per year for various types of equipment used in systems for the removal of residual heat of the reactor (RHR) and the cooling water system for the reactor compartment (CCW).

Figure 2.13 shows an example of data approximation by a linear model with the ageing starting point.

It should be noted that the analysis is quite limited in time, uses samples from different stations and does not take into account the possible impact of maintenance strategy on the failure rate.

Examples of more detailed analysis of operational data on reliability can be found in Ref. 35, 36, 50. In Ref. 36, the data used for diesel generators and other safety equipment obtained in PSA for nuclear power units of the Gösgen nuclear power plant are discussed as an example. A set of virtual data whose structure is similar to reliability data reliability for PSA of French nuclear power plants was used in Ref. 35. Reliability data, used for

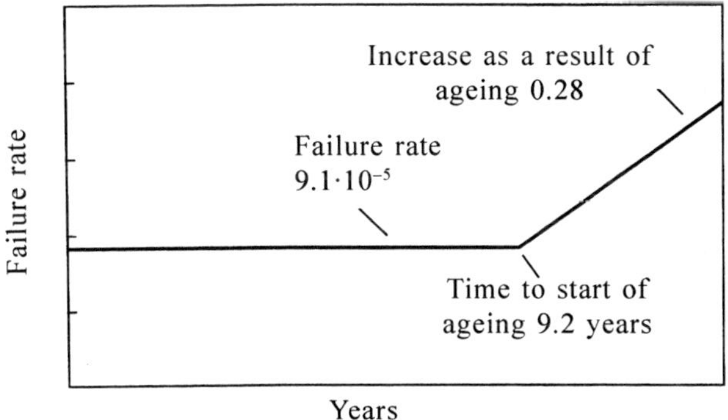

2.13 Failure rate for a pump of the CCW system – linear model with a starting point for ageing of 9.2 years.

Table 2.3 Relative increase of the rate of failure per annum, %

Equipment	RHR system	CCW system
Pumps	30.7%	17%
Electric drive fittings	0.2%	11%
Check valves	1.7%	–
Manual valves	21.9%	–
Heat exchangers	1.6%	8%
Pressure sensors	–	3%

PSA of German plants are analysed in Ref. 50.

Processed information about failures and operating time of equipment is used in all these cases in one of the formats needed to calculate the reliability indices:

• operating time to failure and/or censoring of observations,
• interval values of intensities of the flow of failures.

Analysis of the data shows that for a large group of equipment included in the PSA model analysis of failure rate trends does not make sense due to lack of data on equipment failures.

For groups of components where available statistics allows for quantitative analysis of reliability there is no increasing trend of the failure rate parameter. In cases where the increasing trend is identified, its nature is close to a linear function with relatively low values of the 'ageing parameter'.

The applied reliability models do not require any additional information about the relationship between failure and ageing. On the other hand, there are difficulties in obtaining other necessary information, such as:

• information on the commissioning date ($T = 0$);
• Information on starting and ending time of observation;

• the date of replacement and reconditioning of equipment;
• strategy for periodic testing and maintenance.

Availability and reliability of the additional data is a major obstacle for the practical application of reliability models of ageing.

Generalised data on the reliability of nuclear plant equipment, taking into account the effects of ageing, are extremely limited. The data set TRIGALEX, proposed in Ref. 49 for the linear model of the failure rate is referred to most frequently. Another example of the generalised data can be Weibull model parameters given in Ref. 33 for various types of safety equipment.

Direct application of the parameters of the generalised data in a specific PSA model is impossible because of the lack of important characteristics affecting the reliability of ageing equipment such as the types and intensity of the load, the strategy of testing and maintenance, etc. The generalised data can be used for sensitivity analysis or Bayesian estimates[51,52].

Thus, the methods of ageing data analysis discussed in this chapter: graphical analysis and non-parametric methods for testing the trend – are easy to implement and do not require special software.

If the original data are accurate, these methods make it possible to obtain the primary data quite quickly and provide information about important attributes such as the homogeneity of the sample, the presence of running-in failure, the level of uncertainty and importance of the estimates.

However, the main factor limiting the use of all these proposed models is the lack of representative and accessible baseline data. It is recommended to choose the simplest two-parameter reliability models and perform detailed analysis when calculating the parameters of the model and extrapolating the results. This approach is applicable mainly to the active elements of safety of the nuclear power plant.

3

Physico–statistical approach:
Procedures using the defect-free model of
structural material

During operation, damage cumulates in the metal of structural components of nuclear power plants, i.e. ageing takes place.

The basic mechanisms of ageing for various components of nuclear power plants with PWR and BWR were given earlier in Tables 1.1 and 1.2.

It should be noted that for the limiting state of the damaged structural element for any ageing mechanism specified in Tables 1 and 2 is determined by the strength criteria.

For example, when erosive–corrosive wear of steam pipes takes place failure occurs when the stresses in the wall of the pipe reach the value equal to the fracture stress.

In radiation embrittlement, the limiting state (for example, of the reactor casing) is determined by the brittle fracture resistance criteria.

In pitting corrosion, stresses concentrate in pits and cause the nucleation and propagation of corrosion fatigue cracks or stress corrosion cracking. Kinetics of development of such cracks is described in the section of the science of strength – fracture mechanics.

Thus, one could argue that to determine the probability of failure of components and of equipment and pipelines of nuclear power plant it is imperative to use the strength models of failure. The physical mechanisms of radiation embrittlement, corrosion, corrosion–erosion wear can be used, if necessary, as auxiliary mechanisms.

The first probabilistic model of fracture developed in the context of the science of strength was the model proposed by A.R. Rzhanitsyn[53,54,etc].

3.1 Probability of failure under random static loading. The method proposed by Rzhanitsyn

The modern science of strength is based on the concept of the acceptable stress–strain state.In accordance with this concept, the strength condition is written as:

$$\left(\begin{array}{c}\text{Stress}-\text{strain state}\\ \text{of structural element}\end{array}\right)\leq\left(\begin{array}{c}\text{Permitted service}\\ \text{characterisics of the}\\ \text{stress}-\text{strain state}\end{array}\right)$$

The simplest example of such a condition could be the formula of the strength of a tensile-loaded bar

$$R=\sigma_T-\frac{N}{F}>0, \qquad\qquad [3.1]$$

where N is tension; F is the cross-sectional area of the bar; σ_T is the tensile strength of the bar material.

In general, the condition [3.2] can be written as:

$$R(x_1,x_2,...x_n)>0, \qquad\qquad [3.2]$$

where x_1, x_2 ,..., x_n are some calculated quantities.

Each of the calculated quantities x_1, x_2 ,..., x_n can deviate from its mean (expected) values, and these deviations can be characterised as a function of their distribution p_z $(x_1$, x_2 ,..., $x_n)$ obtained statistically or on the basis of theoretical considerations. From this function, one can go to the distribution curve R using the formula[52]

$$p_R=\int\limits_{-\infty}^{\infty}\underbrace{\int...\int}_{[(n-1)\text{times}]}\frac{\partial R}{\partial x_1}p_z\left(x_1,x_2,...,x_n\right)dx_2dx_3,...,dx_n. \qquad [3.3]$$

If the law of distribution of p_g is normal and the function R $(x_1$, x_2,...,$x_n)$ is linear or if the function R $(x_1$, x_2,...,$x_n)$ is non-linear but the variance of the distribution of p_R is so small that within the square root of variances, multiplied by a small number (two or three) the function R with sufficient accuracy can be replaced by a linear function, the curve p_R will be expressed by the normal distribution

$$p_R=\frac{1}{\sqrt{D_R}}\Phi'\frac{R-m_R}{\sqrt{D_R}}=\frac{1}{\sqrt{2\pi D_R}}e^{-\frac{(R-m_R)^2}{2D}}. \qquad [3.4]$$

where m_R and D_R are determined by replacing u by R [52].

It remains to determine the probability of default of inequality [3.2] or, equivalently, the probability of fulfilling the fracture condition

$$R(x_1,x_2...,x_n)<0, \qquad\qquad [3.5]$$

Knowing the curve p_R, this can be done very easily. It is enough to integrate it from minus infinity to zero, i.e., determine the ordinate of the integrated distribution curve R for the values $R = 0$:

$$V = \int_{-\infty}^{0} p_R dR = P_R(0),$$ [3.6]

Here V is the probability of fracture.

For the normal law of distribution of R formula [3.6] has the form according to Ref. 24

$$V = \frac{1}{2} + \Phi\left(-\frac{m_R}{\sqrt{D_R}}\right) = \frac{1}{2} - \Phi\frac{m_R}{\sqrt{D_R}} = \frac{1}{2} - \int_{0}^{\frac{m_R}{\sqrt{D_R}}} e^{-\frac{u^2}{2}} du.$$ [3.7]

The value $m_R / \sqrt{D_R}$ is denoted by γ and termed the safety characteristic

$$\gamma = m_R / \sqrt{D_R} = 1 / A_R,$$ [3.8]

A_R here denotes the coefficient of variation of the indestructibility function R.

The probability of failure is then expressed by the formula

$$V = 1/2 - \Phi(\gamma),$$ [3.9]

i.e., it is functionally linked to only one value, namely safety characteristic γ. The safety can be defined as the ratio of the standard of the strength function to its expected value. In comparison with quantity V, it has the advantage that it can be expressed in ordinary cases by a simple numeric value of the order of 1–5 and not by a very small fraction like V. As an example, some values of V as a function of γ are given in Table 3.1.

In many cases, the indestructibility function can be expressed by the following simple formula:

$$R = r - q,$$ [3.10]

Here r is the strength of the installation, measured in some units of the scale x, for example, in kg/cm² of the strength of the material of the structure; q is the load on the installation, measured in units of the same scale x, for example, in kg/cm², the stress in a dangerous cross section caused by external forces.

If the distribution functions r and q are known and can be represented with sufficient accuracy by the distribution normal law, the distribution curve of R is also normal with the centre

Table 3.1 Values of V as a function of γ

V	0.1	0.01	0.001	0.0001	$3.2 \cdot 10^{-5}$	$3 \cdot 10^{-6}$	$2.9 \cdot 10^{-7}$
γ	1.28	2.32	3.15	3.77	4.00	4.50	5.00

$$m_R = m_r - m_q$$

[3.11]

and dispersion

$$D_R = D_r - 2D_{rq} + D_q.$$

[3.12]

Usually, the strength r and the load q can be considered independent, random variables. Then $D_{rq} = 0$, and formula [3.12] is simplified:

$$D_R = D_r + D_q.$$

[3.13]

The safety characteristic in this case is:

$$\gamma = \frac{m_r - m_q}{\sqrt{D_r - 2D_{rq} + D_q}}.$$

[3.14]

Dispersion D_{rq} is not equal to zero only in the presence of a correlation between strength and load. This relationship is generally absent or has a negligible value. In this case, the safety characteristic can be determined by a simple formula

$$\gamma = \frac{m_r - m_q}{\sqrt{D_r + D_q}}.$$

[3.15]

In the general case of the arbitrary distribution functions r and q, the probability of failure V can be determined by the formula

$$p_R = \int_{-\infty}^{\infty} p(R+q, q)\,dq;$$

[3.16]

$$V = P_R(0) = \int_{-\infty}^{\infty} p_R\,dR = \int_{-\infty}^{\infty}\int_{-\infty}^{\infty} p(R+q, q)\,dq\,dR = \int_{-\infty}^{\infty}\int_{-\infty}^{r=q} p(r, q)\,dr\,dq.$$

Thus, the distribution function $p(r, q)$ should be integrated in the plane r, q over the area lying below the line $r = q$ (Fig. 3.1).

In practice, it is easy to do this numerically, as values of $p(r, q)$ in all directions decrease rather quickly.

An example of calculation of steel structures

In steel structures, fracture stresses are represented by yield strength σ_T. The permissible stress in structures $[\sigma]$ is defined by the safety factor which in various regulatory documents may have different meanings, such as for steel St3 it is 1400 kgf/cm² and then increases to 1600 kgf/cm². It was assumed that in the former case, the safety factor with respect to the minimum yield strength was 2200 kgf/cm²: 2200/1400 = 1.57 and 2200/1600 = 1.37, respectively.

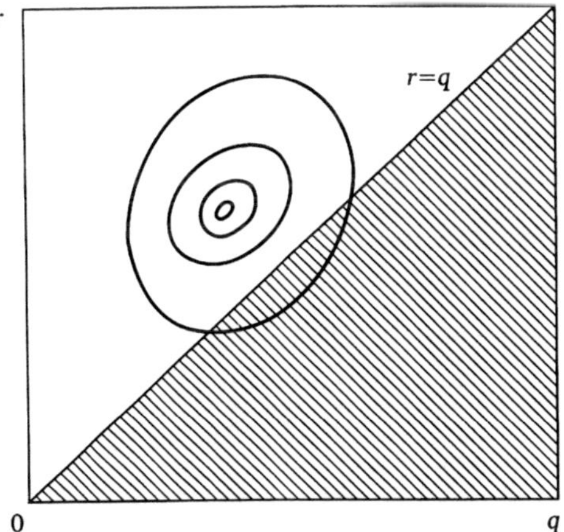

3.1 Determining the probability of failure at an arbitrary distribution function of strength and load.

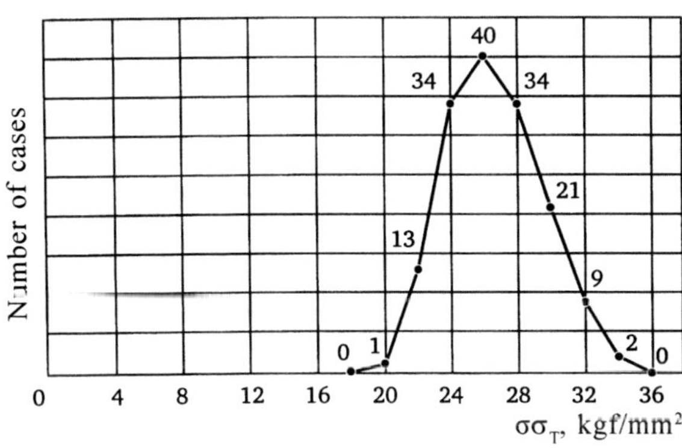

3.2 Distribution curve of yield strength of steel grade St. 3.

In the above approach, the safety factor should be calculated on the basis of the mean expected yield strength and not the minimum yield strength. The mean expected yield strength is determined on the basis of the actual distribution curve of yield strength for grade St. 3 steel (Fig. 3.2). Processing this curve, the following is obtained for yield strength:

$$m_{\sigma_T} = 2663 \text{ kgf/cm}^2, \quad \sqrt{D_{\sigma_T}} = 284 \text{ kgf/cm}^2.$$

The coefficient of variation of yield strength is thus

$$A_r = A_{\sigma_T} - 284/2663 = 0.106$$

Safety factors in relation to the expected yield strength will be: $2663/1400 = 1.90$ and $2663/1600 = 1.67$, respectively.

In this case, the permissible stress is taken as the expected stress in a dangerous cross section at a given load.

We take the coefficient of variation of the load $A_q = 0.100$, ie., approximately the same as for the material strength σ_T. Then the safety characteristics and probability of failure for both cases can be defined by the formulas:

$$\text{at}\,[\sigma]=1400\ \text{kgf/cm}^2 \quad \gamma=\frac{1.90-1.0}{\sqrt{0.106^2\cdot1.90^2+0.1^2}}=4;\ V=0.000032;$$

$$\text{at}\,[\sigma]=1400\ \text{kgf/cm}^2 \quad \gamma=\frac{1.67-1.0}{\sqrt{0.106^2\cdot1.67^2+0.1^2}}=3.3;\ V=0.00045.$$

This implies that an increase in permissible stress to 200 kgf/cm² increases in this case the probability of failure by approximately 14 times. However, this probability still remains sufficiently small.

3.2 Probability of failure under cyclic loading causing fatigue of constructional materials

Under cyclic loading the basic process of aging is the cumulation of fatigue damage. Fatigue of structural steels is determined by the characteristics of fatigue which can be represented in the form of the fatigue curve (Fig. 3.3).

If the conditions of cyclic loading are stationary (i.e., constant loading amplitude), then the probability of failure can be determined using the approach described above in section 3.1.

Let the endurance limit of a component be a random quantity distributed normally with parameters $\overline{\sigma}_{-\lg}$, $s_{\sigma-\lg}$ ($\overline{\sigma}_{-\lg}$ is the endurance limit of components, $s_{\sigma-\lg}$ is the standard deviation, the ratio $\upsilon_{\sigma-\lg}=s_{\sigma-\lg}/\overline{\sigma}_{-\lg}$ is the variation coefficient).

The distribution function of stress amplitudes is characterised by the values: $\overline{\sigma}_a$ – the average amplitude of the stress cycle; $s_{\sigma a}$ – standard

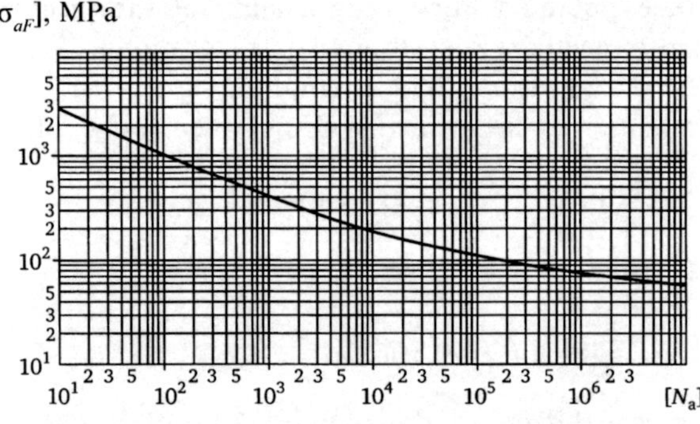

3.3   Fatigue curve of type 0Cr18Ni10Ti steel.

deviation σ_a; $\upsilon_{\sigma a}$ – the coefficient of variation of stress amplitude.

Using arguments similar to the previous case leads to:

$$u_p = \frac{M}{S_M} = -\frac{1-n}{\sqrt{\upsilon_{\sigma_2}^2 + n\upsilon_{s_{-\lg}}^2}}, \qquad [3.17]$$

where $n = \overline{\sigma}_a / \overline{\sigma}_{-\lg}$ is the ratio of the average stress amplitude in the component to the average value of the endurance limit, $\upsilon_{\sigma-\lg}$ is the coefficient of variation of the endurance limit.

Using the known quantile u_p, the failure probability P is calculated from normal distribution tables.

For other distribution laws failure probability can be determined from the relation:

$$P = P\left[(z-y)<0\right] = \int_0^{\infty} f_y(y) \int_0^y f(z)dzdy, \qquad [3.18]$$

where y and z can be substituted by the variables from [3.17] for cyclic loading.

In non-stationary random cyclic loading the problem is more complicated. A solution to this problem based on the linear hypothesis of damage cummulation is described in Ref. 55.

The task of taking fatigue damage into account is solved in the framework of the Automated control system of residual life (SAKOR)[5], installed at a number of nuclear power plants with VVER-1000 reactors.

The advantage of the above-described methods for determining the relative probability of failure is their simplicity and clear physical meaning of the models. Their drawback is that they neglect the real state of the structure, particularly its defectiveness.

4

Physico–statistical approach taking defects into account and using binomial distribution

4.1 Key elements of the behaviour of structures with crack-type defects

4.1.1 Critical and allowable defect sizes

The critical size of the defect is the size at which instantaneous failure of the structure takes place.

The permissible size of a discontinuity during operation should not exceed the value that is equal to the critical size reduced by the appropriate safety factors.

For example, a pipeline is usually characterised by the viscous state of the metal. The defect can be schematised as shown in Fig. 4.1.

For cylindrical shells with discontinuities oriented in the circumferential direction the critical defect sizes a_c and c_c can be determined by the formula[5]:

$$\sigma_b = \frac{2R\left\{2\sin\left[0.5\left(\pi - n_a y n_\varphi x - \pi\frac{\sigma_m}{R}\right)\right]\right\} - n_a y \sin(n_\varphi x)}{\pi}, \qquad [4.1]$$

where $y = a / s$ is relative depth of the crack of the critical size; a is the crack depth in mm; s is the wall thickness of the cylinder in mm; $x=C/R_c$ is the half length of the crack of the critical size; C is the half crack length in mm; R_c is the radius of the cylinder in mm; R is the fracture criterion in the viscous region; $R_{p0.2}$ is the yield strength of the material of the cylinder in MPa; σ_b is bending stress; n_a is the safety factor in the depth of the crack;

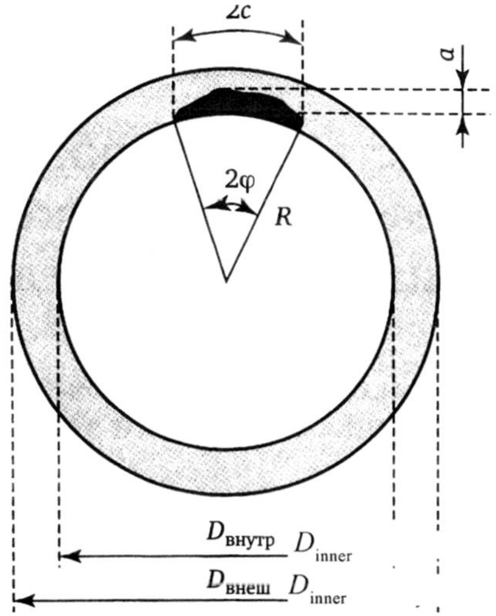

4.1   Schematization of a discontinuity.

$n_a = 1$; n_φ is the safety factor along the crack length, $n_\varphi = 1$.

4.1.2 Growth of the discontinuity under cyclic loading

The development (growth) of a discontinuity with the original sizes a_0 and c_0 under the influence of N cycles can be determined using the equation:

$$\frac{da}{dN} = C\left(\frac{dK_1}{\sqrt{1-R}}\right)^m ,$$

[4.2]

where C, m are the constants depending on material and operating conditions; R is the stress ratio, for a pressure cylinder it is equal to 0; ΔK_1 is the range of the stress intensity factor.

The stress intensity factor for a nonuniform stress distribution in the vicinity of the crack is determined by the equation[5]:

$$K_1 = Y\sigma_1(a/1000)^{0.5},$$

where

$$Y = [2 - 0.82(a/c)]/\{1 - [0.89 - 0.57(a/c)^{0.5}]^3(a/c)^{1.5}\}^{3.25},$$
$$\sigma_1 = 0.61\ \sigma^A + 0.39\ \sigma^B +$$
$$+[0.11\ (a/c) - 0.28(a/s)(1 - (a/c)^{0.5})](\sigma^A - \sigma^B);$$

σ^A is the stress at the crack tip; σ^B is the stress on the surface of the component at the crack root.

For a particular case $Y = 1.12\sqrt{\pi}$. Integrating the expression [4.2] we get:

$$N = \int_{a_0}^{a_f} \frac{1}{C\left(\dfrac{\Delta K_1}{\sqrt{1-R}}\right)^m}\, da.$$

[4.3]

Substituting the previous expression into expression [4.3] and solving it with respect to the finite size of the crack a_f, we can determine the increase in the crack length Δa_N under the influence of N load cycles.

Usually, by defect we mean a discontinuity larger than the permissible size in service. For brevity, any crack-type discontinuity in metals will be sometimes called a defect.

4.1.3　Growth of discontinuities in static loading in a corrosive environment

The crack growth rate under static loading in a corrosive environment can be described by an equation of the following type:

$$da/d\tau = C\tau\,(\Delta K)^m,$$

[4.4]

where C_τ and m_τ are the constants depending on material properties and the environment; a is crack length, τ is time.

In Ref. 29, the crack growth rate is described by the following equation:

$$da/dt = f(a)z,$$

[4.5]

where a is crack length, t is time, $f(a)$ is the function of crack length a, and z is a statistical random variable which takes into account a random error.

One of the simplest forms of equation [4.5] is as follows:

$$da/dt = Ca^\lambda z,$$

[4.6]

where C and λ are constants. Logarithmic transformation of both sides of equation [4.6] leads to the following:

$$\log(da/dt) = \log C + \lambda \log a + \log z.$$

[4.7]

The constants C and λ can be obtained from the linear relationship between the variables $\log a$ and $\log (da/dt)$.

Damage for each corrosive factor is determined from[56]

$$a_{cd} = \int_0^t \frac{d\tau}{\tau_{cor}},$$

[4.8]

where the a_{cd} is the damage from the corrosive factors at time τ; τ is the time to nucleation of defects in the corrosive environment.

4.2 Methods of determining failure probability using a binomial distribution

4.2.1 Methods of determining the probability of failure on the basis of computer program MAVR-1.1

Probabilistic analysis of fracture of a section of a component based on the methodology proposed by A.A. Tutnov[57] is performed using the computer program MAVR-1.1[58,59]. In this method it is assumed that:

– the predicted depth distribution of calculated cracks is described by the density distribution $p(a)$, which is determined by the distribution of defects detected by monitoring or by the conditional distribution of the calculated probability $p_{cond}(a)$ of cracks and the probability of detection of defects $p_d(a)$:

$$p(a) = \frac{p_{cond}(a)}{p_d(a) \int\limits_0^s \frac{p_{cond}(a)}{p_d(a)}\, da}, \qquad [4.9]$$

and for the predicted distribution of the lengths of the calculated cracks l:

$$p(l) = p_{cond}(l); \qquad [4.10]$$

– fatigue growth of cracks is described by the Paris equation;
– distribution of the probability k of finding defects ($k = 1, 2, 3,...$) along the length of a heat exchanger pipe is described by the Poisson distribution:

$$p(k) = \mu_0^k \exp[-\mu_0]/k!, \qquad [4.11]$$

where μ_0 is the mathematical expectation of the number of cracks in the reference area (e.g. the length of a section of the pipe);

– distribution of critical crack dimensions $F^{a_{cr}}(a, t)$ and $F^{l_{cr}}(a, t)$ can be determined on the basis of various criteria, for example, two criteria used in Ref. 59 for heat exchanger pipes of steam generators:

1) occurrence of stress corrosion cracking;
2) onset of the limiting plastic state (ductile fracture).

In this case, the stress intensity factors for the Paris equation are calculated using quasi-elastic stresses, but in the conditions of stress corrosion cracking and formation of a plastic hinge the actual stresses are taken into account.

Each area of the heat exchanger pipe is divided into Q reference areas. In the presence of k cracks in the reference area the probability of failure of the investigated section l of the heat exchanger pipe P_{1jL} by the j-th failure mechanism for calculation cracks oriented in the direction L ($L = 1$ – axial, $L = 2$ – circumferential) is defined as follows:

$$P_{jL}(t) = 1 - [1 - P^1_{1jL}(t)]^{kQ}, \quad j = 1, 2, \qquad [4.12]$$

where the index j meets the criterion of failure (j = 1 – corrosion cracking, J = 2 – ductile fracture).

The probability of failure $P_{1jL}^1(t)$ in the presence of a crack at time t for the case of continuous cracks is replaced by $P_{TjL}^1(t)$ and for the case of large-scale destruction by $P_{crjL}^1(t)$. These probabilities are defined as follows:

$$P_{T1jL}^1(t) = \int_0^s P_1(a,t)F_{1jL}^{a_{cr}}(a,t)\,da; \qquad [4.13]$$

$$P_{cr1jL}^1(t) = \int_0^{2\pi R} P_1(l,t)F_{1jL}^{l_{cr}}(l,t)\,dl. \qquad [4.14]$$

The distribution of critical crack dimensions at the appropriate fracture criterion is defined as follows:

$$P_{11L}^{a_{cr}}(a,t) = F_{11L}^{K_{1SCC}}(K_{1SCC},t); \qquad [4.15]$$

$$P_{12L}^{a_{cr}}(a,t) = F_{12L}^{R_{Po2}}(K_{po2},t). \qquad [4.16]$$

The distributions of critical crack lengths calculated by the appropriate failure criterion $F_{1jL}^{l_{cr}}(l,t)$ are defined similarly to [4.15] and [4.16].

The probability of failure of heat exchanger pipes for each criterion j in the time interval t is determined by the summation of the probabilities of failure of all sections of the pipe according to this criterion. The total probability of failure of pipes according to two criteria for the failure time t is defined as:

$$P(t) = P_1(t) + [1 - P_1(t)]P_2(t). \qquad [4.17]$$

The size of leaks in a vessel or pressure pipeline for a given length and orientation of the continuous crack is determined in two ways:
– realistic calculation and
– conservative calculation.

Realistic calculation is reduced to determining the equivalent area of opening of the crack edges on the basis of the size of the continuous crack.

Conservative calculation is based on conservative assumptions that the size of a hole is equal to the size of the local zone of concentration of elastic energy in the vicinity of the crack.

When analysing the efficiency of heat exchange pipes of steam generators it is assumed that the number of damaged heat exchanger pipes (problems with the steam generator pipes of nuclear power plants are described in section 9.1) can be described by the binomial distribution, provided the heat exchanger tubes are in the same service conditions and the same medium of the secondary circuit and that the failure of one pipe has no effect on the failure of other pipes. If the fracture of pipes depends on their location, then for each group of pipes the probability of failure in time period t for i heat exchange pipes from the considered number n $P_n^i(t)$ is determined by the formula:

$$P_n^i(t) = C_n^i [P(t)]^i [1 - P(t)]^{n-i},$$ [4.18]

where $P(t)$ is the probability of failure of one of heat exchanger pipes over time interval t, assumed to be the same for a group of heat exchanger pipes and defined by [4.18], and C_n^i is binomial coefficient defined as

$$C_n^i = \frac{n!}{i!(n-i)!}.$$ [4.19]

In this case, the total likelihood of leaks or large-scale destruction in n heat exchanger pipes is

$$P_n = \sum_{i=1}^{n} P_n^i.$$ [4.20]

Reliability indices of pipes are determined on the basis on the failure rate and transition to the limiting state. The probability and rate of large-scale destruction (limiting state) and of formation of leaks (failure) are related as follows:

$$\lambda(t) = \frac{dP(t)/dt}{1 - P(t)}.$$ [4.21]

The calculations were carried out to estimate the probability of formation of continuous cracks and, consequently, the occurrence of leaks, as well as the probability of large-scale destruction for different variants of setting the probability of detecting defects and the accuracy of their size. The permissibility of the investigated variants is assessed by the reliability criterion.

4.2.2 Development of methodology based on MAVR-1.1 program

The development of a probabilistic methodology described in section 4.2.1 led to a more sophisticated and efficient method for analysis of the reliability of equipment and pipelines of the reactor installation[56,58–60].

Key provisions of this technology are as follows:

– the rate of of failures and limiting states associated with the propagation of defects in metal components and piping under pressure in the primary and secondary circuits are determined using the probabilistic methods of fracture mechanics;

– reliability (reliability, durability, availability coefficient) of equipment and pipelines under pressure of the primary and secondary circuits, defined by failure rate and limiting state, including the failure rates obtained on the basis of propagation of defects in the metal of elements;

– the level of reliability of a reactor depends on the reliability of equipment and pressure pipelines of the primary and secondary circuits;

– the reliability criterion for equipment and piping is based on the requirements of normative and technical documentation, domestic and international operating experience of similar plants based on the safe operation of the reactor core and nuclear power plants in general;

– satisfying the reliability criterion of test equipment and pipelines if the determining factor is the propagation of defects in metal parts. This is achieved by issuing recommendations for the characteristics of metal inspection and, if necessary, for their design and working conditions. Analysis of reliability of equipment and pipelines of VVER reactors based on probabilistic methods of fracture mechanics consists of four main steps:

– analysis of input data;

– calculation of the probability of destruction;

– definition of reliability;

– preparation of recommendations for achieving the necessary reliability.

To determine the reliability of equipment or piping it is necessary to construct the physical and mechanical model of destruction of their elements. This model, which includes all the stages of fracture (nucleation, fatigue growth, the critical growth of defects under corrosion and mechanical effects), is called the *unified model of fracture*[61]. The idea of this model is as follows. Before operation, an element of a composite piece of equipment may contain defects that were missed during testing of metal and also nuclei of future defects may be found. In service, parts of equipment or piping are subjected to thermal, mechanical and corrosive influences. This may be accompanied by the nucleation of new defects and propagation of existing defects from the start of operation. The ratio of the rates of these processes may differ depending on the service conditions of equipment or piping. Two cases are of particular interest here. In the first case the nucleation of new defects is much slower than the process of propagation of the already existing defects. In this case, in determining, for exmple, SG reliability indices, the nucleation of new defects can be neglected. This case requires mandatory testing and validation, as commonly used in design analysis. In the second case, the rate of nucleation of new defects exceeds the rate of propagation of existing defects and this leads to merging of the nucleated and existing defects. Nucleation of new defects in this case is of fundamental importance for determining the reliability of equipment or pipeline. Thus, the damage cumulation in a section of equipment or piping can lead to an increase in both the number and size of the existing defects. In addition, the operating rules and regulations define the extent and frequency of testing of metal which in turn lead to the identification of defects during the plant preventive maintenance. This reduces the number of dangerous defects in components of equipment or pipeline.

The physical–mechanical models of faults due destruction of parts of equipment or pipeline have been developed using the terms defined below.

The composite unit – part of equipment or pipeline (e.g., body, primary coolant header, heat exchanger pipes, seal assembly of the primary and secondary circuits, tube units, etc). The composite unit is divided into elements.

Element – a component part into which the composite unit is divided to analyze growth of defects. The element is divided into cells.

Cell – part of the metal volume into which the element for the analysis of defect nucleation is divided.

Determining the reliability of equipment and pipelines includes the following procedures:

– decomposition of the considered composite unit into elements for the analysis of propagation of defects;

– dividing the elements into cells for the analysis of defect nucleation;
– formation of the data for the element in question;

– formation of the data for the cell in question;
– uncertainty analysis of initial data;
– determination of the probability of nucleation of defects by the fatigue-mechanical and corrosion mechanisms in the investigated cell;
– adjustment of the size distribution of defects for the given elements as a result of metal inspection and analysis of nucleation of new defects, analysis of growth of defects;
– adjustment of the distribution of the number of defects for these elements based on metal inspection, analysis of nucleation of new defects; analysis of the nucleation of new defects based on the cumulation of fatigue, mechanical and corrosion damage;
– analysis of propagation of corrosion fatigue defects;
– distribution of defects on the basis of the mechanisms of their critical growth depending on the conditions in which the given element works;
– analysis of the critical growth of defects using the brittle fracture criterion;
– analysis of the critical growth of defects using the ductile fracture criterion;
– analysis of the critical growth of defects using the elastic plastic fracture criterion;
– determination of equivalent size D_0 on the basis of the size of continuous defects;
– determination of the probability of leaks by equivalent D_0 in this element with the uncertainties of input data taken into account;
– determination of the probability of large-scale destruction of the element in question, taking into account the uncertainty of initial data;
– determination of the probability of leaks by equivalent D_0 for the considered composite unit of equipment or piping;
– determination of the probability of large-scale destruction for the considered composite unit of equipment or piping;
– determination of the failure intensity of the composite unit element;

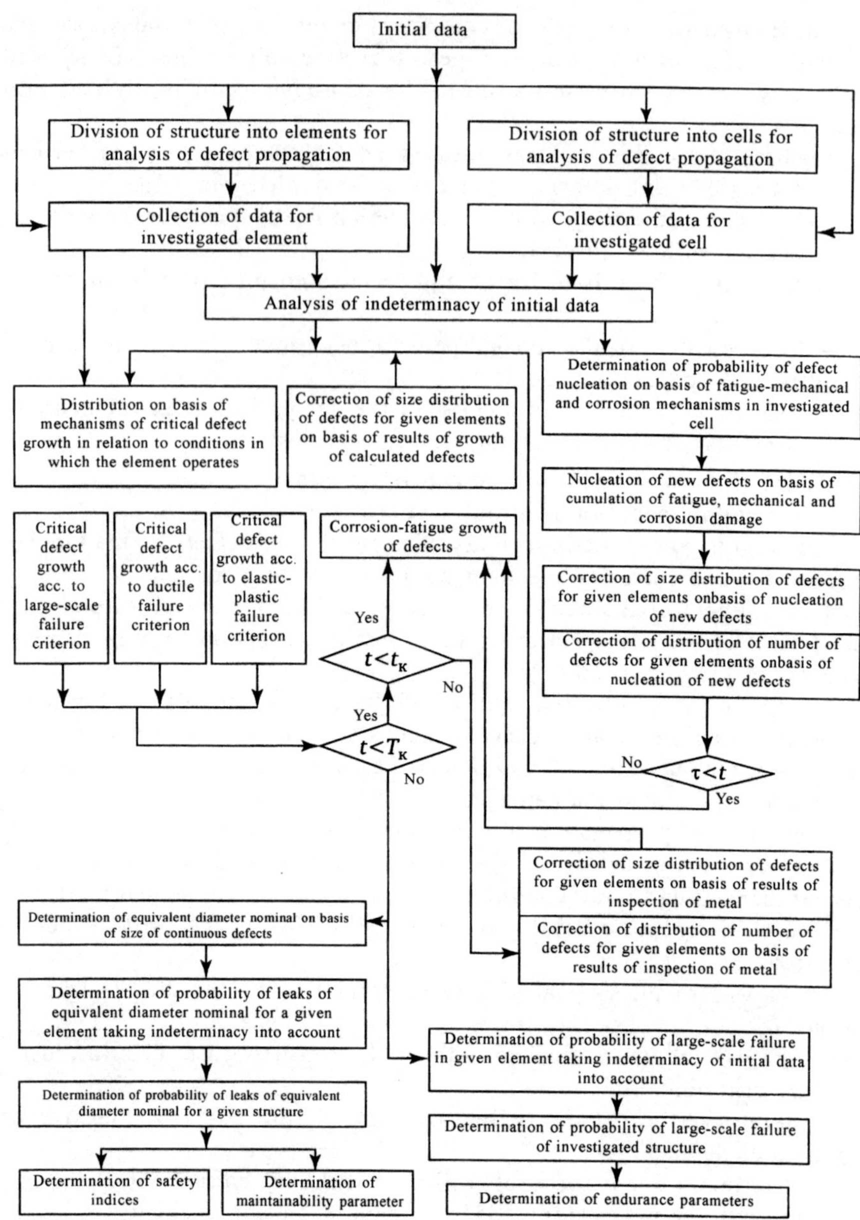

4.2 Enlarged flow diagram of the definition of the rate of failure and the transition to the limiting state.

– determination of the intensity of the transition of the element of the composite unit of equipment or piping to the limiting state;

– determination of the reliability of equipment or pipeline.

The relationships between the above procedures are shown in Figure 4.2.

To construct a probabilistic model of nucleation of new defects, four problems should be solved: on the basis of operating experience and experimental studies to identify factors that affect defect nucleation; construct mathematical models of the influence of factors on defect nucleation, define the probability of nucleation of a defect under the effect of the identified factors; a mathematical model that combines nucleation of all defects.

Analysis of operating experience and experimental studies has shown that nucleation of defects is a complex multifactorial process. In many cases, defect nucleation was of an individual character as it was associated with individual violations of manufacturing and installation technology or with individual violations of the water-chemical regime. The model of nucleation of defects is constructed taking into account factors that can be determined for several power units. These factors differ for pearlitic and austenitic steels.

Moreover, the following factors can be defined for *pearlitic steels*: mechanical fatigue damage cumulation (low-cycle fatigue); vibration exposure (high-cycle fatigue), hydrogen embrittlement.

For *austenitic steels*: mechanical fatigue damage cumulation (low-cycle fatigue); vibration exposure (high-cycle fatigue), oxygen and chloride cracking, formation of defects near copper deposits.

The mathematical model of nucleation of defects is based on the damage as a characteristic of damage cumulation, leading to the formation of defects. In this case, the criterion of defect nucleation is the condition

$$a_d = 1.   [4.22]$$

Regulatory approaches are used to determine mechanical fatigue damage. In this case, a_{mp} is determined by the formula

$$a_{m.d} = \sum_{i=1}^{1} \frac{N_i}{[N_{0i}]},   [4.23]$$

where $[N_{0i}]$ is the permissible number of cycles at the i-th type of the calculation regime of change of stress; N_i is the number of cycles at the i-th type of the calculation regime of stress changes.

The damage for every corrosive factor is determined by the formula

$$a_{c.d} = \int_{0}^{t} \frac{d\tau}{\tau_{corr}},   [4.25]$$

where τ_{corr} is the time to defect nucleation under the effect of corrosion.

It is assumed that the combined effects of mechanical and corrosive factors lead to damage which is determined by the non-linear summation of damage from each factor[62]. In this case, prior to non-linear summation of damage from each factor, the factors are ranked in the order from high to low, i.e. the first number a_{d1}, is assigned the highest value of all the failure rates. Accordingly, the damage from the joint action of all factors is given by:

$$a_d = \frac{a_{d1}}{(1-a_{d2})/(1-a_{d3})/(1-a_{d4})/...},$$
[4.25]

where a_{d1}, a_{d2}, a_{d3}, a_{d4} is ranked damage by each factor.

Thus, formulas [4.22] and [4.25] determine the conditions under which the defects nucleate as a result of the combined action of mechanical and corrosive factors.

The ranked damage from the corrosive factors in formula [4.25] is expressed on the basis of the time to nuleaction of defects in accordance with [4.24]. The damage from mechanical factors is defined in accordance with [4.23]. It is assumed that the characteristics included in the formula for determining the failure rate from the effect of mechanical and corrosive factors are random variables and are described by statistical distributions. Based on the above, it is possible to determine the probability of nucleation defects anywhere in the structure. This probability, in accordance with the criterion of nucleation of defects [4.22] is defined as follows

$$P_n = P(a_d \geq 1).$$
[4.26]

where P_s is the the probability of nucleation of a defect anywhere in the structure.

Given the large number of characteristics with the corresponding statistical distributions and the non-linear dependences which define the damage, the probability of nucleation is estimated by statistical simulation (Monte Carlo)[63,64].

To solve the problem of determining the number of incipient defects in relation to the operating time, the composite unit sample is divided into elements on the grounds of features of the structure, taking into account the values of the loading parameters: pressure and temperature, uniformity (homogeneity) of the stress–strain state, the number of defects and material properties. Elements are divided into cells assuming that each cell may contain a nucleus for a future defect. Given that a defect may form in each cell, the size of this cell should correspond to the size of the incipient defect. In general, the metal of the structure is divided into cubic elements and the area of one face of this element is equal to the area corresponding to the sensitivity of inspection of metal P_{sens}. In this case, different cells within one element are exposed to different effects of different factors influencing defect nucleation. Therefore, to analyze the number of incipient defects, the cells are placed in groups existing in similar conditions during operation, for example, in layers over the wall thickness of the structure. The probability of formation of a single defect in any cell of each group of cells is determined by the formula [4.26].

Suppose the i-th element of a structure is divided into i cells. Cells are arranged in jj groups with L^i_{jj} being the number of cells in each group. The probability of nucleation of a defect in any cell $P^i_{jj}(\tau_1)$ at time τ_1 is determined for each group.

To determine the number of nucleated defects in the jj-th group of cells, it is assumed that the nucleation of defects in the group is described by a binomial distribution, i.e. the cells are independent and $P^i_{jj}(\tau_1)$ is the same for all cells of one group

$$P^i_{jj}(k/n) = C^k_n [P^i_{jj}(\tau_1)]^k [1-P^i_{jj}(\tau_1)]^{n-k},$$

[4.27]

where $P^i_{jj}(k/n)(\tau_1)$ is the probability of nucleation of defects in the jj-th group of the i-th element at time τ_1; $n = L^i_{jj}$ is the number of cells in the jj-th group, C^k_n is the binomial coefficient, defined as

$$C^k_n = \frac{n!}{k!(n-k)!}.$$

[4.28]

Based on the properties of the binomial distribution [4.27], the mathematical expectation of nucleated defects in the jj-th group of the i-th element at time τ is equal to

$$v^i_{jj}(\tau_1) = L^i_{jj} \times P^i_{jj}(\tau_1).$$

[4.29]

where $v^i_{jj}(\tau_1)$ is the mathematical expectation of the number of nucleated defects in the jj-th group of i-th element at time τ_1.

Then for the i-th element the mathematical expectation of the number of nucleated defects at time τ is defined as follows:

$$v^i_n(\tau_1) = \sum_{ii=1}^{jj} v^i_{jj}(\tau_1).$$

[4.30]

where $v^i_n(\tau_1)$ is the expected number of nucleated defects in the i-th element at time τ_1.

In this case, the mathematical expectation of the number of nucleated defects at time interval $\Delta\tau_1$ is

$$v^i_n(\Delta\tau_1) = v^i_n(\tau_1) \times v^i_n(\tau_{i-1}).$$

[4.31]

Given that the nucleated defects are equally likely to be found in any cell of a group of cells, it may be assumed that defects nucleate both far away from each other and also in the neighbouring cells. Also, if a defect formed in a cell away from other cells, the size of the nucleated defect corresponds to the sensitivity of inspection of metal, that is, for example, in accordance with Ref. 65

$$a_{0.05} = 0.54\sqrt{F_{sens}}\,; \, l_{0.05} = 5a0.05.$$

[4.32]

If the defects formed in the neighboring cells, it is assumed that these defects have merged into a single defect. The average size of nucleated and merged defects over time $\Delta\tau_1$, taking into account Ref. 64, is defined as follows:

$$F_{0.5}=k_{0.5}^i(\Delta\tau_i)\times F_{\text{sens}};\; a_{0.5}=0.54\sqrt{F_{0.5}};\; F_{0.5}=5la_{0.5}, \qquad [4.33]$$

where $K_{0.5}^i(\Delta\tau_1)$ is a parameter corresponding to the average size of the nucleated defect and depends on the partition of the element into the cells, on the number of these cells and on the likelihood of nucleation of a defect in the i-th element. Assuming that the depth and length of nucleated defects are described by, for example, a log-normal distribution, the formulas [4.32]) and [4.33] are used to determine the mathematical expectations $M_3[a^i(\Delta\tau_1)]>M_3[l^i(\Delta\tau_1)]$ and dispersion $D_3[a^i(\Delta\tau_1)]$, $D_3[l^i(\Delta\tau_1)]$ of these distributions.

When carrying out adjustment of distributions of depth, length and the number of defects in a structure due to the nucleation of new defects over time $\Delta\tau_1$, two cases should be considered:
– the nucleated defects are far away from the defects present in the structure;
– new defects form in the immediate vicinity of the defects already present in the structure. This result in merger of these defects.

For the first case, the depth distribution of defects is adjusted as follows

$$p_i\left[a^i(\tau_1)\right]=\frac{v^i(\tau_{1-i})\times p_i\left[a^i(\tau_{l-1})\right]+v_n^i(\Delta\tau_1)\times p_n\left[a^i(\Delta\tau_1)\right]}{v^i(\tau_1)}, \qquad [4.34]$$

where $p_i\left[a^i(\tau_1)\right], p_i\left[a^i(\tau_{l-1})\right]$ are the the densities of distribution of the depths of the defects existing in the i-th element of a composite unit, respectively, at time τ_1, and $(\tau_{l-1}); p_n\left[a^i(\Delta\tau_1)\right]$ is the density of distribution of the depth of nucleated defects in the i-th element of the structure for time period $\Delta\tau_1$; $v^i(\tau_1)$ is the the mathematical expectation of the number of defects in the i-th element of the structure at time τ_1, after accounting for the defects nucleated by the formula

$$v_n^i(\tau_1)=v_n^i(\tau_{l-i})+v_n^i(\Delta\tau_1). \qquad [4.35]$$

The procedure for correction of the length distribution of defects is identical with that used for correction of the depth distribution of defects. The distribution of the number of defects is always corrected at the nucleation of defects irrespective of any conditions and is reduced to calculation of the mathematical expectation using formula [4.35].

For the second case, new defects nucleate in the immediate vicinity of the defects present in the structure and this leads to their merger. We consider two schemes of merger of the nucleated defects with the defects existing in the structure:

– In the first scheme, defects formed in the cells located in the surface or subsurface layers of metal structures merge with the existing defects. This increases the length of defects in the structure;

– In the second scheme, the defects formed in the cells located in the inner layers of metal of the structure merge with the existing defects resulting in an increase of the depth of defects in the structure. For the first scheme of merger of defects only the distribution of the length of the defects is corrected and this is accompanied by the 'shift' of the distribution to the right by the value of mathematical expectation of the length of the nucleated defects and by 'stretching' in relation to the mathematical expectation on the axis of the length of the defects $\{D_i\,[l^i(\tau_1)]/D_i[l^i(\tau_{l-1})]\}^{0.5}$ times.

In the second scheme only the depth of defects is corrected and this is accompanied by a 'shift' of the distribution to the right by the value of the mathematical expectation of the depth of nucleated defects and by 'stretching' in relation to the mathematical expectation on the axis of the depth of defects in $\{D_i\,[a^i(\tau_1)]/D_i[a^i(\tau_{l-1})]\}^{0.5}$ times.

Fatigue growth of defects is possible if the range of the stress intensity coefficients is not lower than $K^i_{min}(\tau_1)$ or $K^i_{max}(\tau_1)$, depending on the environment in which the structure operates. This should be taken into account in the mathematical expectation and dispersion of the range of stress intensity factors by rejecting the values below the threshold, followed by normalisation of the distribution obtained. In addition, the values of the stress intensity factor differ for different points of the front of the calculated cracks (defects), for example, points corresponding to the maximum depth and the maximum length of the defects. Therefore, for these points defining the direction of fatigue growth of the defect into depth and length will be different and the corresponding mathematical expectations and variance of the stress intensity factors will also differ.

In this case, adjustment of the distribution of the depth and length of defects due to their corrosion-fatigue growth in the time interval $\Delta\tau$, provided no new defects nucleate during this period, is accompanied by:

– 'shift' of distributions to the right by increasing the value of the mathematical expectations of the depth or length of defects;

– 'stretching' in relation to the mathematical expectation on the axis of the depth or length of defects, respectively, $\{D[a^i\,(\tau_1)]\,/\,D_i\,[a^i\,(\tau_{l-1})]\}^{0.5}$ and $\{D[l^i(\tau_1)]/D_i[l^i(\tau_{l-1})]\}^{0.5}$ times.

The stress intensity factors $K^i_1\,[\sigma^i(\tau_1),\,a^i(\tau_{l-1}),\,l^i(\tau_{l-1})]$, $K^i_1\,[\sigma^i(\tau_{l-1}),\,a^i(\tau_{l-1}),\,l^i(\tau_{l-1})]$ and, correspondingly, the range of stress intensity factors $\Delta K^i_1\,(\Delta\tau_1)$ depend on the size (width and depth) of defects, which in the framework of this approach are random variables and, therefore, they also are random variables. The statistical distribution of the stress intensity factors can be obtained numerically if the appropriate distributions of depths and lengths of the defects are known. The distributions are characterised by the corresponding mathematical expectations and variances.

Therefore, the mathematical expectation and variance of the range of the stress intensity coefficients over time $\Delta\tau_1$ are

$$M[\Delta K_l^i(\Delta\tau_l)] = M[K_l^i(\tau_l)] - M[K_l^i(\tau_{l-1})];$$ [4.36]

$$D[\Delta K_l^i(\Delta\tau_l)] = D[K_l^i(\tau_l)] - D[\Delta K_l^i(\tau_{l-1})].$$ [4.37]

It should be noted that the growth of fatigue defects is possible if the range of the stress intensity factors is not lower than $K_{\min}^i(\tau_1)$ or $K_{\max}^i(\tau_1)$, depending on the environment in which the structure works. This should be taken into account in the determination of the mathematical expectation and variance of the amplitude of stress intensity factors by rejecting the values below the threshold, followed by normalisation of the distribution obtained. In addition, the values of the stress intensity factor are different for different points of the front of the calculated cracks (defects), for example, points corresponding to the maximum depth and the maximum length of the defects. Therefore, the corresponding mathematical expectation and variance of stress intensity factors [4.36], [4.37] for these points, which determine the direction of the growth of fatigue defects in the depth and length, will be different.

Correction of the distributions of depth and length of defects due to their corrosion-fatigue growth during time interval $\Delta\tau_1$, provided that no new defects form during this time, is reduced to increase of the mathematical expectation

$$M[a^i(\tau_l)] = M_i\left[a^i(\tau_{l-1})\right] + M[\Delta a^i(\Delta\tau_l)]$$ [4.38]

and an increase in the variance

$$D[a^i(\tau_l)] = D_i\left[a^i(\tau_{l-1})\right] + D[\Delta a^i(\Delta\tau_l)].$$ [4.39]

The length distribution of defects is corrected by the same procedure as the depth of defects.

Thus, the correction of the distributions of depth and length of defects due to their corrosion-fatigue growth during time period $\Delta\tau_1$, provided no new defects form during this time, is accompanied by:

– 'shift' of distributions to the right by the value of the mathematical expectations of the depth or length of defects;

– 'stretching' in relation to the mathematical expectation of the axis of depth or length of defects, respectively, $\{D[a^i(\tau_l)]/D$ and $[a^i(\tau_{l-1})]\}^{0.5}$ and $\{D[l^i(\tau_l)]/D$ and $[l^i(\tau_{l-1})]\}^{0.5}$ times.

Inspection may detect defects which do not meet requirements on the quality of metal. In the next stage these defects are repaired. Therefore, it is necessary to adjust the distribution of sizes (width and depth) and the number of defects in the metal structure[57]. The following main characteristics of inspection of metal should be taken into account in correcting these distributions:

– frequency of inspection of metal (τ_{ins});
– requirements for the quality of metal: the permissible defect size (permissible depth (a^i_{per}) and permissible length (l^i_{per}), permissible number of defects (k^i_{rep});
– effectiveness of the means of inspection of metal characterised by the dependence of the probability of detecting defects on the size of the defects ($P_{ins}(a^i)$, $P_{ins}(l^i)$);
– The duration of inspection of metal and subsequent repair ($\Delta\tau_{ins}$).

In this case, adjustment of the size distribution and number of defects in the metal structure due to metal inspection and subsequent repair is different in the initial stage of operation (e.g. at the stage of preoperational monitoring) and in subsequent stages of operation (e.g. during in-service inspection). At the initial stage of operation during testing of metal defects are detected and repaired in accordance with the effectiveness of inspection. In this case, the missed defects remain in the metal of the structure. This means that the following situation forms in analysis at the initial stage of operation[57]:

– initial distributions of the depth and length of the calculated defect are the distributions plotted on the basis of the defects identified in testing of metal $p_0(a^i)$, $p_0(l^i)$;
– projected distributions of the depth and length of the calculated defect are determined on the basis of missed defects by means of probability of detection, for example, in accordance with Ref. 57 the predicted depth distribution of calculated defects was determined as follows:

$$p_d(a^i)=\frac{p_0(a^i)C_i}{P_{ins}(a^i)} \ \text{at } a^i < a^i_{per};$$

$$p_d(a^i)=\left[\frac{p_0(a^i)}{P_{ins}(a^i)} - p_0(a^i)\right]C_i \ \text{at } a^i \geq a^i_{per},$$

where

$$C_i = \int_0^{a^i_{per}} \frac{p_0(a^i)}{P_{ins}(a^i)}\,da^i + \int_0^{a^i_{per}} \frac{p_0(a^i)}{P_{ins}(a^i)p_0(a^i)}\,da^i. \qquad [4.40]$$

The mathematical expectation of the number of defects in the i-th element of the structure in accordance with Ref. 57 is

$$v^i_d(\tau_0) = C_i \times v^i_0(\tau_0), \qquad [4.41]$$

where $v^i_0(\tau_0)$ is the mathematical expectation of the number of projected defects in the i-th element after inspection of metal t the initial period of operation of the facilities. The graphical layout of correction of the depth distribution of defects in inspection of metal at the initial period of operation of the facilities in shown in Fig. 4.3.

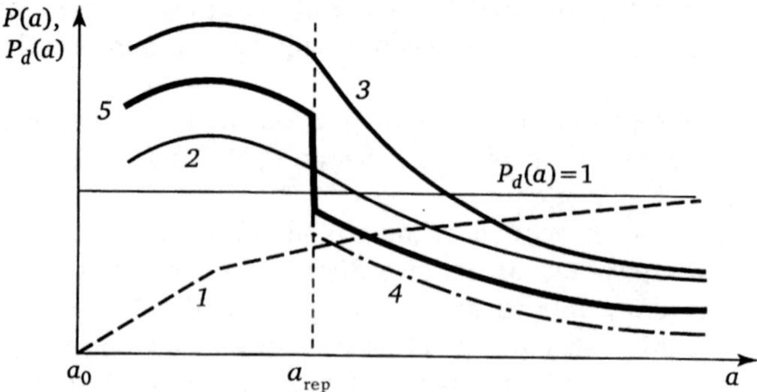

4.3 Correction of the depth distribution of defects in inspection of metal in the initial period of operation of the structure. 1 – dependence of the probability of detection on crack depth; 2 – distribution of the depth of the detected cracks prior to repair; 3 – irregular distribution of defects based on the likelihood of their non-detection; 4 – the distribution with defect repair taken into account; 5 – projected distribution of the normalized crack depth after repair.

In subsequent stages of operation inspection of metal also identifies defects which are then repaired in accordance with the efficiency of inspection devices. However, the following situation arises as a result of calculation analysis in subsequent stages taking into account the results of analysis at the initial stage of operation:

– the initial distribution of the size and number of calculated defects is the predicted distribution of the size and number of calculated defects identified at the initial stage of operation and adjusted due to the nucleation of new defects and corrosion fatigue growth of defects at the time of operational inspection $p_d\,[a^i(\tau_i)]$, $p_d\,[l^i(\tau_i)]$, $v_d^i\,(\tau_i)$;

– projected distributions of the size and number of defects calculated after inspecting and repairing the metal are determined based on the application procedure for adjusting the projected distributions of the size and number of defects prior to inspecting the metal.

If it is assumed that inspection of metal starts at time τ_1, the projected depth distributions of calculated defects after testing of metal and repairs are defined as follows

$$p_d\left[a^i\left(\tau_i+\tau_{\text{ins}}\right)\right]=\frac{p_d\left[a^i(\tau_i)\right]}{C_i}\ \text{ at } a^i\geq a^i_{\text{per}};$$

$$p_d\left[a^i\left(\tau_i+\Delta\tau_{\text{ins}}\right)\right]=\frac{p_d\left[a^i(\tau_i)\right]-p_{\text{ins}}(a^i)\times P_d\left[a^i(\tau_i)\right]}{C_i}\ \text{ at } a^i\geq a^i_{\text{per}},$$

$$C_i=\int_0^{a^i_{\text{per}}}p_d\left[a^i\left(\tau_i\right)\right]+\int_0^{a^i_{\text{per}}}\left\{p_d\left[a^i\left(\tau_i\right)\right]-p_{\text{ins}}(a^i)\times p_d\left[a^i\left(\tau_i\right)\right]\right\}da^i. \qquad [4.42]$$

In this case, the mathematical expectation of the number of projected defects in the *i*-th element of the structure is determined from:

$$v_d^i (\tau_d + \Delta\tau_{ins}) = C_i \times v_d^i (\tau_d),$$ [4.43]

where $v_d^i (\tau_1)$ is the mathematical expectation of the number of projected defects in the *i*-th element after metal inspection of metal in the consecutive operating periods of the facilities.

Analysis of the fracture mechanisms[66–69] shows that the critical crack growth can occur as a result of brittle, ductile or elastic–plastic fracture.

The conditions defining the scope of these fracture mechanisms, in accordance with Ref. 69, may be described as follow:

– zone of brittle fracture

$$0 \leq \left(\frac{K_{IC}^i}{R_F^i}\right)^2 l^{-1} < 1.2,$$ [4.44]

where $R_F^i = (R_m^i + R_{p0,2}^i)/2$ is the the flow stress;

– zone of ductile fracture

$$1.2 \leq \left(\frac{K_{IC}^i}{R_F^i}\right)^2 l^{-1} < 7.0;$$ [4.45]

– zone of elastic-plastic fracture

$$\left(\frac{K_{IC}^i}{R_F^i}\right)^2 l^{-1} \geq 7.0.$$ [4.46]

There is no critical crack growth leading to the formation of a continuous crack in the zone of brittle fracture if

$$K_I^i(\tau_1) \leq K_{IC}^i(\tau_1).$$ [4.47]

Critical crack growth, leading to large-scale destruction, does not occur if

$$K_I^i(\tau_1) \leq K_C^i(\tau_1).$$ [4.48]

In the zone of elastic–plastic fracture there is no critical crack growth, leading to the formation of continuous cracks, according to Ref. 69, if the point with coordinates $K_r^i(\tau_1)$, $S_r^i(\tau_1)$ is within the area bounded by the curve (Fig. 4.4)

$$K_r^i(\tau_1) = S_r^i(\tau_1) \times \left\{\frac{8}{\pi^2} \times \ln\left[\sec\left(\frac{\pi}{2} \times S_r^i(\tau_1)\right)\right]\right\}^{-1/2},$$ [4.49]

where

$$K_r^i(\tau_1) = K_I^i(\tau_1) / K_{IC}^i(\tau_1), \quad S_r^i(\tau_1) - \sigma_{eq}^i(\tau_1) / R_{p0.2}^i(\tau_1); \quad \sigma_{eq}^i(\tau_1)$$

is the equivalent stress, defined by the von Mises criterion[70].

Critical crack growth, leading to large-scale failure, does not

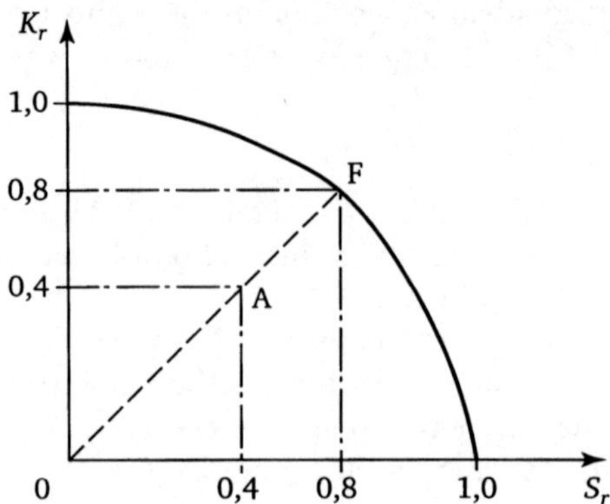

4.4 The integrity–failure diagram: A – a point that reflects the work of construction; F is the critical point.

occur if the point with coordinates $K_r^i(\tau_1), S_r^i(\tau_1)$ is situated inside the region bounded by the curve described by the expression similar to [4.49]. In this case, $K_r^i(\tau_1)$ and $S_r^i(\tau_1)$ are calculated differently: $K_r^i(\tau_1) = K_r^i(\tau\lambda) / K_C^i(\tau_1);\ S_r^i(\tau_1) = \sigma_{eq}^i(\tau_1) / R_F^i(\tau_1).$

There is no critical crack growth in the zone of ductile fracture leading to the formation of a continuous crack (the yielding of the bridge between the crack and the back surface) if[69]

$$\sigma^i(\tau_1) < k_F^i(\tau_1) R_{p0.2}^i,\qquad\qquad [4.50]$$

where $k_F^i(\tau_1)$ is the attenuation coefficient for the i-th element at time τ_1, (determined in accordance with Ref. 69); $\sigma^i(\tau_1)$ are the total tensile stresses in the i-th element at time τ_1.

According to Ref. 69, critical crack growth, leading to large-scale destruction (plastic instability in the cross section with a crack), does not occur if

$$\sigma^i(\tau_1) < k_F^i(\tau_1) R_F^i \eta^i(\tau_1),\qquad\qquad [4.51]$$

where $\eta^i(\tau_1) = \dfrac{R_m^i}{R_p^i} - \left(\dfrac{R_m^i}{R_F^i} - 1\right)(1 - k_F(\tau_1))^{0.2}.$

In determining the probability of structural failure, it is necessary to define:

– conditional probability of the formation of a through-wall defect and large-scale destruction from a single defect in the element under the condition of occurrence of a calculated event CE_k;

– conditional probability of a through-wall defect and large-scale destruction by the effect of several defects in the element under the condition of calculated events CE_k;

– the probability of formation of through-wall defects and large-scale destruction of an element, taking into account the probability of occurrence of calculated events;

– complete probability of formation of through-wall defects and large-scale of destruction of the structure.

When analyzing the formation of through-wall defects, three reasons for their occurrence are considered:

– corrosion–mechanical nucleation of new defects;

– corrosion fatigue growth of surface and subsurface defects;

– critical growth of surface and subsurface defects. In this case, the conditional probability of the formation of through-wall defects in the i-th element at time τ, when the calculated event CE_k takes place due to corrosion-mechanical nucleation and corrosion-fatigue defect growth, is:

$$p^i_{1cy/CE_k}(\tau_1)=\int_s^\infty p_d[a^i(\tau_1)]da^i,\qquad\qquad [4.52]$$

and that of critical growth of the defect is

$$P^i_{1ck/CE_k}(\tau_1)=\int_0^s p_a[a^i(\tau_1)]F\left[a^i_{cr}(\tau_1)\right]da^i,\qquad\qquad [4.53]$$

where $F[a^i_{cr}(\tau_1)]$ is the distribution of critical defects for the i-th element of the structure at the time τ_1, defined for each area of failure on the basis of the relevant critical characteristics for the formation of through-wall defects.

The conditional probability of formation of a through-wall which takes into account three reasons for its formation is:

$$p^i_{1cont/CE_2}(\tau_1)=p^i_{1/CE_k}(\tau_1)+p^i_{1cy/CE_k}(\tau_1)\left[1-p^i_{1cont/CE_k}(\tau_1)\right],\qquad\qquad [4.54]$$

The through-wall defect formed can be either stable or unstable, i.e. it can continue to grow and lead to large-scale destruction. The conditional probability of the resulting defect being unstable and can lead to large-scale destruction is

$$p^i_{1unst/CE_k}(\tau_1)=\int_0^{2\pi R} p_d\left[l^i(\tau_1)\right]F\left[l^i_{cr}(\tau_1)\right]dl^i,\qquad\qquad [4.55]$$

where $F\left[l^i_{cr}(\tau_1)\right]$ is the distribution function of the critical defect length for the i-th element of the structure at the time τ_1, defined for each area of destruction on the basis of the relevant critical features for large-scale destruction.

If the probability of formation of a through-wall defect and the probability of its propagation to large-scale failure are known, then the conditional probability of large-scale destruction in the presence of a single defect in the i-th element at time τ_1, when a calculated event CE_k took place, is determined as follows

$$P^i_{1cr/CE_k}(\tau_l) = P^i_{1cont/CE_k}(\tau_l)P^i_{1unst/CE_k}(\tau_l).$$

[4.56]

The conditional probability of a leak through a through-wall defect in the i-th element at time τ when when a calculated event CE_k took place is

$$P^i_{1leak/CE_k}(\tau_l) = P^i_{1cont/CE_k}(\tau_l)\left[1 - P^i_{1unst/CE_k}(\tau_l)\right].$$

[4.57]

The conditional probability of a leak through a through-wall defect with the length in the range $l^i_j \leq l^i(\tau_l) \leq l^i_{j+1}$, is determined on the basis of equtions [4.55] and [4.53]

$$P^i_{1leak/CE_k}\left[l^i_j \leq l^i(\tau_l) \leq l^i_{j+1}\right] = P^i_{1cont/CE_k}(\tau_l)\int_{l^i_j}^{l_{j+1}} p_d\left[l^i(\tau_l)\right]\left\{1 - F\left[l^i_{cr}(\tau_l)\right]\right\} dl^i.$$

[4.58]

The volume of metal of the composite unit is divided into Q reference volumes. In the presence of n defects in the reference volume the conditional probability of formation of through-wall defects or large-scale destruction of the i-th element at time τ_l, when a calculated event CE_k took place[57,61]:

$$P^i_{n/CE_k}(\tau_l) = 1 - \left[1 - P^i_{1/CE_k}(\tau_l)\right]^{nQ},$$

[4.59]

where $P^i_{1/CE_k}(\tau_l)$ is the conditional probability of the formation of through-walldefects or large-scale destruction in the presence of a defect in the i-th element at time τ_l when a calculated event CE_k takes place.

The conditional probability of failure (formation of through-wall defects or large-scale destruction) of i-th element of the structure when a calculated event CE_k takes place for calculated defects is defined as follows:

$$P^i_{CE_k}(\tau_l) = \sum_n P^i_{n/CE_k}(\tau_l)P^i_n(\tau_l),$$

[4.60]

where $P^i_{n/CE_k}(\tau_l)$ is the conditional probability of formation of a through-wall defect or large-scale destruction in the i-th element at time τ_l when a calculated event CE_k takes place; $P^i_n(\tau_l)$ is the probability of presence in the reference volume of the i-th element of n defects defined by, for example, the Poisson law.

Determination of the total probability of failure (formation of through-wall defects or large-scale destruction) of an element of the composite unit is reduced to the summation of products of conditional probability of failure by the probability of occurrence of the appropriate calculated events:

$$P^i(\tau_l) = \sum_{k=1}^{N_{PC}} P^i_{n/CE_k}(\tau_l)P_{CE_k},$$

(4.61)

where N_{CE} is the number of calculated events CD_k.

Determination of the total probability of failure (formation of open defects, or large-scale destruction) of the structure is reduced to the summation of the total probabilities of failure of elements in accordance with the structural scheme designed in the initial stage of analysis:

$$P(\tau_I) = 1 - \prod_{i=1}^{I}\left(1 - P^i(\tau_I)\right), \qquad\qquad [4.62]$$

where I is the the number of elements that make up the structural model of the structure.

To determine the reliability parameters, it is necessary to determine the following distribution functions:

– To determine the indicators of reliability and suitability for repair– distribution functions of time to failure

$$[4.63]$$

where $P_T^i\,[l^i(\tau) \leq l_{per}^i]$ is the probability determined using [4.58] and [4.61], of the formation of stable through-wall defects in the i-th element which can be repaired $F_I^i(\tau) = P_T^i\,[l^i(\tau) \leq l_{per}^i] + P_{unst}^i\left[a^i(\tau) > a_{per}^i\right]\left\{1 - P_T^i\left[l^i - P_T^i(\tau) \leq l_{per}^i\right]\right\}$,

– to determine the parameters of durability – the distribution function of time to transition to the limiting state.

$$F_{II}^i(\tau) = P_{cr}^i + P_T^i(\tau)\left[l^i(\tau) > l_{per}^i\right]\left[1 - P_{cr}^i(\tau)\right], \qquad\qquad [4.64]$$

where $P_T^i\,[l^i(\tau) > l_{per}^i]$ is the probability, determined using [4.58] and [4.61], of the formation of stable through-wall defects in the i-th element which can not be repaired; $P_{cr}^i(\tau)$ is the probability of large-scale destruction of the i-th element determined by the formula [4.61].

If the functions of distribution of time to failure $F_I^i(\tau)$ and to transition to the limit state $F_{II}^i(\tau)$ are determined, then the corresponding rates of failure $\lambda_I^i(\tau)$ and the transition to the limit state $\lambda_{II}^i(\tau)$ are defined in the same manner[71,72]:

$$\lambda_I^i(\tau) = \frac{dF_I^i(\tau)}{d\tau}\,/\left[1 - F_I^i(\tau)\right]; \qquad\qquad [4.65]$$

$$\lambda_{II}^i(\tau) = \frac{dF_{II}^i(\tau)}{d\tau}\,/\left[1 - F_{II}^i(\tau)\right]. \qquad\qquad [4.66]$$

The flow diagram of equipment or piping has usually the form of a series connection of elements for which the failure rate [4.65] and transition to the limiting state are determined [4.66]. Therefore, the reliability indices of equipment or piping are defined as follows:

– mean service life to failure

$$T_0 = \int\limits_0^\infty \left\{ \exp\left[-\int\limits_0^t \sum_{i=1}^l \lambda_1^i(\tau)d\tau \right] \right\} dt;$$

[4.67]

– γ-percentile lifetime T_γ from the equation:

$$\gamma/100 = \exp\left[-\int\limits_0^{T_\gamma} \sum_{i=1}^l \lambda_1^i(\tau)\,d\tau \right].$$

[4.68]

The indicators of life-cycle for the series connection of elements are:
– the average resource till write-off (full)

$$T_{P_{mean}} = \int\limits_0^\infty \left\{ \exp\left[-\int\limits_0^t \sum_{i=1}^l \lambda_1^i(\tau)d\tau \right] \right\} dt;$$

[4.69]

– γ-percentile resource till write-off (full) T_{P_γ} from the equation:

$$\gamma/100 = \exp\left[-\int\limits_0^{T_{P_\gamma}} \sum_{i=1}^l \lambda_{\mathrm{II}}^i(\tau)d\tau \right].$$

[4.70]

In analyzing the availability and suitability for repair it was assumed that during reconditioning of any element other elements work. The indicator of suitability for repair [71] for series connection of elements – the average reconditioning time is equal

$$T_{P_{mean}} = T_0 \sum_{i=1}^l \left[T_{B_{mean}} \lambda_i^i(\tau) \right].$$

[4.71]

In this case, the availability factor K_r (complex value) is defined as follows

$$K_r = \prod_{i=1}^l \frac{T_0^i}{T_0^i + T_{B_{mean}}^i},$$

[4.72]

Here T_0 is the mean time to failure of i-th element of a composite unit; $T_{B_{mean}}$ is the average (mean) recovery time of i-th element of a composite unit.

Formulas [4.67], [4.68], [4.71], [4.72] take into reliability indicators up to the first failure. The mean reliability indicators of the elements to failure for the reconditioning process with increasing failure intensity are determined using the formulas given in Ref. 71.

4.2.3 Criterion values of failure probability, based on the procedure described in section 4.2.2

The reliability criteria are selected by comparing the reliability indices obtained by creating a new design or when upgrading an old one, or when determining the reliability of the elements of the existing energy unit, with indicators of reliability (reliability level) in previous instances (or equivalent).

It is assumed that the reliability criterion is satisfied if reliable operation of the elements ensures:

– safe operation of the core and nuclear power plant in general;
– transfer of the element to the safe state when a failure occurs (the principle of safe failure);
– the required value of the readiness coefficient;
– optimum cost of repairs related to failure.

Based on the first principle, acceptable level of reliability is determined on the basis of the PSA data the corresponding generalised IAEA data. At the present time the databases in Russia and the European Community contain information regarding the reliability.

For nuclear power plants these reliability indices ensure safety of the reactor core, i.e. in accordance with domestic and foreign safety requirements the probability of core damage is less than 10^{-5} per reactor per year. These reliability indices, which characterise the reliability of components and systems, comply with such a (acceptable) level of reliability at which the safe operation of nuclear installations and NPP is ensured as a whole.

The second principle of design and regulatory documentation is used to determine the permissible level of reliability in terms of transfer of an item into a safe state when a failure occurs in it.

The third principle is based on the availability factor of nuclear power plant and the reactor, the availability factor of the elements should satisfy the required availability factor for nuclear power plant.

The fourth principle relates the allowable reliability indices and the required factor of economic efficiency of operation of the element, reactor installation the nuclear power plant as a whole.

If the resulting level of reliability (and/or failure rate and the limiting state) of equipment does not meet the reliability criterion, then based on the reliability criterion, it can be improved (and/or reduced to the failure rate and limiting state), for example, due to: changes of periodicity of metal inspection, design requirements for flaw inspection with regard to the allowable defect sizes with the corresponding allowable number of defects or more accurate determination of its service life.

5

Physico–statistical models based on the residual defectiveness of structural materials

It is known that defects form in the material of structures during production and service It is also known[12, etc.] that there is always a finite probability (in many cases, large) of not detecting defects during inspection. In this connection, it can be argued that defects can remain in the structure undetected after manufacture of a component, its inspection and repair of detected defects.

The set of the defects not detected by NDT methods and remaining in the material of the structure after manufacture of a component, inspection and repair of identified defects is termed *residual defectiveness*.

It is clear that for ensuring the strength and residual life of the structure the residual defectiveness is the most important characteristic of the given structure. Any prediction of strength, reliability and service life of the structure without taking residual defectiveness into account will be inaccurate.

The following are the basic concepts, definitions and laws of the formation of residual defectiveness, and some results of its assessment for the real structures of nuclear power plants of different types are also presented.

5.1 Regularities of the formation, detection and omission of defects during non-destructive testing

5.1.1 Formation of defects in the metal of structural elements

Depending on their size, the discontinuities (defects) can be divided into three groups: submicroscopic (comparable to the size of atoms), microscopic (comparable to the size of grains in the metal) and macroscopic (comparable to the size of structural elements).

The number of submicroscopic defects in the metal (also called lattice defects – dislocations, vacancies, etc.) is very large. In a 1 cm^2 section there are 10^8–10^{12} dislocations.

Microscopic defects are associated with the process of production of ingots, pressure working, manufacturing of semi-finished products. They are mainly micropores, non-metallic inclusions, mikrotears, etc.

The number of microscopic defects is much smaller than that of the submicroscopic ones, but is still large. There can be several defects on 1 cm^2.

Macroscopic discontinuities are usually characteristic of welded joints. The probability of starting operation of a structure with a macroscopic defect in the base metal is very small, possibly 3–5 orders of magnitude smaller than the probability of occurrence of defects in welded joints (this does not apply to cast components).

As an example, the characteristic defects in a critical component of the nuclear reactor – steam generators (SG) – immediately after manufacture are shown below. In Ref. 74 macrodefectiveness was evaluated as the number of defects per 10 m of the weld γ_d^L per 1 t of deposited metal γ_d^m. In studying the manufacturing technology of SGs made of steel 22K quantitative data for γ_d^L and γ_d^m were obtained, depending on the type of welding and the size of the welded structural element (Fig. 5.1).

Obviously, the number of defects according to the standards for the manufacture of unit length or mass ranges from about 1 to 1000.

In general, it can be argued that the number of defects in the structure decreases with increasing defect size (Fig. 5.2a). In this case, it can be assumed that for a structure weighing several tons the curve tends to infinity as the size of defects tends to zero.

The curve in Fig. 5.2a can be expressed as equations, the simplest of which is:

$$N = Aa^{-n},\qquad\qquad [5.1]$$

where a is the characteristic size of the defect, such as depth (in the direction of the wall of the pressurevessel); A, n are the constants for the given structure, steel grade and manufacturing technology.

The following exponential equation can also be used

$$N(a) = \lambda \exp [-\lambda a],$$

where λ is a constant.

A shortcoming of this equation is that it does not adequately describe the number of small defects.

The curve in Fig. 5.2a can be restricted on the right-hand side. For pressure vessels and piping this restriction is the wall thickness s (see Fig. 5.2b).

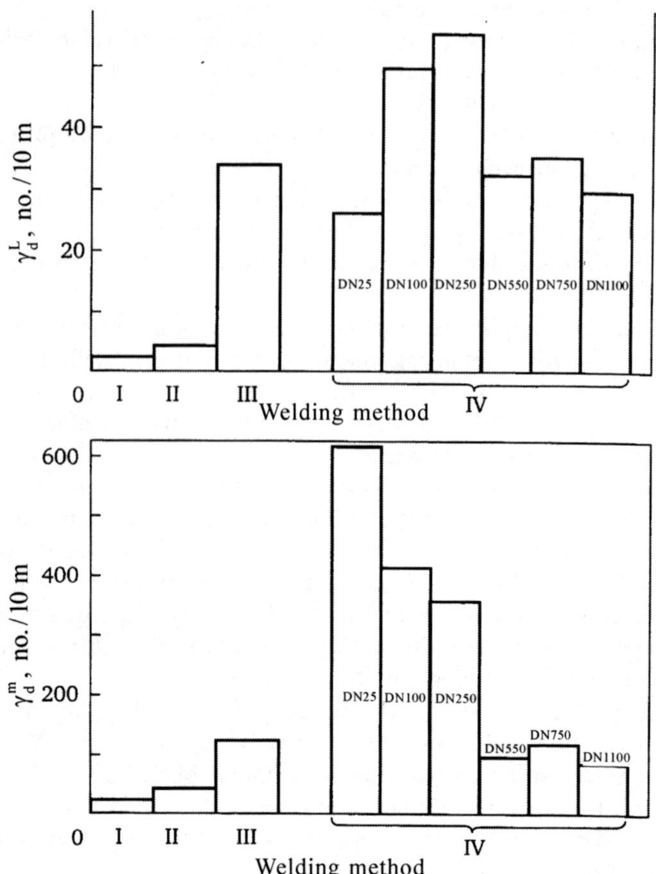

5.1   Diagrams of specific defects in welds in steam generator casings (I–III) and nozzle welds (IV) under different types of welding ised: I, II – respectively, electroslag, automatic submerged arc; III, IV - manual electric welding.

In conducting successful hydraulic tests, if an $a_{cr} < s$, the dependence N (a) corresponds to that shown Fig. 5.2c.

The dependence N (a) also affects repair according to the results of flaw inspection (Fig. 5.3). The value a_0 in Fig. 5.3 corresponds to the rejection size of the defect.

The curves N (a) (see Fig. 5.2) can be called curves that characterise real defects in a construction. Curve 2 in Fig. 5.3 corresponds to the residual defectiveness remaining in the structure after inspection and repair of real defects. For brevity, the term 'residual real defectiveness after inspection and repair' will be replaced by the term 'residual defectiveness'.

The actual initial (prior to inspection) defectiveness and residual defectiveness should distinguish between themselves and also should be

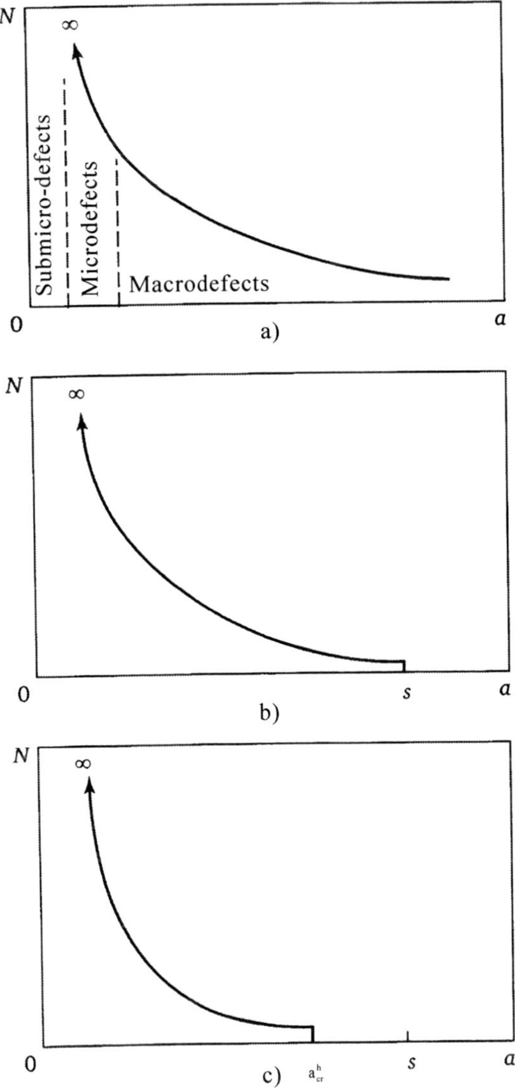

5.2   Dependence of the number of defects *N* on the size *a*: a) general view, b) pressure vessel with wall thickness *s*; c) pressure vessel after the hydraulic test.

distinguished them from the defectiveness identified in inspection, which in most cases is characterised by the curve shown in Fig. 5.4.

Obviously, if all the defects in the structure with the size $a > a_0$ would be detected with 100% certainty, i.e. all defects with $a > a_0$ would be detected and repaired, then the curves of the detected and residual defectiveness would correspond to curves 2 and 3 in Fig. 5.5. Moreover, the number of defects with the size $a > a_0$ after inspection and repair would be equal to 0.

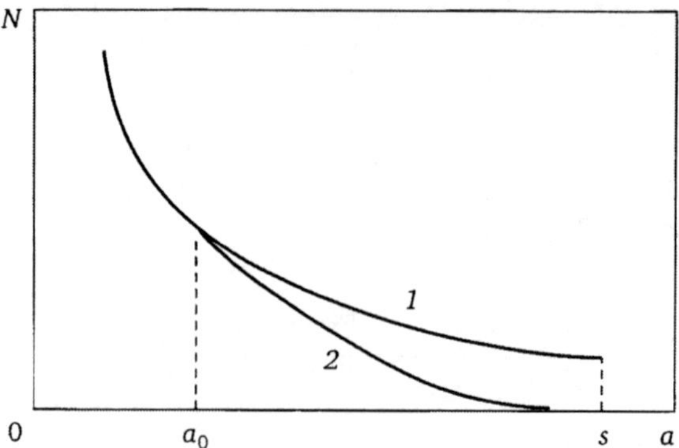

5.3 Dependence of the number of defects on their size before flaw inspection (1) and after testing and repair (2).

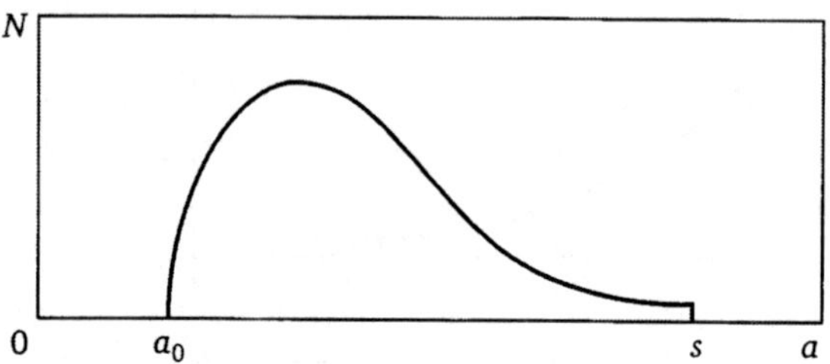

5.4   Defects detected in inspection.

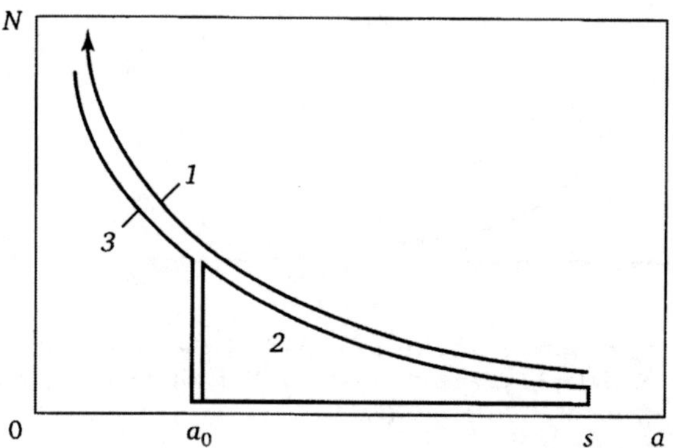

5.5 Initial defectiveness (1), defects detected at 100% detectability of inspection (2) and residual defectiveness (3).

Unfortunately, in most cases, and possibly in all cases, the inspection using NDT methods does not achiev 100% reliability of quality control.

5.1.2 Probabilistic relationships of inspection. Detectability of inspection

Probabilistic relationships of defect detection have been studied insufficiently. Even the PISC programme in which one of the principal goals was to study the detectability of defects, could not cover all aspects of the problem (despite the duration of studies for almost 15 years with the participation of 16 Western countries including the USA and Japan). In Russia, the pattern of detection of defects (including the probabilistic aspects) was studied by A.K. Gurvich, V.G. Shcherbinsky, N.P. Aleshin and others. In VNIIAES this problem has been studied since 1978

To understand the laws of probabilistic relationships of defect detection, it is desirable to describe more accurately concepts such as detectability of the control and the probability of detecting a defect.

The notion of detectability of inspection and related terms has been analysed most extensively by V.N. Volchenko[75]. He uses the term comparative detectability, understanding it as the objectivity of inspection, i.e. degree of coincidence of signals relating to quality with the actual values of its indicators. It is assumed that the real value of quality indicators is assessed by a reference method. When inspecting by NDT methods, the reference method can be metallographic examination of defects in layer cutting of components (i.e. in fracturing components.)

The relative detectability can be quantified by the probability of making error-free decisions when evaluating the quality of the object or a batch of components. Thus, the quantitative characteristic of the reliability of inspection is associated with the norms of defects in contrast to the detection of defects. Detection of defects is the probability of detecting a discontinuity on the basis of the specified parameter of the discontinuity.

An important indicator of the detectability of inspection is the reproducibility of its results. This characteristic can be defined as the frequency of coincidence of test results in different conditions. Reproducibility may be a particular characteristic of detectability.

In terms of the strength nd lie the real reliability unlike the relative detectability of inspection serves as the controlling characteristic of the detectability of inspection. Real reliability of non-destructive inspection can be defined as the degree of coincidence of the NDT results with the real characteristics of discontinuities in the structure.

The opportunity to systematically investigate the real detectability of NDT appeared after the development of manufacturing technology of test specimens with artificial discontinuities to fully simulate the discontinuity of technological and operational nature.

5.1.3 Methods for studying the detectability of non-destructive testing

To compare different inspection methods, an apriori initial inspection method can be developed and compared with the test method. The initial method should give the largest amount of information about the discontinuities. In cases where the initial method gives complete information on the discontinuities it can also be referred to as the reference method.

Before the advent of technologies for production of test samples, the only apparently reference method which provided complete information about the real defectiveness of the structure was the method of cutting layers from a structure, combined with metallographic studies (method of opening the defects).

The method of opening the defects is costly and time consuming. Also, in this case there is a likelihood of not detecting cracks,. Therefore, it is sometimes useful in assessing the relative detectability to apply the most economical and widespread NDT method as the initial method. In some cases, the initial method can be the same as the test method but with different sensitivity of the apparatus and different configuration.

New possibilities in the study of the detectability of NDT are offered by the test specimens with artificial, i.e. specifically embedded, hidden discontinuities. In this case, defects of almost any type, size, location and orientation can be formed in the sample. With the well-developed manufacturing technology of test samples it is not necessary to use the reference and initial inspection methods, since the information about discontinuities is known in advance.

Studies using the test samples also have drawbacks, since in practice it is not always possible to produce defects in accordance with the requirements. In addition, the test samples may contain unplanned defects due to imperfect welding. In this case, information about the real defectiveness can be obtained on the basis of design data on the discontinuities as well as the NDT results obtained by different methods (e.g., ultrasonic and radiographic) and different NDT inspectors.

The reliability of NDT can also be estimated by the analysis of testing results. In this case it is not necessary to use the the initial inspection method and preliminary information about the discontinuities in the structure. This method of analysis is described below. The calculation method also has disadvantages.

The most efficient methods of determining detectability are, apparently, the methods based on the application of the test samples. In addition, the same test samples can be used to assess the qualifications of NDT inspectors, as trainers, to assess the effectiveness of the NDT methods and means.

The ASME rules of (XI) have legalised the use of test samples for evaluation of inspection means and methods as well as to assess the skill level of NDT inspectors. The application of test samples is prescribed by the draft European standard for certification in the field of nondestructive testing (ENIQ). The technology used for manufacturing test samples is described in detail in Ref. 12. There is a detailed description of the test samples prepared for study of the detectability of the inspection of key elements of the primary circuits of the VVER and RBMK reactors.

5.1.4 Experimental study of detectability of non-destructive testing using test samples

Some of the results obtained in various programmes relating to the safety of elements of the operating nuclear power plant reactors are described below. The results are selected in such a way as to reflect the possibility of solving the following problems: study of the detection of defects under given inspection conditions, comparative analysis of different methods of inspection, investigation of inspection technology to improve its efficiency; study the effect of the 'human factor' on the testing results. All of these problems were solved with respect to the conditions of VVER and RBMK reactors.

Full-scale test samples with hidden artificial plane and volume defects such as cracks, incomplete penetration, pores, slag inclusions, were produced for major items of equipment and pipelines of the primary circuit of the nuclear reactors.

Cracks and other defects of a given size and orientation were made in the welds in specific areas. The term 'full-scale' reflects the fact on the 1:1 scale the sample corresponds approximately to the selected element of

5.6 Test specimen of the main circulation circuit DN 800 of the RBMK-100 reactor.

nuclear power plant equipment. For example, Fig. 5.6 shows a test sample of the main circulation line (MCL) of the RBMK-1000 reactor made of the same grade of steel and electrodes as the piping in nuclear power plants. The pipe size (diameter and wall thickness) are also consistent with MCL. This test sample has two welds in which 26 plane and volume defects were made.

The detectability of inspection was assessed by defect detection probability (DDP).

Study of detection of defects at the given constant conditions of inspection

These studies can be illustrated by the example of inspection of a test sample of the element of the reactor vessel into which incomplete fusion defects between the cladding and the base metal were introduced.

A crack-like defect was located on the fusion boundary and was sometimes difficult to detect. Ultrasound testing was the main inspection method of incomplete fusion defects. However, it is fraught with considerable difficulties which are caused by various conditions of propagation of ultrasound in austenitic and pearlitic metal, as well as in the fusion zone between them. In this regard, detection of such defects is difficult and associated with many objective and subjective factors such as qualifications, fatigue, psychological and physical condition of operators, etc.

The aim of the study was to evaluate the detectability of incomplete fusion defects between the corrosion-resisting cladding and the base metal of the casing by different ultrasound operators in conditions close to the actual conditions. To assess the detectability of incomplete fusion defects between the corrosion-resisting cladding and the base metal, it is necessary to know accurately in advance the position, size, shape and nature of the defects. For this purpose, a special test-sample with the size of $720{\times}960{\times}100$ mm, made of steel 15Kh2NMFA, was produced. The sample surface was clad with two corrosion-resisting layers in accordance with the manufacturing technology of the VVER-1000 reactor. The material of the first layer was steel 07Kh25N13, the second layer was made of 04Kh20N13G2B steel.

The coordinates of the location of discontinuities and their size were chosen randomly using a random number generator. The area of the discontinuities was chosen so as to have fewer defects and more acceptable size equal to 20 mm^2 in accordance with the PK1514-72.[73] Defects with the area of 12, 20, 29, 63, 113 mm^2 were produced in the test sample. The number of discontinuities of different standard sizes was: first 12, second 12, third 16, fourth 10, fifth 12. A total of 62 defects were made. Four defects were located in the zone close to the edge of the plate where detection is difficult, so they were excluded from consideration.

The test sample was subjected to verification by VNIIAES ultrasonic flaw inspection highly skilled operators invited from different nuclear power plants. The DUK66PM flaw detector owned by one of the stations was used. Inspection was carried out according to the rules valid for irradiating the fusion zone of the casing with the corrosion-resisting cladding on the side of the base metal with a direct searching probe with a frequency of 2.5 MHz.

Operators were not given special inspection conditions, they were asked to implement ispection in accordance with the same requirements as when working on the reactor vessel at their plants. The length of service of operator No. 1 was 2 years; operator No. 2 5 years; operator No. 3 2 years; operator No. 4 6 years.

The operators recorded all discontinuities and classified them to acceptable, with an area of up to 20 mm², and unacceptable, i.e., with a large area.

As a result, the operator No. 1 detected 3 out of 9 inspected discontinuities with the area of 12 mm² which equalled 33,3% of the 12 discontinuities with the area of 20 mm² the operator identified as 3, i.e. 26%. Of the 15 embedded defects with an area of 29 mm² 6 were detected and 3 of them were rated as acceptable, i.e. having an area less than 20 mm². The overall detection rate was 40%, but with the correct evaluation taken into account (3 defects) it was 20%. The total detectablity of the defects with the area of 63 mm² was 50% (5 out of 10 embedded defects), but in this case two defects were evaluated as acceptable, i.e., having an area smaller than 20 mm². Therefore, based on correct assessment, the detectability was 30%. Of the 12 defects with the area of 113 mm² 8 defects were found, which accounted for 66.6%, but with such large defects 2 were rated as acceptable and, consequently, the detectability was 50%. From the total of 58 defects 25 were detected, i.e. 43.1%, but with the correct evaluation of the permissibility of defect detection was 34.5%. Similar results were obtained by all other operators (Table 5.1).

The results can be represented graphically (Fig. 5.7).

The horizontal axis in Fig. 5.7 gives the size of the defect in the form of its area S, the vertical axis is the value that can be interpreted as the probability of non-detection of a defect with size S:

$$P_{nod} = 1 - N_{det}/N_{emb} = 1 - R_{d.d},$$

where N_{emb} is the number of defects embedded in a sample of this size; N_{det} is the number of detected defects of the same size; $P_{d.d}$ is the value which can be viewed as the probability of defect detection.

These results indicate the probability of non-detection of a defect at ultrasonic flaw inspection is relatively high and decreases with increasing size of the inspected defect.

Differences in the curves X_1–X_4 reveal the influence of the subjective factor on the inspection results. The curve Σ describes the overall result of

Table 5.1 Detection of incomplete fusion defects in a plate of steel 15Kh2MFA with a cladding when using a test sample

Defect size, mm²	Number of defects	Operator No. 1		Operator No. 2		Operator No. 3		Operator No. 4	
		Number	%	Number	%	Number	%	Number	%
12	9	3	33.3	1	11.1	-	-	-	-
20	12	3	25	2	16.6	(1)	-	-	-
29	15	3 (6)	20 (40)	2(4)	13.3	1	6.6	3	20
63	10	3 (6)	33 (50)	4	40	3	30	3	30
113	12	3 (8)	50 (66.6)	9	75	4	33.3	7	58.3
Average detectability		34.5 (43.1)%		31 (34.5)%		26.4%		35%	
Detection of unacceptable defects		35.2 (44)%		44 (50)%		26.4%		35%	

Note: 1. Values in parentheses indicate the number of identified defects of this size, and detection without taking into account estimates of the size of defects. 2. Operator No. 3 classified a defect with area of 20 mm² as unacceptable, i.e. having a large area. 3. The time spent by operators in inspection: operator No. 1 2 h 30 min; operator No. 2 2 hr 40 min; operator No. 3 1 hr 15 min; operator No. 4 2 h 05 min.

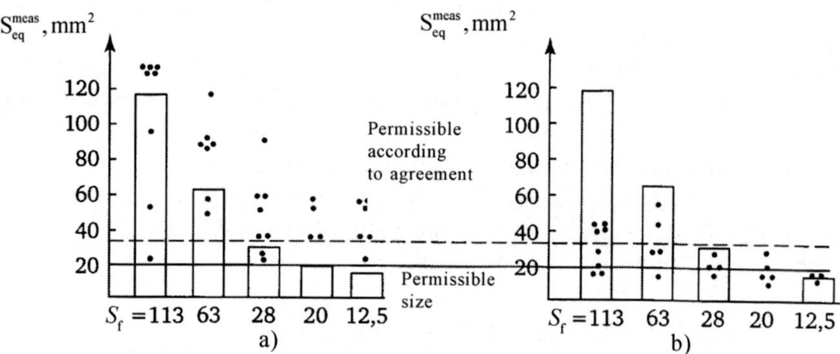

5.8 Results of inspection of the test sample with incomplete fusion on the side of the base metal (a) and the cladding (b).

control of four NDT inspectors.

In a second experiment on the same test sample, but with a different composition of NDT inspectors, the results of the inspection on the side of the base metal and on the cladding side were compared (Fig. 5.8). It is evident that inspection on the side of the base metal leads to excessive rejection of defects.

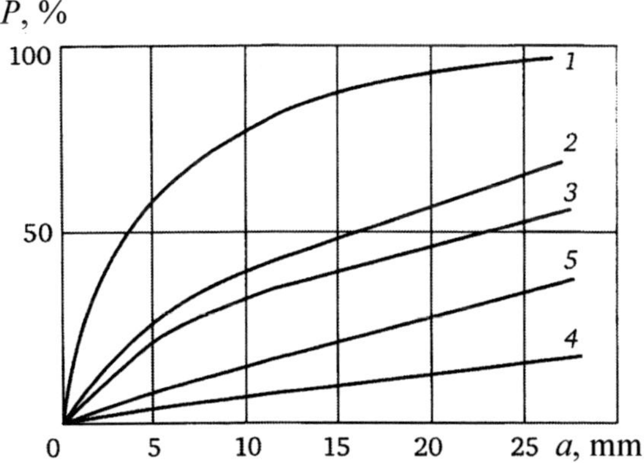

5.9 Dependence of the detectability of defects in inspection using different techniques, on the length of the defect: 1) method No. 1 as used by developers; 2) method No. 2 as used by developers; 3) method No. 2 as used by nuclear power plants; 4) inspection in 1988 in a nuclear power plant; 5) inspection in 1991 in a nuclear power.

Comparison of different inspection methods

In this case, the results of inspection carried out by different methods are plotted in the same coordinate system: the probability of defect detection – the defect size. Figure 5.9 shows the results obtained by five different inspection methods, and the curves 1–3 were obtained using test samples from a pipeline with a diameter of 55 mm made of austenitic steel. The test sample of the piping DN 500 was made of standard welding technology for shells of pipes made of 08Kh18N9T steel with EA-400/10T welding electrodes 4 mm in diameter (Fig. 5.8).

Curve 1 (see Fig. 5.9) was obtained by the method developed by NPO NIKIMT. The weld was inspected out by the dual-frequency method in a ADMT-21UB flaw detector with a converter with the prism angle of 50° at frequencies of 1.2–1.8 MHz and with a transducer with the prism angle of 40° at the same frequencies. Sensitivity setting for inspection with the converter with the prism angle of 50° was performed in a 3.5; 2.2 mm

Table 5.2 Smallest detectable level in inspection with a tranducer with the prism angle of 50°

Depth mm	Point defects	Defect length
0–10	Signal from the notch 2 dB	Signal from notch 4 dB
10–24	Direct beam: the signal from the notch 8 dB	Reflected beam: the signal from the notch 2 dB
24–34	Signal from the notch 8 dB	Signal from the notch 2 dB

Table 5.3 The detection of the total number of defects in the control of different methods

Parameter	Method No. 1	Method No. 2	Method No. 3
N_{det}	11	5	4
K_d	0.81	0.38	0.307
K_{ex}	-	3.6	1.92

Note. N_{det} is the number of detected defects coincident on the coordinate with the certificate of the sample; K_d is the coefficient of defect detection having the form of the ratio of the number of detected defects, which coincide with the certificate for the sample, to the number of defects embedded in the sample, K_{ex} is the coefficient of excessive rejection in inspection.

levelwith a once- reflected beam. Sensitivity setting for inspection with the converter with the prism angle of 40° was performed in a 3.5; 2.0 mm notch in the base metal in the direct beam. The lowest level recorded in inspection with the transducer with a prism angle of 50° is shown in Table 5.2.

The lowest level recorded in inspection with the transducer with the prism angle of 40° corresponded to a signal from a level of 8 dB.

The inspection No. 2 (Fig. 5.9, curves 2 and 3) was developed by one of the institutes servicing thermal power engineering. Curves 2 and 3 were obtained by the method No. 2 by different groups of NDT inspectors: curve 2 by the developers of the method, curve 3 by the NDT inspectors from one of the nuclear power plants.

The method No. 2 is described in the document MU 34-70-023-86. The acoustic customised transducer produced a slowly diverging acoustic beam of the shear wave with the main parameters: frequency f = 1.3 MHz; input angle α_0 = 48°, width of the angular capture A = 7, the amplitude of the echo signal at 6 dB from the cylindrical side reflector with diameter of 6 mm SOP No. 2.

The weld was inspected with a standard flaw detector UD 2-12 with an acoustic beam reflected once from the inner surface of the sample from twosides of the weld.

Curves 4 and 5 (see Fig. 5.9), obtained by the computational method,[12] integrally reflect the state of inspection of pipelines DN 500 at the nuclear power plant in 1988 and 1991, respectively.

In assessing the detectability of defects by various techniques it was found that the slag and cracks, as well as the total number of defects are detected more efficiently by the method No. 1 (Table 5.3).

Additional studies were subsequently carried out using an automated ultrasound inspection system 'Sumiad' together with X-ray inspection.

Automated ultrasonic inspection was performed by standard technology using the Sumiad automated inspection system developed by Technatom (Spain) with printout of the test results. The general view of the apparatus, mounted on a test sample, is shown in Fig. 5.10. Sensitivity was set in

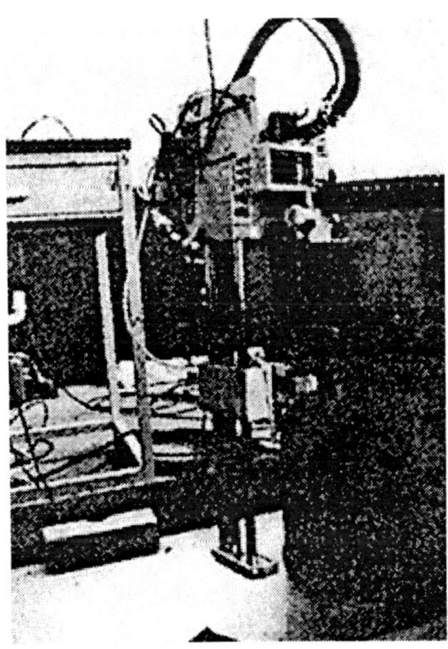

5.10 Inspection of the test sample DN 500 in automated ultrasonic equipment Sumiad (Spain).

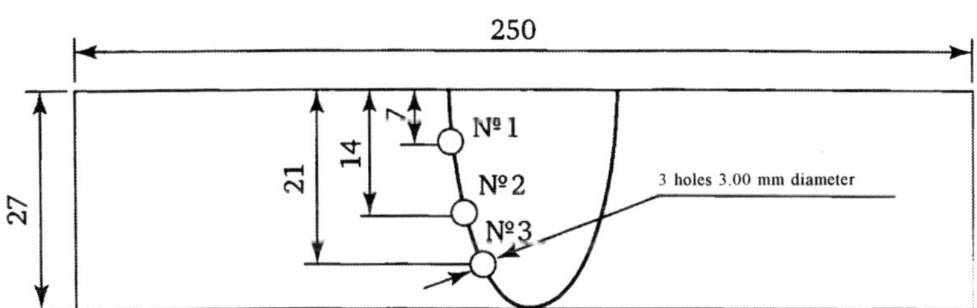

5.11 The standard sample of the enterprise.

accordance with the methodology PNAEG-7-032-91 on the basis of a signal from the cylindrical reflectors 1, 2 and 3 in the SPS (standard plant sample) with a diameter of 3 mm (Fig. 5.11). Inspection conducted longitudinal wave transducers SEW-45-F2 and SEW-60-F2 produced by Yuzhtekhenergo with the ultrasonic beam at angles of 45 and 60°.

Manual inspection was carried out by experts of VNIIAES using transducers of two types:

• SEW-45-F2, SEW-60-F2 (longitudinal wave transducers with a frequency of 2 MHz with angles of entry 45 and 60° respectively, the manufacturer Yuzhtekhenergo (L'viv);

• PEP60-2.5, PEP45-2.5 (longitudinal wave transducers with a frequency of 2.5 MHz with input angles of 45 and 60° respectively, manufacturer N.E. Bauman Moscow State Technical University).

Serial flaw detector UD-2-12 was used for inspection. The discontinuities whose amplitude of the echo signal exceeded the recording level were recorded.

Radiographic inspection of the test sample was carried out by specialists of the Expert Centre for Technical Diagnostics of Metals (MIKIMT) at three angles to the normal to the circumference of the joint: 0, +20 and −20° (sensitivity 0.3 mm).

The results of all experiments were used to prepare charts of defects in the weld test specimen. The following factors were taken into account when analyzing the results:

- inaccuracy in the coordinates of the defect location introduced during sample preparation;
- measurement inaccuracy in radiographic inspection, for example, if the radiographs were taken with overlapping of the sector size, the inaccuracy in the range of ± 17 mm was also made in the coordinates of defects on the overall chart;
- measurement inaccuracy in the coordinates of the detected discontinuities in manual and automatic ultrasonic testing, for example, the effect of the wall thickness of the sample.

In addition to the defects mentioned in the producer certificate, additional discontinuities can form in the welded joint of the test sample. The positive conclusion on the existence of such discontinuities was made if indications from discontinuities in the near coordinates were recorded by at least two different methods of ultrasonic testing and/or radiography.

The detectability of discontinuities was assessed using the coefficient of overall detection of discontinuities (i.e., the ratio of the number of detected discontinuities from the number taken into consideration to the number of discontinuities taken into account).

Table 5.4 and Fig. 5.12 show the results of detection of different types

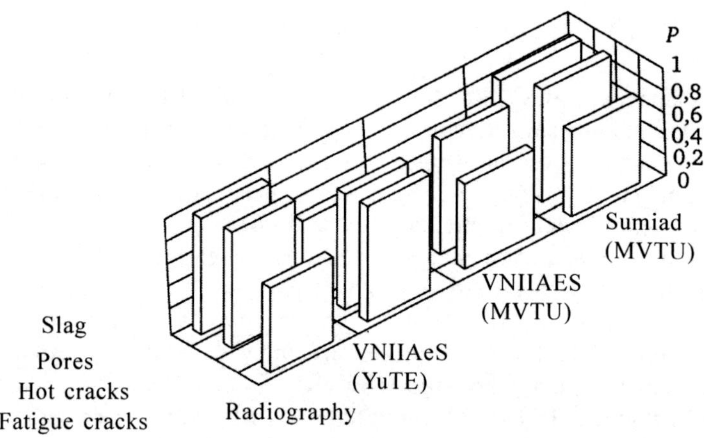

5.12 Detection of discontinuities of different types using different methods.

Table 5.4 Detection of embedded discontinuities using different techniques (weld No. 1)

Method	Pore 3×1	Slag 5×5×20	TU 5×5	Slag 3×3×15	Pore 10	TU 5×10	TG 15
Radiography	+	+	+	+	+	+	-
VNIIAES (YuTe)	-	+	+	-	-	+	+
VNIIAES (Bauman)	-	+	+	-	+	+	+
Sumiad (YuTE)	+	+	-	-	+	+	+

Methods	Slag 3×3×20	TG 15-17	Pore 3×10	TI 5×5	Slag 3×3×20	TI 20	Total
Radiography	+	+	-	+	+	-	10
VNIIAES (JT)	+	+	+	-	-	+	9
VNIIAES (Bauman)	-	+	+	+	-	+	9
Sumiad (YuTE)	+	+	+	+	+	+	11

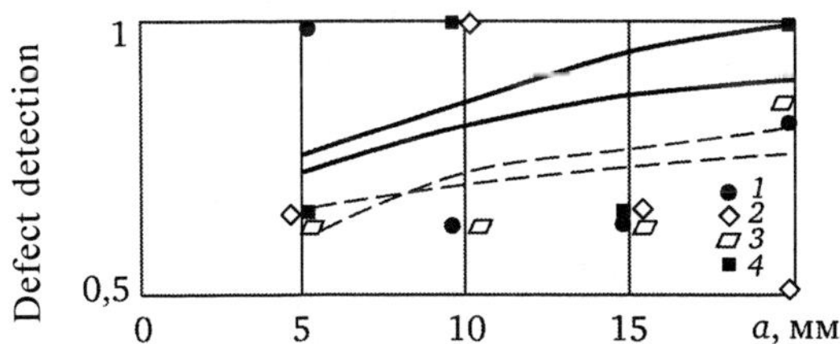

5.13 Dependence of the detection of discontinuities on their length in inspection by different methods: 1) radiography; 2) VNIIAES (YuTE) 3) VNIIAES (Bauman Institute), 4) Sumiad (YuTe).

of defects. Table 5.5 and Fig. 5.14 summarise the inspection results. From these results it follows that the highest total detectability of cracks, pores and slag inclusions was observed in inspection by Sumiad equipment where the overall detectability coefficient was $K_d = 0.8$. The excessive rejection coefficients, listed in Table 5.5, are tentative and require further verification. It is also noteworthy that, according to radiographic inspection, all the detected discontinuities which were represented by cracks, were in fact identified as slag. This can play a negative role in assessing the risk of a defect.

Table 5.5 Integrated assessment of the detection of discontinuities

Parameter	Radiography	VNIIAES (Bauman)	VNIIAES (YuTe)	SUMIAD (YuTe)
N_{det}	15	19	15	20
K_d	0.60	0.76	0.60	0.80
K_p	0	1	4	4

The dependence of the detectability of discontinuities on the length of the defect was estimated as follows. The data on the length of the defects were represented by the data from the test sample certificate, and investigations were carried out on discontinuities corresponding to those embedded in the preparation of the test sample and confirmed by at least two different methods of ultrasonic testing and/or radiography. The dependences obtained are shown in Fig. 5.13.

The detectability of defects using different inspection methods of was also analysed for the main circulation line DN 800 of the RBMK-1000 reactor.

A test sample (see Fig. 5.6 and 5.14), made of three coils of pipe DN800 was used in this case. The pipe material was steel 22K, clad with an austenitic deposit (in this case, imported steel KREZELSO was used). The test sample contained two welded joints: repair and assembly. In this book, attention is only given to the results of inspection of the assembly weld No. 1 in plane and volume defects were produced (incomplete fusion of the weld root, incomplete fusion with the edge, slag and pores). Assembly weld No. 1 (Fig. 5.15) is based on standard technology of manual arc welding with

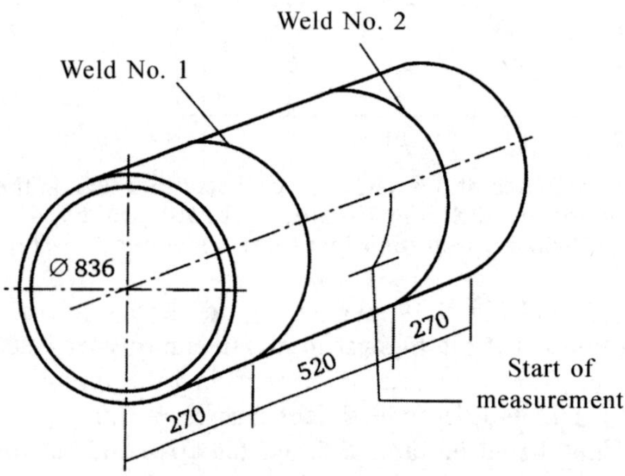

5.14 Diagram of the test sample DN800 of the main circulation pipeline of the RBMK-1000 reactor.

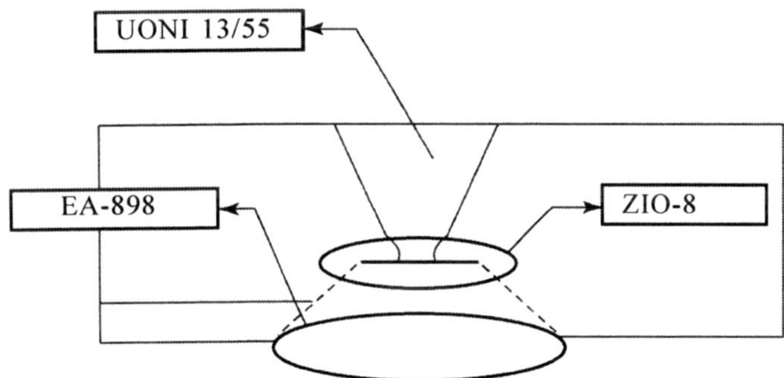

5.15   Assembly weld No. 1 (cross section).

electrodes UONI 13/55 with the transition layer welded with electrodes ZIO-8 and EA 898 (cladding layer).

The welded joint was described by a confirmed certificate showing the coordinates, size and type of defects. Due to confidentiality of the data for the coordinates of the location of the defects each defect was assigned a number.

Manual ultrasonic testing was carried out as 'a blind' test, i.e., NDT inspectors were not aware of the number and coordinates of the location of defects. Serial flaw detectors UD2-12 and UDTs-107 were used, together with Priz transducers with the entry angles of 0, 50 and 65°. Sensitivity was adjusted using the SO-2 reference sample in accordance with GOST 14782-86. Regular NDT inspectors of four nuclear power plants with RBMK reactors carried our inspection.

Automatic ultrasonic testing was performed as 'a blind' test using Tomoscan equipment, followed by data analysis and comparison of reports at the TomoLuis workstation (development of the Swedish company TRC).

The transverse welded joint was inspected by the non-parallel TOFD-method with scanning in the clockwise direction. Focusing was carried out on 2/3 of wall thickness. Additionally, to cover the entire volume of the welded joint, the echo-pulse method on longitudinal (45 and 60°) and transverse (45 and 60°) waves was used.

Radiographic inspection of the sample was carried out in three directions to the normal of the weld 0, +13 and −13° (Fig. 5.16).

A defect was regarded as detected if the coordinate X_f, mm, recorded in the report, corresponded to the interval

$$X_m - (a_m/2) - \Delta X_m < X_f < X_m + (a_m/2) + \Delta X_m,$$

where X_m is the mean detection coordinate according to the certificate; $a_m/2$ is the average length of the defect according to the certificate; ΔX_m is the confidence interval of location X_m of the defect on the certificate.

The inspection results indicate that most discontinuities were detected

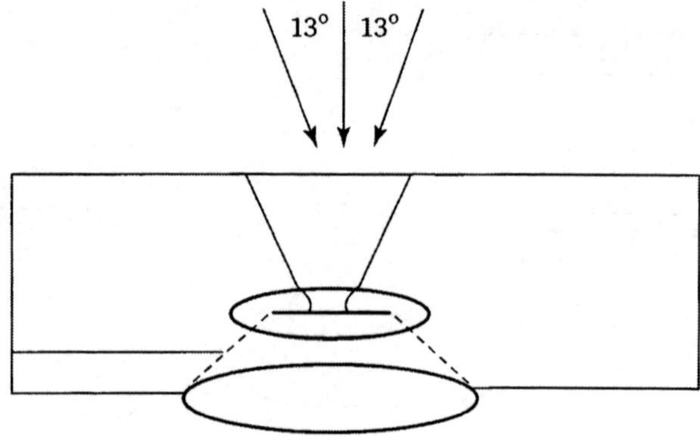

5.16   Diagram of radiographic inspection of the assembly weld No. 1 in test sample DN 800.

by automatic ultrasonic testing and radiography. However, the number of detected defects can not be regarded as a complete characteristic of detectability since it does not take into account the size and, hence, the importance of the detected discontinuities. In this case, radiography did not detect two defects.

The probability of detecting defects of the given lengthis determined by the formula:

$$P = n_{deti} \, / \, N_{embi,}$$

where n_{deti} is the number of detected defects with the given length; N_{embi} is number of embedded defects with this length.

The results calculated by this formula are presented in Table 5.6 and Fig. 5.17. The calculations were performed for radiography and automated ultrasonic testing (NDT inspector Nos. 12, 18 and 11 who found the maximum, minimum and average number of defects, respectively).

As seen in Fig. 5.17, the highest detectability was recorded for automatic ultrasonic testing. Combined application of TOFD and the pulse-echo technique in automatic ultrasonic testing allows conclusions to be drawn about the type of detected defect. In the conclusion, drawn up by TRC experts, the given type of defects is identical with the certificate data of the test sample and is confirmed by radiography.

Investigation of inspection methods to enhance their effectiveness

In this case, the effect of various elements of inspection methods (inspection duration, scanning method, frequency, etc.). As an example, Fig. 5.7 and 5.18 presents the data on the impact on repeated inspection on the detectability of defects.

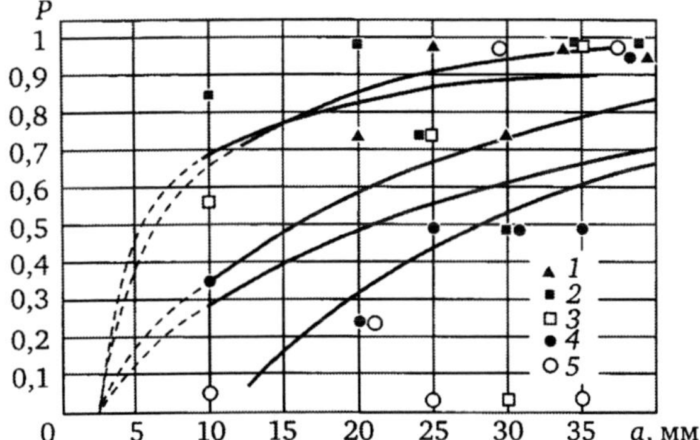

5.17   Dependence of the probability of detecting a defect on the linear size of defects: 1) automatic ultrasonic inspection; 2) radiography, 3) ultrasonic inspection, No. 12, 4) ultrasonic inspection, No. 11, 5) ultrasonic inspection, No. 18.

Table 5.6   Probability of detection of defects of different length

Methods	To 15	From 15 to 20	From 20 to 25	From 25 to 30	From 30 to 35	From 35 to 40
Radiography	0.86	1	0.75	0.5	1	1
Automatic ultrasonic	0.84	0.75	1	0.75	1	1
Ultrasonic, No. 12	0.58	0.5	0.75	0	1	1
Ultrasonic, No. 18	0.04	0.25	0	1	0	1
Ultrasonic, No. 11	0.37	0.25	0.5	0.5	0.5	1

Curve Σ (see Fig. 5.7) reflects the total result of inspection carried out by four NDT inspectors. It is seen that the curve Σ lies significantly below each of the curves obtained by NDT inspectors X_1-X_4. This means that the defectiveness of the component can be reduced by a factor of 2 or more only by organisation of repeated inspection by different NDT inspectors.

Figure 5.18 shows that the effective number of inspections is 6. Subsequent inspections ($7^{th}-10^{th}$) do not lead to any further detection of defects in the inspected sample (in this case, pipes 800 mm in diameter, made of steel 22K representing the model of the main circulation line of the RBMK reactor were inspected). Each inspection was performed by manual ultrasonic testing by different NDT inspectors. Curve 2 in Fig. 5.18 reflects the individual results of each NDT inspector.

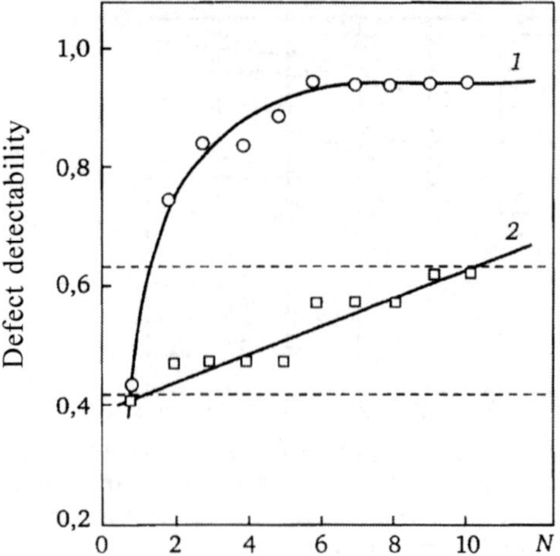

5.18   Dependence of the detectability of defects on the number of inspections (1) carried by different NDT operators (2).

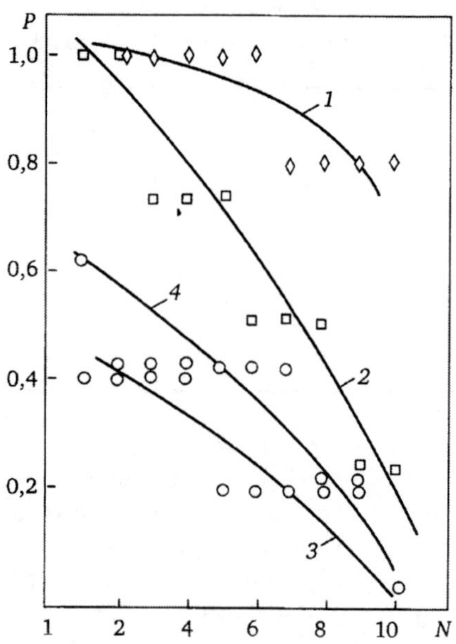

5.19   Dependence of the detectability of defects on defect type and serial number of NDT inspector *N* (the test sample DN 800).

Investigation of the influence of the 'human factor' on test results

The influence of the 'human factor' on the test results can be seen from Fig. 5.5, where the curves X_1-X_4, obtained by different NDT inspectors under the same experimental conditions, significantly differ.

Another experiment involved ten NDT inspectors (Fig. 5.19). The analyzed the test sample (DN800) contained four types of defects (lack of penetration, incomplete fusion, slag, pores). The NDT inspectors were ranked based on the results of inspection of one type of defect (the best was assigned '1', the worst '10'). It turned out that the nature of the ranking does not depend on the type of inspected defect. The best NDT inspector shows the best results for all types of defects. The worst case shows the corresponding results also for all kinds of defects. Thus, the 'human factor' in flaw inspection in this experiment was stable and a major factor influencing the quality of inspection.

To conclude this section it can be noted that the described inspection results obtained in assessing the detectability of NDIS, used in nuclear power plants, indicates the great importance of these studies for predicting and ensuring the required service life, reliability and safe operation of pressure vessels and piping of nuclear power reactors. The next section presents the main results obtained in the framework of PISC, which provide a more comprehensive view of this important issue.

5.1.5 Results obtained in the PISC programme

The research program to investigate the reliability of structural elements working under pressure in nuclear power plants of American manufacture (Programme for the Inspection of Steel Components, PISC) was initiated by the Commission on the strength of reactor casing (USA) and was carried out during 20 years in the Commission of the European Community (CEC) and the OSCE with the participation of 18 countries. The programme was implemented in three phases. After completion of the PISC-I programme, the PISC-II programme was developed and implemented, and this was followed by the PISC-III programme. The results of these studies are presented in accordance with the publications Ref. 76–83.

The aim of the PISC-I programme was to study the reliability of determination of the linear dimensions of the defects by ultrasonic testing, recommended for NDIS of equipment of nuclear power plants by the ASME (XI) rules. The technique is based on 50% DAC calibration and is recommended for manual ultrasonic testing of welds and the heat-affected zone of the reactor vessels from the outside.

Two plates and one plate with a pipe branch with diameter of 18 inches with artificial defects (Fig. 5.20) were proposed for inspection. All defects were introduced during welding, and after ultrasonic testing they were cut at the Research Centre of the CEC in Ispra (Italy). Several test samples were also transferred for further investigation by the defect opening method at IPA (Stuttgart) and other organisations (a total of six organisations).

The results of non-destructive and destructive testing were processed in order to obtain the following results:
- probability of detection (POD);

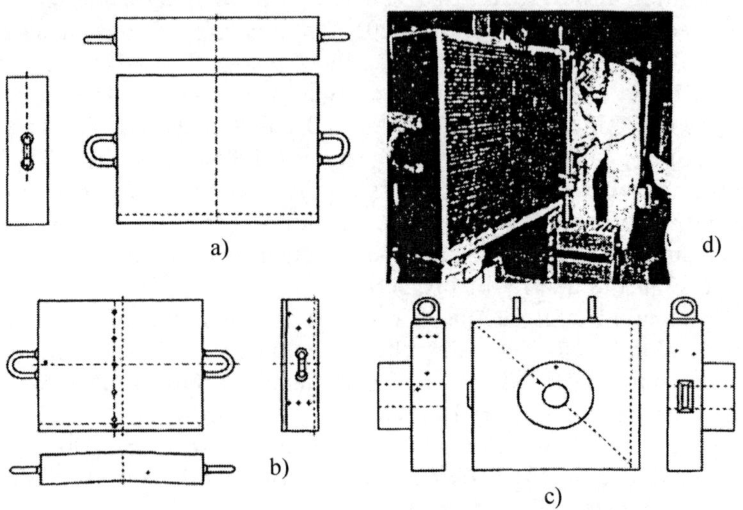

5.20 Test samples for PISC-I: a) plate no. 51/52; b) plate no. 51/53; c) view of the sample, d) nozzle 204.

- quality and errors in determining the location of the defect:
- quality and errors in determining the linear dimensions of the defects;
- probability of making correct or incorrect decisions on the basis of the inspection results accordance with the ASME (XI) rules.

The main results obtained within the framework of PISC-I are shown in Fig. 5.21. These results indicate that the reliability of NDT is unacceptable for nuclear engineering. It was shown experimentally that the probability of making wrong decisions in the analysis of test results is very high. In some cases, for example, for chains of defects it is actually equal to 100%.

In connection with the results, methods alternative to the ASME (XI) code were introduced to the PISC-I programme. The results obtained by additional methods were significantly better, but it was decided to continue research under the programme PISC-II.

The purpose of the PISC-II programme was the following:
- assessment of the effectiveness of various alternative methods of inspecting the elements of reactors in operation;
- define procedures acceptable for input inspection, preoperational and operational inspection;
- bring to the attention of supervisors the results of research and develop normative and technical documents (rules, standards) on the basis of these results.

One of the generalised results of studies of detection of defects produced by PISC-II is shown in Fig. 5.22. It can be seen that the detection of fatigue cracks and chains of fissures in the investigations has been greatly

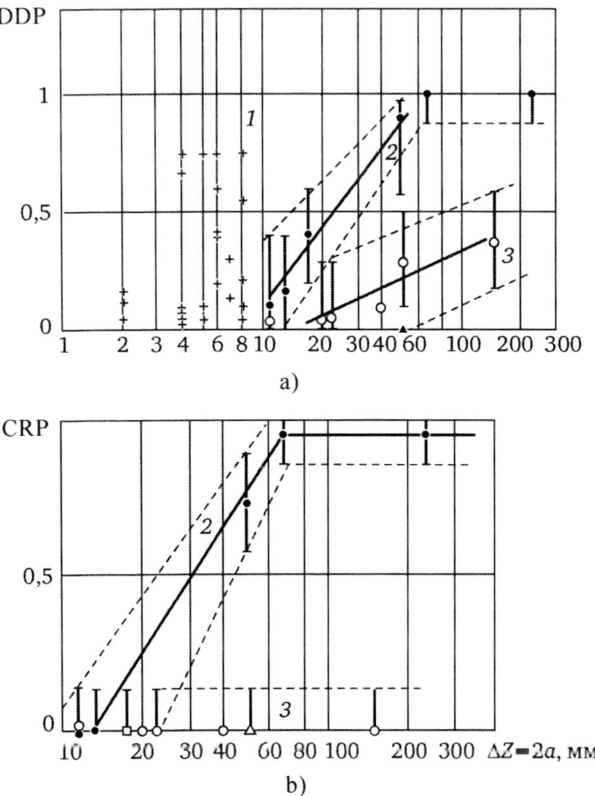

5.21 The main results of PISC-I: a) defect detection probability (DDP), depending on the size, b) correct rejection probability (CRP) of defects depending on their size; + – permissible defects; • – unacceptable defects (vertical cracks); o – chains of defects; □, Δ – difficult to detect defects because of their location; 1) bulk defect; 2) crack-type defect, 3 – chains defects.

improved, but not enough to consider the results satisfactory.

Following the PISC-II programme, the PISC-III programme was developed. This programme covers a wider range of problems associated with the reliability of NDIS. Eight areas of work were proposed:

Direction 1 (real deactivated structural elements) provided the study of reliability of inspection for the decommissioning of damaged elements of the reactor structures.

Direction 2 (inspection on a full-scale model of the reactor pressure vessel) is a direct continuation of PISC-I, PISC-II programmes but this time the study was as close as possible to the real conditions. The main volume of work was conducted at the Institute for Testing and Materials in Stuttgart (Germany). The full-scale model of the reactor provided an opportunity to test the robotic arms for NDIS of reactor casings.

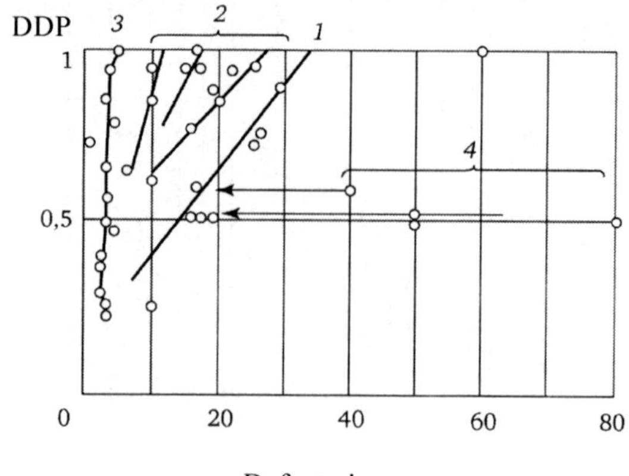

5.22   Defect detection probability (DDP) by the ASME method using 20% DAC, depending on the size of defects in the direction of wall thickness: 1) fatigue cracks, 2) volume–planar defects or hot cracks; 3) volume defects, including flat-bottom drilled holes 3 mm in diameter and gauge holes 9.5 mm in diameter, 4) chains of defects.

Direction 3 (nozzles and composite welds) focused on the inspection of nozzles and places for joining pipelines to the casing. Since the shell is made of a pearlitic steel and piping of an austenitic steel, the programme included research on test samples containing defects in the composite welds.

Direction 4 (inspection of austenitic steels) provided the study of detection of stress corrosion cracking with specific features of the structure of austenitic steels taken into account.

Direction 5 (inspection of steam generator tubes) takes into account the specific issues of eddy current testing of steam generator tubes. Investigations were carried out on both the tubes cut out from steam generator damaged in operation and containing real defects, and the tubes containing artificial defects.

Direction 6 (mathematical modelling of NDT) included a range of studies related to the mathematical processing of signals and obtaining the maximum information about defects from the inspection results. Mathematical models and computer programs provided by 16 organisations from eight countries were considered.

Direction 7 (human factor) included studies of the role of the human factor in inspection and its influence on the inspection results.

Direction 8 (improving the legal and technical instruments in the field of NDT of the member countries of the PISC programme) included the following activities:

- informing the technical committees, involved in developing codes and standards, about the results of PISC;
- critical analysis of national and international technical documents

by the PISC members;
- preparation of technical reports on issues of rules and standards and transfer them to the authorities responsible for the development of normative documents.

The main results of these studies are summarised concisely below. The results of studies of the reliability of the linear dimensions of discontinuities in the direction of the wall thickness in inspection on the full-scale model of the reactor vessel are described in Fig. 5.23. As can be seen, the error in determining the depth of the discontinuity may reach significant values, both in the direction of understating the size and in the direction of overstatement. Figure 5.24 and 5.25 show defects No. 8 and No. 9 (see Fig. 5.23) for which the errors in detection were the highest.

The results of the study of the probability of detecting defects in austenitic steel are shown in Fig. 5.26. 5.27, and the test samples on which

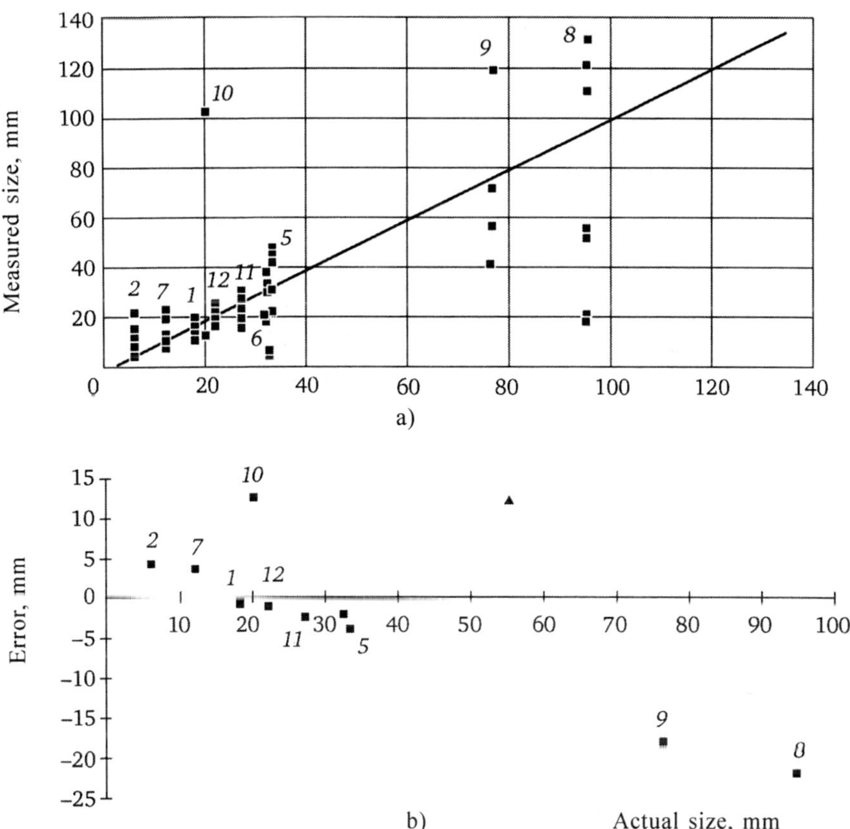

5.23 Variation of the size of defects (except for defects Nos. 3 and 4) according to the results of inspection by all groups of NDT inspectors (a) and dependence of the error in determining the size of defects in the direction of the wall thickness of the reactor vessel on the actual size of the defect (b).

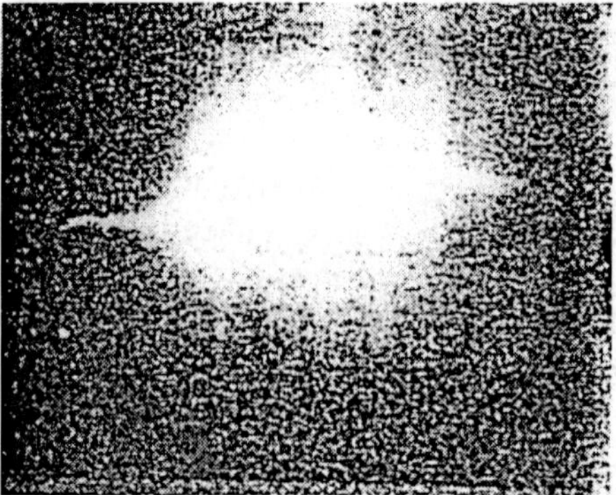

5.24 Trepan of defect No. 8. Result of radiographic inspection (for Fig. 5.23).

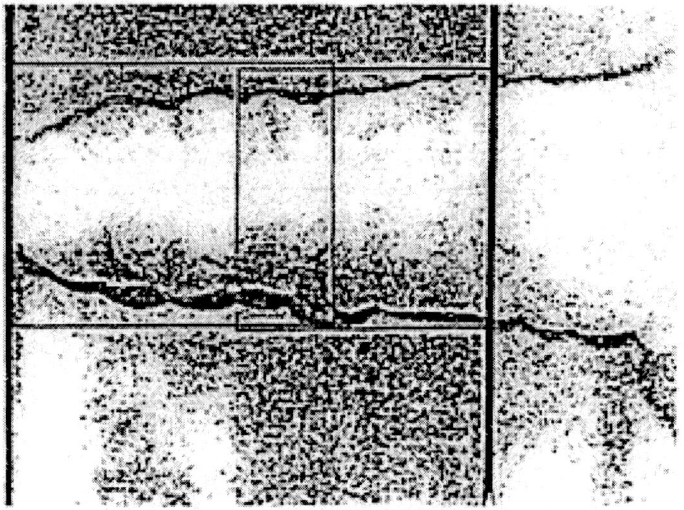

5.25   Macrograph of defect No. 9 (for Fig. 5.23)

the results were obtained are presented in Fig. 5.28. The samples contained both intergranular stress corrosion (ISC) cracks, fatigue cracks and also the cracks produced by the electrospark method. The data presented in Fig. 5.26 and 5.27 show that that the ISC cracks are more difficult to detect than other cracks. Even for the cracks that make up 50–70% of wall thickness the probability of non-detection is very high.

Information on the detectability of defects, obtained in studies in direction 3, is shown in the diagrams in Fig. 5.27 and 5.28.

5.1.6 Reasons for non-detection of defects

The detectability of defects is affected by a large number of factors,

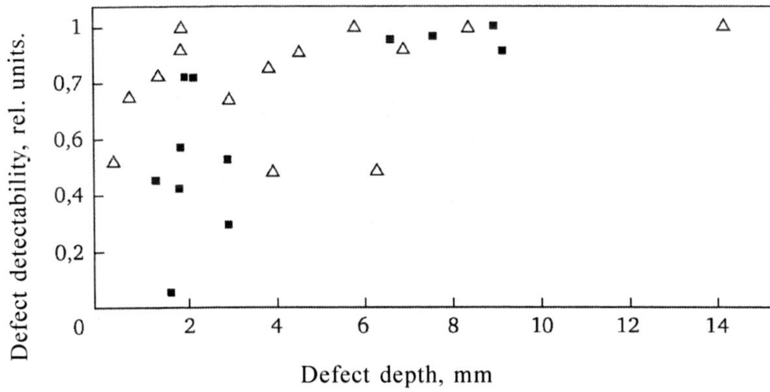

5.26   Dependence of the detectability of the defects on their depth in the direction of wall thickness in austenitic steel samples: ■ – intercrystalline corrosion cracks; Δ – other cracks.

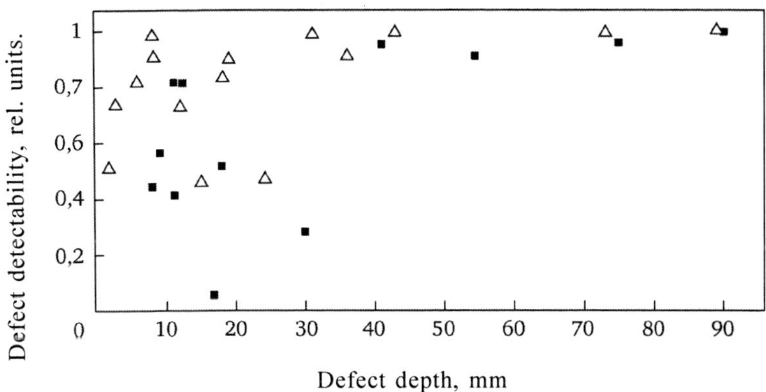

5.27   Dependence of the detectability of the defects on their depth in the direction of wall thickness in austenitic steel samples crack (defect depth is expressed in percent of wall thickness): ■ – intercrystalline corrosion cracks; Δ – other cracks.

including the physical features of NDT methods and materials in which the defect is situated, the defect type, the characteristics of tools and inspection techniques, the environment and inspection conditions, individual characteristics of NDT inspectors.

Despite numerous studies into the causes of insufficient reliability of quality inspection carried out both in Russia and abroad, only qualitative or semiquantitative description of the effect of various factors in reducing the reliability of inspection can be provided at the present time.

There are various classifications of factors affecting the reliability of inspection. V.G. Shcherbinsky and N.P. Aleshin[84] attribute the insufficient reliability of inspection to the errors of inspection, including: systematic, random, and misses.

5.28   The experimental test specimens of austenitic steel (for Fig. 5.26 and 5.27), PISC-III programme.

Systematic errors are caused by factors operating in the same manner in inspection under the same conditions or they change the results if the inspection conditions change.

Random errors lead to the spread of the results of repeated measurements with respect to some mean value.

Misses constitute gross errors associated either with the behaviour of the operator or with undetected hardware failure.

The authors of this book considered the effect of these errors in conjunction with the state of inspection means, methods, and the type and orientation of the defect.

V.N. Volchenko[75] analyzed the reliability of quality inspection and singled out the errors of the first (excessive rejection) and second kind (insufficient rejection). In this case, reliability D is estimated by

$$D = 1-\chi_\alpha-\chi_\beta,$$
[5.2]

where χ_α is the fraction of excessive rejection, χ_β the fraction of insufficient rejection.

In the PISC programme it was also attempted to isolate the impact on reliability of such factors as the physical method of inspection, inspection methods, excessive rejection and insufficient rejection, 'human factor', the type of defect, material.

The dependence of the detectability of discontinuities in:
- physical methods of inspection;
- type and orientation of the defect;
- inspection methods;

- material;
- type of weld;
- human factor;

is discussed below.

The effect of material on detectability is shown in Fig. 5.29. Comparative analysis of the detection of discontinuities in relation to the physical method of inspection was carried out in Ref. 82–86 and others. Figure 5.30 shows the results of the pilot study of detection of discontinuities by three inspection methods, depending on their linear size a.[86] The best detectability was obtained for ultrasonic testing, the worst for radiography (RG).

Ultrasound (US), radiography and magnetic power (MP) inspection methods were compared in Ref. 84 (Fig. 5.31)). In this case, the best detection was shown by US inspection, the worst by MP inspection. The

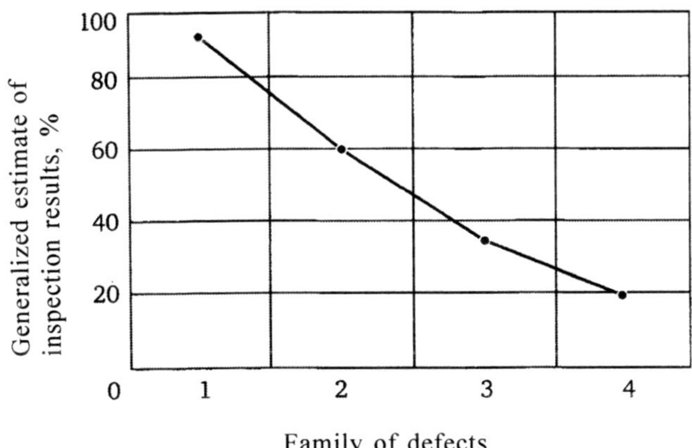

5.29 Effect of material in which the defect is situated: 1) forged steel, 2) fusion zone of the base metal with the weld; 3) clad austenitic stainless steel, 4) weld.

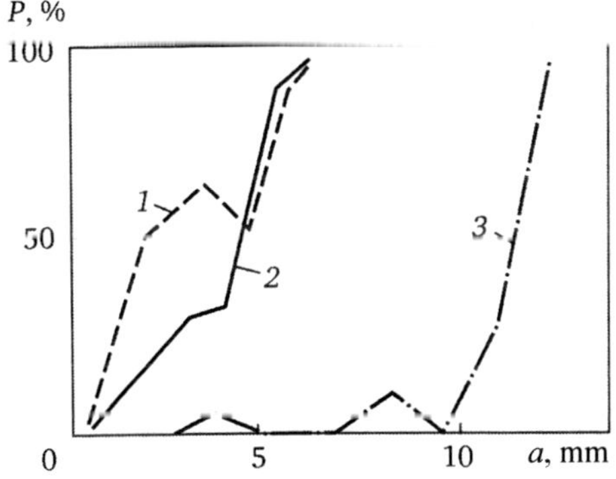

5.30 Comparison of three methods of detection of discontinuities: 1) ultrasound; 2) dye penetrant, 3) radiography.

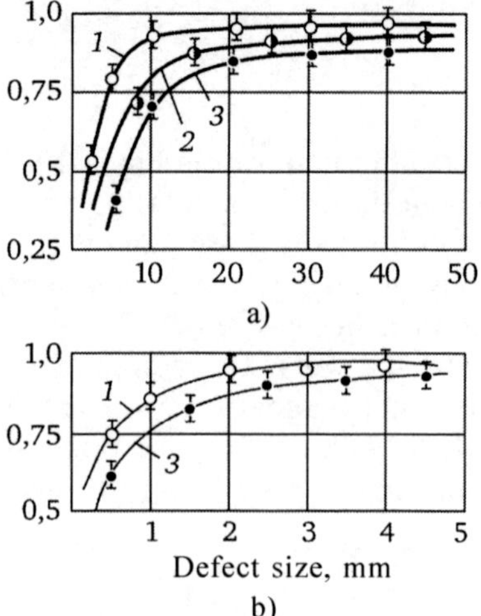

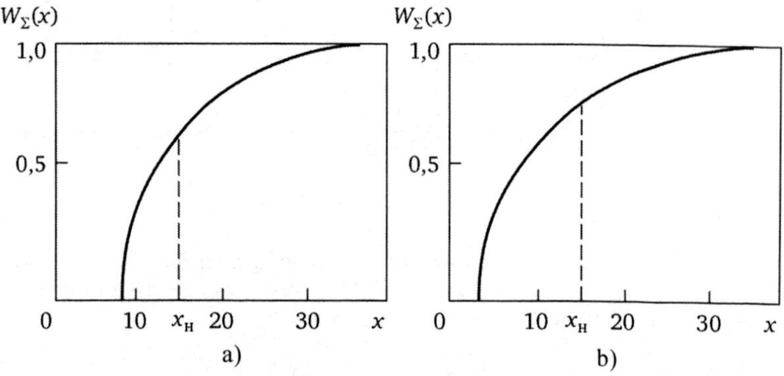

5.31 Detection of incomplete fusion and cracks (a), pores and slag inclusions (b) by different methods: 1) ultrasound, 2) radiographic, 3) magnetic powder.

5.32 Total detectability $W_\Sigma(x)$ of defects of various sizes in butt 40 mm thick by different methods: a) ultrasonic, $W_\Sigma(x) = 1-\exp[-0.17(x-9)]$ b) gamma inspection, $W_\Sigma(x) = 1 - \exp[-0.12(x-6)]$.

results were obtained in the Surguttruboprovodstroy Trust in inspection of welded joints of pipelines with a diameter of 820–1020 mm.

The best results obtained in US inspection as compared with RG inspection were also reported in Ref. 75 (Fig. 5.34).

Table 5.7 shows the results of a study to assess the relative information content of different methods of testing welded joints in pipelines with a diameter of 50–500 mm with a wall thickness of 3–10 mm. After inspection (more than 1000 joints) the weld metal was machined layer by layer and all defects were measured.[87] Table 5.6 shows that radiographic provides greater accuracy for bulk defects (90%), and the ultrasonic method – for

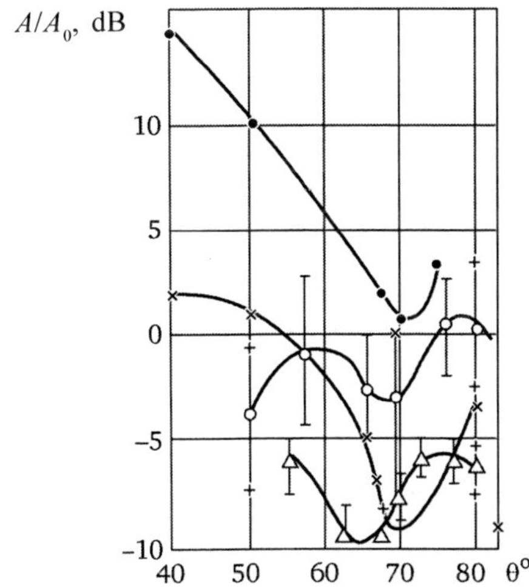

5.33 Amplitude of the echo signal from the angle of contact with the defect:
• – slit 3×2 mm deep; x – 3×2 mm notch; o – cracks 1–3 mm deep;
Δ – pitting corrosion, depth 0.8–2mm.

the planar defects with a small opening (90%).

Higher detectability of defects using ultrasonic inspection (as compared to RG) was recorded also in the studies in the PISC-III programme, although RG was used only in isolated cases. The results shown in Fig. 5.13 and 5.17 indicate that the ultrasonic method is somewhat better than RG, if applied automated ultrasonic testing (Technatom equipment and TRC installation are used). Manual ultrasonic testing showed worse results compared with the results obtained by automated ultrasonic testing as well as RG.

Estimates of the detectability of defects such as hot cracks by the dye penetrant method showed low detectability of defects. The total detectability of the defects is shown in Fig. 5.32.

The above results have already shown that detectability depends strongly on the type and size of the discontinuity. Thus, for a planar discontinuity its orientation with respect to the ultrasonic wave and the direction of radiation is important. Figure 5.33 shows the dependence of the echo signal from the angle of contact with the defect.[88]

The detectability is strongly by the volume of defects. Cracks under the effect of compressive stresses are especially difficult to detect. V.G. Shcherbinsky and N.P. Aleshin[84] showed the effect of the size of the defect, the type and method of inspection on the accuracy of detection. Figure 5.34 shows the dependence of the true d_s and equivalent d_{eq} dimensions of the real defects in the vertical plane. For bulky defects with $2 < d_s < 8$ mm, this dependence is approximated by $d_{eq} = 2d_s^{0,58}$, suggesting that the true size of bulk defects can be determined on the calibration curve. Measurement of planar defects with a single probe gives greater

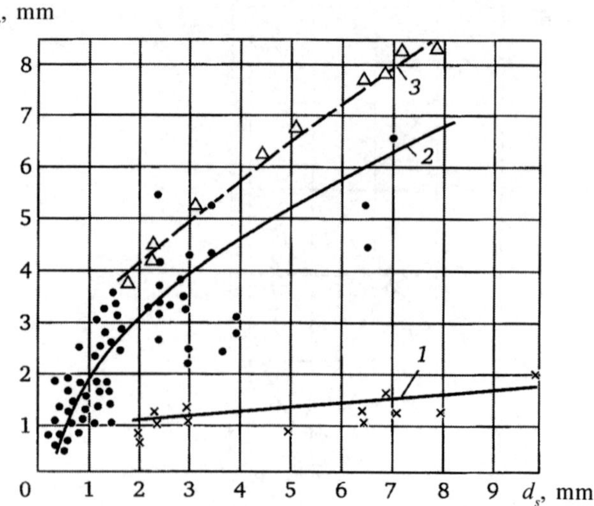

5.34   Correlation between the equivalent diameter d_e and its true size in the cross section d_s: 1) cracks in inspection with a single piezoelectric transducer, 2) volume defects, 3) crack in tandem inspection.

Table 5.7 Information content of methods and the reliability of detection of defects in welded joints of pipelines

Inspection methd	Identification of defects on the basis of type and appearance	Information content			Reliability of detection compared with opening,%	
		Minimum detected sizes of internal defects, mm			Bulk defects (pores, slag)	Planar defects (cracks, narrow incomplete fusion, etc.)
		Height,	Length	Width		
Radiography	Good	0.1-0.2	0.1-0.2	0.1-0.2	85-90	65-75
Gammagraphy	Satisfactory	0.2-0.3	0.2-0.5	0.3-0.5	75-85	60-70
Ultrasonic testing	Good for thickness 40 mm	0.8	0.5	10.5	65-70	85-90
Colour flaw test*	Satisfactory	0.3-0.4	0.5-1	0.5-1	75-80	55-65
Eddy current testing*	Cannot be distinguished	2-3	5-10	2-3	30-40	45-55

*Colour flaw test and eddy current testing for joints of pipelines with diameter 57–108 mm, thickness 3.52–4.5 mm

Table 5.8   A generalized estimate of detection of discontinuities by various physical inspection methods

The type of defect	UI	RG	MP	ECI	DP
Fatigue cracks under compressive stresses	Bad	Bad	Bad	Bad	Bad
Opened cracks	Good	Poor	Poor	Satisfactory	Satisfactory
Planar discontinuity type fusions	Good	Fair	Fair	Good	Good

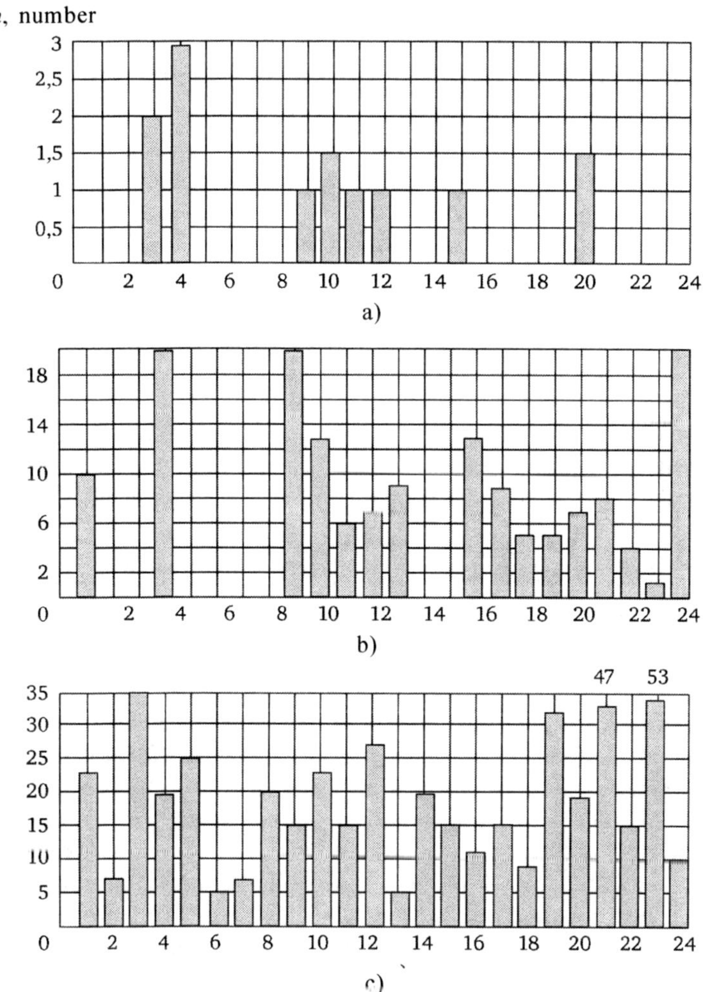

5.36   Results of decoding 26 gammagraphs by 24 radiographs with the service life of 0.5 to 10 years in radiographic testing: a) incomplete fusion, b) pores, c) slag inclusions.

accuracy in assessing their equivalent area. The use of the 'tandem' scheme (curve 3) provides better results.

The dependence of the detectability of defects on the material is summarised in Fig. 5.33. The graph was plotted using the results of the third direction of the PISC-III programme. In general, it can be assumed that the detectability of defects decreases with increasing heterogeneity of the micro-and macrostructure of the material and with increasing grain size. The detectability of defects is adversely affected by structural heterogeneity. Increased curvature of the surface of the component and more complicated shape of the welded joint and of the structure reduce the reliability of NDT.

Table 5.8 shows the generalised rough estimates of detectability of discontinuities in relation to the physical inspection method and type of defect.

Apparently, the strongest influence on the reliability of inspection is exerted by the human factor. Studies carried out by A.K. Gurvich, V.F. Luk'yanov, B. Margitroid and others have shown that the results of inspection are significantly influenced by the skill, work experience, long breaks in work, psychophysiological factors (stability of attention, temperament, etc.), social factors, incentives to work, working conditions, and the moral quality of NDT inspectors. The scatter in the results of inspection under laboratory conditions because of the influence of the human factor reaches tens of percent. Radiographic testing is also characterised by a significant difference between the test results obtained by different operators working under the same conditions (Fig. 5.35)[85].

In some cases, due to dishonest attitude or lack of incentive to work the detectability of defects can fall to zero even when using efficient inspection methods.

5.2 Residual defects as the most important characteristic of the state of the structure. Methods of determination

As shown in the previous section, in NDT there is always a finite probability of sending to service a structural element with a defect of the metal. The set of all missed defects sent to service due to imperfection of the methods and means of NDT is called the residual defectiveness. If he missed (undetected) defect reaches the critical size, the whole structure or its element can be destroyed. Since no information is available about the missed defect (the defect was not detected by the NDT methods), fracture occurs suddenly. Consequently, the question of obtaining information about the characteristics of the residual defects is a crucial issue in assessing the strength and operating life of structural elements.

5.2.1 Mathematical approximation of the detection of discontinuities, depending on their size

V.N. Volchenko[75] derived the following equation for the curve of detection of defects

$$W = 1 - \exp \lambda\,(x - x_0), \qquad [5.3]$$

where x_0 is the boundary smallest detectable defect size depending on the sensitivity of the inspection method; λ is a constant.

A. Zimmer used the equation of the same type. However, he used it for the probability of missing the defect P

$$P = \varepsilon + (1 - \varepsilon)\,\exp\,(-\gamma a), \qquad [5.4]$$

where ε and γ are constants.

Constant ε indicated that for any type of inspection there is always a probability ε of missing the defect as a result of operator's mistake. According to Ref. 86, the value $\varepsilon = 0.005$ was determined on the basis of a survey of a large number of experts – NDT inspectors.

In Ref. 89, etc. the probability of defect detection was described using the equation

$$W = (1-\varepsilon) - (1-\varepsilon)\,\exp\,[-\alpha\,(a-a_0)], \text{ for } a > a_0, \qquad [5.5]$$
$$W \equiv 0, \text{ at } a < a_0,$$

where ε is a constant characterising the fundamental limitations of this inspection method; α is a constant characterising the detectability of defects depending on their size; a_0 is constant related to the sensitivity of the inspection method.

Given the finite size of the vessel walls and pressure pipelines, commensurate with the size of defects, the constant ε can be omitted, and the influence of the human factor or of instrumentation and methodological shortcomings will be taken into account by the coefficient α. In this case, equation [5.5] transforms into equation [5.3].

The effect on detectability of two linear dimensions of the defect: the depth in the direction of the surface a and length c, was described in Ref. 90 by the equation

$$W = 1 - \exp\,[-\alpha(a-a_0)\,/a/c]. \qquad [5.6]$$

Equation [5.6] is valid in the region $a \geq a_0$, $c \geq a$.

There are also other equations for describing the detection of defects[29] but the curves obtained for these dependences are close to the curves obtained by the above-mentioned equations.

5.2.2 Quantitative assessment of residual defectiveness

If the function of initial defectiveness N_{in} (a, c) is known, and the distribution function of the defects detected in inspection is $N_{det}(a, c)$, the residual defectiveness N_{res} can be defined as

$$N_{res}(a, c) = N_{in}(a, c) - N_{det}(a, c). \qquad [5.7]$$

The number of detected defects depends on the initial defectiveness $N_{in}(a,c)$ and on the reliability of inspection that can be characterised by the probability of detection of defects $W(a, c)$.

$$N_{det}(a, c) = N_{in}(a, c)\, W(a, c). \qquad [5.8]$$

Substituting the last expression in equation [5.8] leads to

$$N_{res}(a, c) = N_{in}(a, c) - N_{in}(a, c)\, W(a, c) \qquad [5.9]$$

or

$$N_{res}(a, c) = N_{in}(a, c)\, [1 - W(a, c)]. \qquad [5.10]$$

Equation [5.10] is valid for the region where $W > 0$. This region is determined by the minimum values of discontinuities a_0, c_0, which can be detected by this inspection method. In the region $(a, c) < (a_0, c_0)$

$$W \equiv 0, \; N_{res} \equiv N_{in}. \qquad [5.11]$$

In a particular case it can be assumed that $N_{in} = A / a^n$, and the expression for W is [5.3], then

$$N_{res}(a) = \frac{A}{a^n} \exp\left[-\alpha(a - a_0)\right]. \qquad [5.12]$$

In repeated inspection by the same method, the number of detected defects is equal to

$$N_{det}^{II}(a) = N_{det}^{I}(a) W(a) \qquad [5.13]$$

or

$$N_{det}^{II}(a) = \frac{A}{a^n}\left(\exp\left[1 - \alpha(a - a_0)\right]\right)\exp\left[-\alpha(a - a_0)\right]; \qquad [5.14]$$

$$N_{res}^{I}(a) = N_{res}^{I}(a) N_{det}^{II}(a); \qquad [5.15]$$

$$N_{ost}^{I}(a) = N_{ost}^{I}\left(1 - \exp\left[1 - \alpha(a - a_0)\right]\right). \qquad [5.16]$$

Converting formula [5.16] for the k-th ispection gives

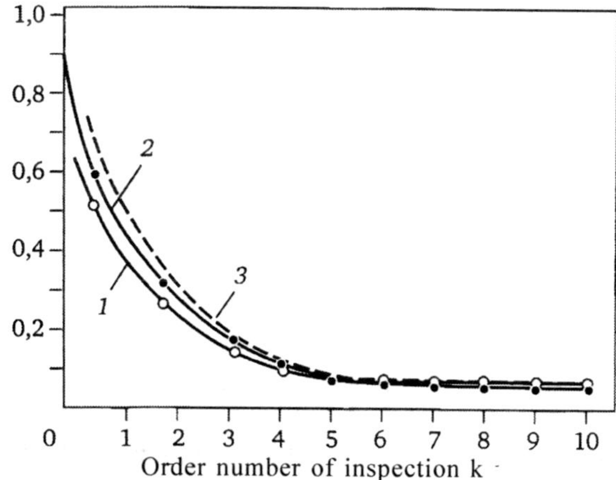

5.36 Residual defectiveness after k inspections; N_{res} is the number of undetected defects in inspection; N_{in} is the total number of defects in the test sample DN800, 1) experimental curve, 2) calculated curve at $\varepsilon = 0$, 3) calculated curve with $\varepsilon = 0{,}05$.

$$(N_{res})^k = N_{res}^{k-1}(1-\exp[-\alpha(a-a_0)]) \qquad [5.17]$$

or

$$(N_{res})^k = \frac{A}{a^n}\exp[-k\alpha(a-a_0)]. \qquad [5.18]$$

Calculations using equation [5.17] yielded the calculated curve (2) (Fig. 5.34). Curve 1 in Fig. 5.36 is the result of an experiment conducted on a test specimen DN800 made of steel 22K.

Figure 5.36 shows the satisfactory agreement between the calculated and experimental curves. The observed discrepancy in the curves for the number of inspections $k > 6$ is of fundamental nature. The theoretical curve 2 tends to 0 with the increase in the number of inspections, whereas the experimental curve tends to 0.05. The value $\varepsilon = 0.05$ is determined by the limited capacity of the inspection method. If the detection of defects is described by equation [5.5], setting $\varepsilon = 0.05$, then there is better agreement between the calculated curve 3 and experimental curves 1 for the number of inspections $k > 5$. Curve (3) was plotted using the modified equation [5.18]:

$$N_{res}^k = N_{in}\left(\exp[-k\alpha(a-a_0)+\varepsilon]\right). \qquad [5.19]$$

Equation [5.19] holds in the region

$$a \geq a_0 - \frac{\ln(1-\varepsilon)}{k}. \qquad [5.20]$$

5.2.3 Assessment of initial and residual defectiveness and detectability of defects using inspection results

Equation [5.8] can be written in the form

$$N_{det}(a) = \frac{A}{a^{-n}}\left\{1 - \exp\left[-\alpha(a - a_0)\right]\right\},$$

[5.21]

(action range for $a \geq a_0$).

Equation [5.21] can be used to predict the results of flaw inspection if the function of detectability of defects $W(a)$ and the initial defectiveness $N_{in}(a)$ are known.

The equations $W(a)$ and $N_{in}(a)$ can be estimated, as shown above, on the basis of analysis of defectiveness at the producer plant in conjunction with the specific technology of production and direct experimental studies of detectability of defects in test samples.

However, in most practical cases the equations $N_{in}(a)$ and $W(a)$ are not known. Equation [5.21] solves the inverse problem: from the known left-hand side of equation [5.21] to determine the right-hand side, i.e. determine the initial defectiveness N_{in} and detectability W.

Indeed, the results of flaw inspection can be used to calculate the function $N_{det}(a)$. This function can be defined as the envelope of the histogram of inspection results (Fig. 5.37). The problem of determining

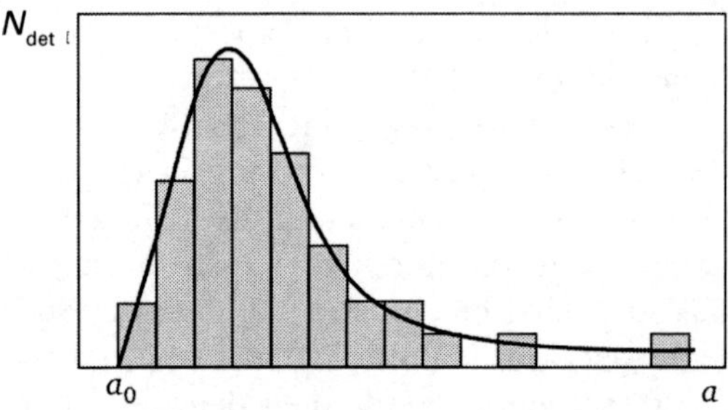

5.37   Curve N_{det} plotted on the basis of inspection results.

Table 5.9   Characteristics of initial defectiveness of equipment and inspection methods for the detection of defects

Equipment	Initial defect		Inspection methods	
	A, mm^{n-1}	n	a_0, mm	α, mm^{-1}
Cladding VVER-1000 reactor casing	1482	2.17	1	0.034
Parent and welds of VVER-1000 MCL	241.7	1.58	2	0.114

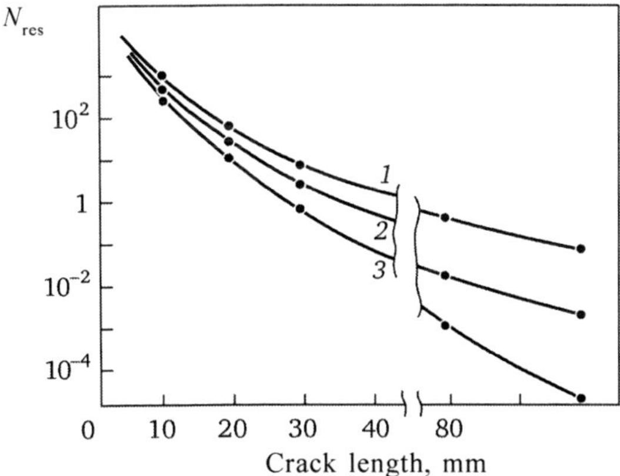

5.38 Number of remaining cracks in the cladding of a pressure vessel: 1) before inspection, 2) after first inspection, 3) after second inspection.

the functions W and N_{in} is reduced to determining the unknown constants A, n and α. The constant a_0 is easy to find from the inspection results.

The coefficient A, n and α can be determined from the condition of maximal proximity of the calculated curve to the experimental one.

Table 5.9 shows the constants A, n and α, obtained by analyzing the results of flaw inspection at nuclear power plants during entry inspection and commissioning works. The cladding of the reactor vessel was inspected by dye penetrant inspection, the pipelines by ultrasonic testing.

Figure 5.38 shows the results of analysis of initial and residual defectiveness in the cladding of the reactor pressure vessel. Curve 1 characterises the initial defectiveness of the cladding after factory inspection in the plant and before inspection at the nuclear power plant. Curves 2 and 3 characterise the residual defectiveness after two inspections at the nuclear power plant.

5.2.4 Possibility of predicting the results of repeated inspection

Generalising equation [5.21] to the case of the k-th inspection gives the equation:[89]

$$(N_{det})^k = Aa^{-n} \exp\left[-\alpha(k-1)(a-a_0)\right]\left\{1-\exp\left[-\alpha(a-a_0)\right]\right\}. \qquad [5.22]$$

Determining the constants A, n, α and a_0 from the results of the first inspection, as described above, the results of subsequent inspections can be predicted.

Comparison of results of repeated inspection of the cladding of the reactor casing during start-up operations by the dye penetrant method with

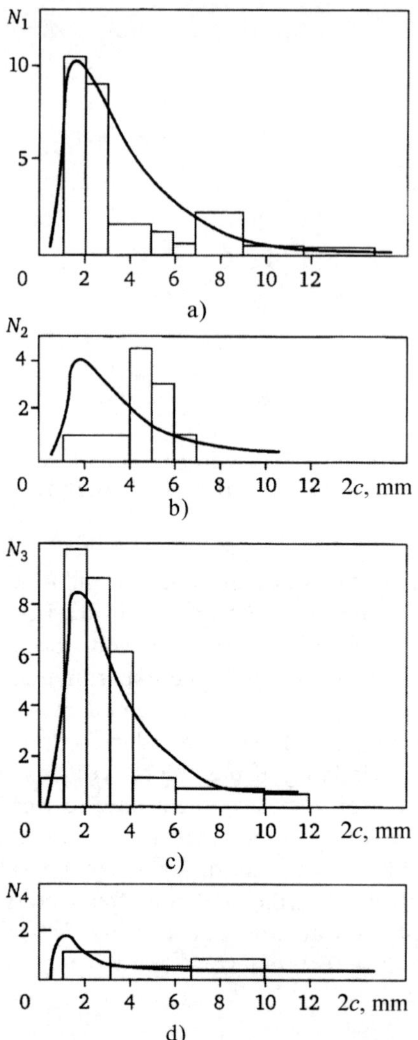

5.39 Experimental histogram and calculated curves of the detection of defects: a) 1st inspection, 2) 2nd inspection, 3) 3rd inspection, 4) 4th inspection.

the results of the forecasts is shown in Fig. 5.39. Figure 5.39a shows the function N_{det} for the cladding of one of the casings of the VVER-1000 reactor constructed on the basis of the inspection results. The following values of the parameters are determined from the condition of maximal proximity of the function [5.21] to the experimental data

$$A = 1.482 \text{ mm}^{1;17}; \ n = 2; \ 17, \ a = 0.034 \text{ mm}^{-1}.$$

Figure 5.39b shows the curve calculated from equation [5.22] for

($k = 1$) and the experimental histogram obtained in the second inspection of the cladding of the casing. Similar curves were also plotted for the third (Fig. 5.39c) and fourth (Fig. 5.39d) inspections. In the calculations it was taken into account that the observed defects were repaired and that the area changed from inspection to inspection.

It is interesting to compare the calculated number of detected cracks N_{tot}^{calc} with the experimental data N_{tot}^{exp}. For the second inspection these values are respectively: 13.5 and 12, for the third inspection 33 and 34, and for the fourth inspection 6, 8 and 5.

Thus, the proposed calculation method gives satisfactory agreement between the calculated curves and the experimental histograms. Essentially, the procedure for predicting the results of subsequent inspections is included in the algorithm for evaluation of the residual defectiveness (see Fig. 5.37), which is also well supported by experiments on the test sample.

5.2.5 Credibility and probabilistic components of residual defectiveness

It has already been mentioned that, in general, the number of defects in a structure decreases with increasing size of defects. It is obvious that there are size ranges where the number of defects is reliably equal to 1, more than 1 or much greater than 1. It is also clear that there is a size range where the defect may or may not be. The size range where the defect (or discontinuity) is present in the structure reliably in the amount equal to or greater than 1, can be called the reliable part of the residual defectiveness. The size range where the defect (or discontinuity) may or may not be is the probabilistic part of the residual defectiveness. The boundary between these regions is formed by defects (discontinuities) with dimensions (a_d, c_d).

The probabilistic part of residual defects, i.e. discontinuities with the size $(a, c) \geq (a_d, c_d)$ is of special interest in terms of strength, reliability and residual life.

The following section is concerned with one of the methods for determining the quantitative characteristics of the probabilistic part of residual defectiveness and examples of their definitions for a number of structural elements of the nuclear reactors. The function of the integrated density of distribution of the probability of existence of discontinuities with dimensions (a, c) is expressed by the equation:

$$P_{a,c}\left[(a,c) \geq (a',c')\right] = \frac{\displaystyle\int_{(a',c')}^{(a,c)_{max}} \int N_{in}(a,c)\,da\,dc}{\displaystyle\int_{(a_\Delta c_\Delta)}^{(a,c)_{max}} \int N_{in}(a,c)\,da\,dc}, \qquad [5.23]$$

where N_{in} (a, c) is the function of defectiveness if inspection is not performed, or the function of residual defectiveness if inspection is carried out, $(a, c)_{max}$ is the maximum possible size of the defects in the structure; a is the size of the defect in the direction of the wall thickness of the vessel, $a_{max} = s$, where s is the wall thickness. For c, the maximum value c_{max} is the value at which the probability of existence of a defect with $c = c_{max}$ is equal to 0. For a circumferential defect c_{max} may be equal to the length of the perimeter of the pipe or the cylindrical part of the pressure vessel.

The denominator in [5.23] has the meaning of the normalisation factor. It is natural to assume that the integral in the denominator is equal to 1 and the boundary values (a_d, c_d) should be sought from the condition:

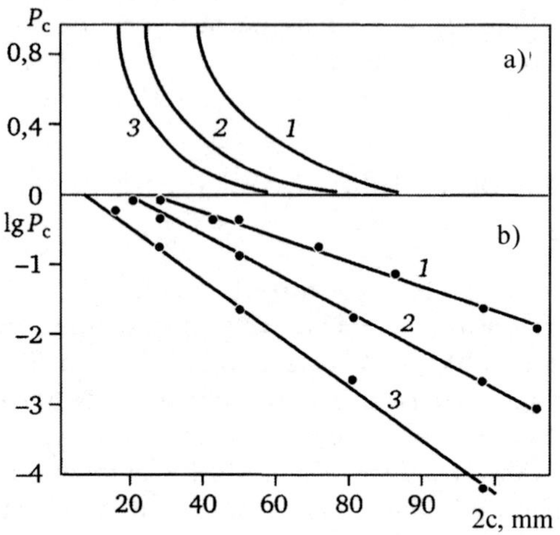

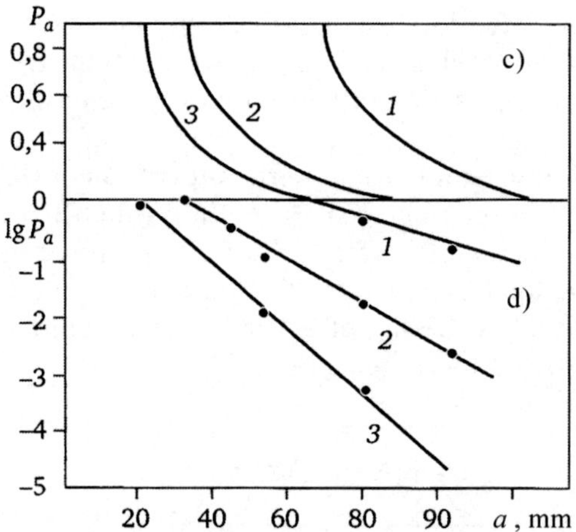

5.40 The probability function of the existence of defects in the cladding of the casing (a, b) and MCP Du850 of the VVER-1000 reactor (c, d): 1) initial state, 2) after input control, 3) after first, second, third inspection (a, b) and first (c, d) inspection.

$$\int\limits_{(a_\Delta c_\Delta)}^{(a,c)_{max}} \int N_{in}(a,c)\, da\, dc = 1.$$

[5.24]

The results of the calculation of function of P_c for the cladding of the reactor pressure vessel, and P_a to the main circulation lines DN850 of the VVER-1000 reactor according to the input inspection and inspection during the start-up period can be seen in Fig. 5.40a and 5.40b, respectively.

Function P_a for the welds of the base metal of the VVER-440 reactor (Fig. 5.40) was determined using the results of the input and in-service inspection of reactors at the Novovoronezh nuclear power plant (NVNPP), as well as data on defectiveness of the reactor casing at the Kola NPP. The maximum size of the defect was accepted regardless of the direction.

The functions P_a and P_c for N_{in} were calculated using the equations $N_{in}(a) = Aa^{-n}$ or $N_{in}(c) = Ac^{-n}$, respectively.

The dependences in Fig. 5.40 in the semilogarithmic scale are well approximated by straight lines. This means that the equations for P_a, P_c can be accurately described by equations such as

$$\left. \begin{aligned} P_a\,(a \geq a^1) &= \exp\left[-\gamma_Q(a^1 - a_\Delta)\right]; \\ P_c\,(c \geq c^1) &= \exp\left[-\gamma_c\,(c^1 - c_\Delta)\right]. \end{aligned} \right\}$$

[5.25]

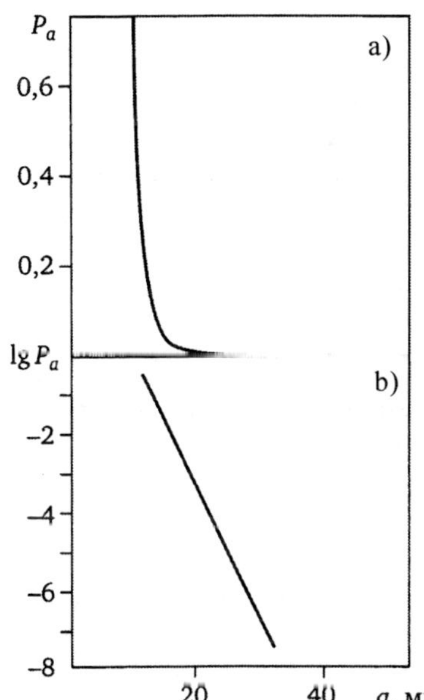

5.41   The probability of detecting in welds and base metal of the VVER-440 reactor defects with depth 2a on the linear (a) and logarithmic (b) scale.

For simplicity, these equations will be written as:

$$P_a = \exp\left[-\gamma_a(a - a_\Delta)\right]; \\ P_c = \exp\left[-\gamma_c(c - c_\Delta)\right].$$

$$[5.26]$$

Constants γ_a and γ_c characterise the rate of decrease of the probability of existence of discontinuities with increase of their size; these constants will be referred to as the probabilistic coefficient of residual defectiveness. The values of a_d and c_d characterise the threshold size of discontinuities; below this size the discontinuities exist with the highest probability and these values can be referred to as the threshold values of the reliably existing discontinuities (defects).

Quantitative estimates of the probabilistic part of residual defectiveness (Fig. 5.40) were made in 1982–1983. All values of a_d and c_d, shown in these figures, lie in the range of unacceptable sizes of discontinuities, if evaluation is carried out in accordance with PK1514 developed for evaluating the quality in the manufacturing stage and valid for the service stage

Later, operational inspection of the above structural elements revealed discontinuities classified as defects according to PK1514 and lying in the range $(a, c) > (a_d, c_d)$. For example, in the 1991 the reactor vessel of block 2 of the Kola Nuclear Power Plant was inspected in service with automated ultrasound installation Reaktortest made in Czechoslovakia. The detected 12 defects, unacceptable according to PK1514, were found in the reliable parts of residual defectiveness in the casing of the VVER-440 reactor. This can be verified this by comparing the geometric characteristics of the detected defects with the values of a_d values in Fig. 5.41. The identified defects were of technological nature and were missed in because of the insufficient of reliability of the plant and input inspections. As will be shown in section 5.3.2, these defects do not influence the design service life and were kept in service without repair according to the analysis of strength and residual life.

Cases like the one described above for the reactor vessel were also detcted for the main pipelines for the VVER-1000 reactors.. They testify to the correctness of the above-described methods of quantitative analysis of residual defectiveness and the reliability of non-destructive testing.

The quantitative characteristics of the probability of residual defectiveness in the cladding and base metal of the reactor vessels and piping of the VVER-1000 reactors are shown in Table 5.10.

In conclusion, it should be noted that these examples of estimation of a_d and c_d comparison of the estimates with the permissible and critical dimensions of discontinuities show that the probabilistic part of residual defectiveness largely determines the strength, longevity, reliability and remaining service life of the investigated structural elements.

Table 5.10   Characteristics of residual defectiveness of equipment and reliability of inspection methods

Equipment	Initial defec-tiveness		Inspection method		Residual defec-tiveness	
	A, mm^{-1}	n	a_0 or c_0, mm	a, mm^{-1}	a_0 or c_0, mm	g, mm^{-1}
Cladding RPV VVER-100	1482	2.17	1	0.034		
Initial state					28	0.041
After entry inspection						0.061
After 1st, 2nd, 3rd inspection						0.088
Base metal and welded joints in MCP of VVER-1000	241.7	1.58	2	0.114		
Initial state					29	0.059
After entry inspection					14	0.150
After first inspection					9.5	0.256
Base metal and welded joints in RPV of VVER-440					10.9	0.841
Pipelines of MCP of VVER-440 after entry, pre-service and in-service inspection	234	1.37	0.5	0.08	6	0.5

5.3 Probabilistic methods for assessing strength and service life taking into account residual defectiveness in structural elements

5.3.1 The general characteristic of methods

The methods described in this section can be used to determine the quantitative characteristics of reliability and gamma-percentile life according to the criteria of fracture resistance of the structural element and the resistance to leakage during operation. In addition, the methods determine the probability of finding a defect (or a group of defects) of a certain size in the process of non-destructive testing in service, as well as to solve some practical problems associated with increased reliability and lower operating costs (the latest technologies are described in the second part of the book).

In general, the determination of the probability of failure, leaks or defects operation comprises the following steps:

- determining the distribution of mechanical properties;

- determination of load distribution;
- determining the actual defect component (residual defects);
- identify future operating conditions;
- identify changes in defect during the operation;
- determination of changes in mechanical properties during the operation;
- determining the mechanism of fracture: brittle, quasi-brittle or ductile;
- determining the probability of failure and its changes during operation.

The distribution laws of mechanical properties, loading, defects, the effects of ageing, cyclic loading, the influence of corrosive environment and fracture mechanisms are taken into account.

The basis of the following methods is to consider the following characteristics of manufacturing techniques and quality control of structural elements (products).

In manufacture, products usually contain technological defects of integrity of metal. Integrity defects may also arise during the operation. Non-destructive testing is performed to identify and eliminate these defects. It is believed that after the non-destructive testing and repairs conducted on the basis of NDT results the product is free from defects identified by NDT. It is assumed that the reliability and safety of the product in operation is guaranteed (see regulatory documents on nuclear energy, for example, Ref. 73, 6, 91).

In fact, there are practically no NDT methods and means in engineering which would guarantee with 100% certainty detection of all defects. Therefore, there is always a certain probability of missing a defect, including critical defects (i.e. propagation of such a defect during operation will lead to product damage or destruction). It is known that in almost all cases, NC there is a substantial likelihood of missing a large defect that substantially exceed the permissible size (see Section 5.1, 5.2). In practice, it turns out that products almost always contain defects after NDT and remoal of the identified defects. These remaining defects ultimately determine the reliability and operating efficiency of the products.

The technical result obtained when using the methods described below can be used for assessing the actual defectiveness of the product after testing and repair of the identified defects and to determine the actual level of reliability and safety of the product before it is destroyed or damaged in service.

5.3.2 Method for determining the probability of failure, the probability of leakage and the probability of existence of hidden defects from the probabilistic part of residual defectiveness

The method can be applied to a specific product (or group of m similar products) for which it is necessary to ensure manufacturing quality, reliability and safety by the given NDT method during inspection by the operator with the knowm skill with subsequent repair of identified defects.

Fracture mechanics methods determine the critical size of defects in the service refime for this product χ_{cr} and the maximum allowable size of defects in service $[\chi]_{d.s}$ (norms defects in the product), defined the valid regulations and / or Technical Instructions (TI) for manufacturing (e.g., in nuclear technology – the statutory procedure[104] described in section 4.1). The characteristic size of the defect χ is, for example, either the linear size of the defect or a combination of linear sizes of the defect, or the defect area, or the defect volume. It should be noted that the defects that determine the quality are the defects whose sizes range from the minimum size of defects that can be detected (search fize) (search) to the size of defects that are allowed in manufacture and larger defects; defects determining reliability – these are defects whose sizes are in the range from rejection in manufacture or service to the acceptable size in service and larger; the defects that define safety – from the allowable in service to the critical size and larger.

The methods are illustrated in Fig. 5.42–5.45.

Fig. 5.42 shows the schematisation of a defect in a pipeline by an ellipse with minor semiaxsis a and major scmiaxis c. Figure 5.43 shows a set of critical size (curve 3), defects with the size allowable in service (curve 2) and the allowable size in manufacture (curve 1). Figure 5.44 is a histogram of defects detected in the component and the curves of initial and residual defectiveness are also shown. The curves of initial and residual

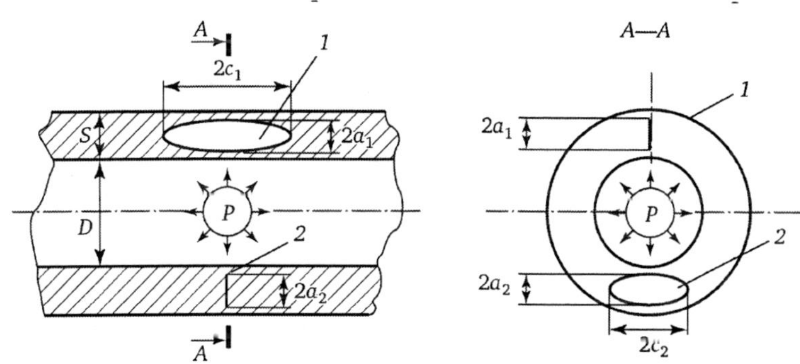

5.42 Schematic representation of the longitudinal and transverse integrity detects of pipeline metal in its longitudinal and transverse sections.

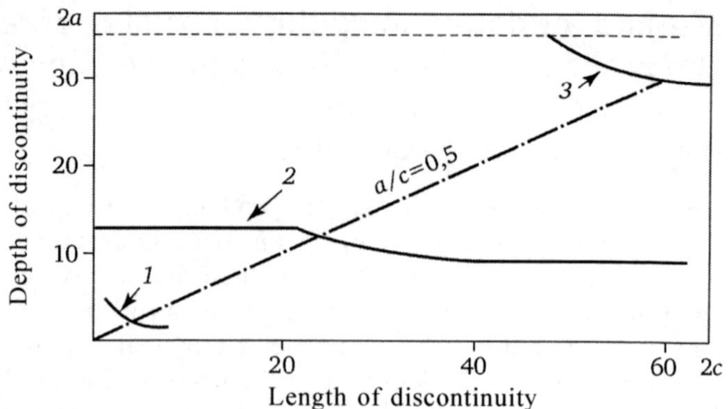

5.43 Set of discontinuities of critical size 3, size permissible in service 2 and permissible in manufacture 1.

defectiveness on the semilogarithmic scale are shown in Fig. 5.45.

The product is inspected by non-destructive testing products using the selected method technical means of inspection (ultrasonic, eddy current, radiographic, or others) as well as by operators of certain qualifications. The discovered defects are then removed.

The inspection results are represented as a histogram in the 'the characteristic size of the defect χ – the number of identified defects of the given size $N_{det.comp.}$' coordinates.

This is followed by determination of the probability of defect detection P_{pdd}, initial defectiveness $N_{in} = f(\chi)$ and residual defectiveness $N_{res} = \varphi(\chi)$ as the difference between N_{in} and N_{det}.

These relations can be determined by various methods.

According to **one of the methods,** the NDT results are presented in the form of analytical expressions.

The structure of the equation which can describe the NDT results, presented in Fig. 5.44, is the following:

$$N_{det}(\chi) = N_{in}(\chi)\, R_{dd}(\chi),$$

where N_{det} is the number of defects detected in inspection in the unit of the characteristic size. If as the characteristic size is represented by the minor axis of the ellipse which schematises the defect, then the unit N_{det} – mm^{-1}; N_{in} is the function of the initial (prior to NDT and repair) defect with the same unit as the N_{det}; P_{dd} is the probability of detecting a defect of given size χ.

The type of functions N_{in} and P_{dd} is defined on the basis of the simplest expression, the minimum number of constants and the correspondence of the physically defined dependence of N_{in} and P_{dd} on χ. In the first approximation the following equations can be used:

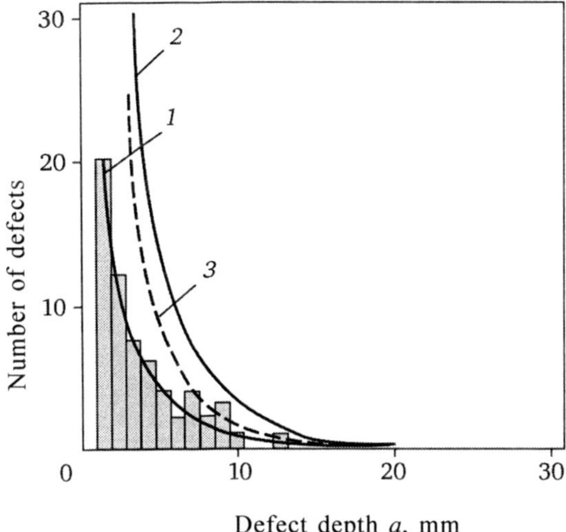

Defect depth a, mm

5.44 Histogram of detected defects 1, initial defectiveness 2 and residual defectiveness after repair of detected defects 3.

$$N_{in} = A\chi^{-n},$$

$$P_{pdd} = 1-(1-\eta)\exp\left[-\alpha(\chi-\chi_0)\right]-\eta;$$

$$N_{det}(\chi) = A\chi^{-n}\left\{(1-\eta)\exp\left[-\alpha(\chi-\chi_0)\right]-\eta\right\};$$

where A, n, α, η, χ_0 are constants.

The numerical values of the constants A, n, α, η are determined from the condition of maximal approximation of equation $N_{det}(\chi)$ to the NDT results presented in the form of a histogram.

The minimum allowable defects size for detection χ_0 is determined in setting flaw inspection equiment used for inspecting the product, or as the minimum size of the defect which can be detected in inspection; in the first approximation η can be taken as equal to 0. This leaves three unknowns, which greatly simplifies the task of their definition.

The constants A, n, α can be determined either by solving a system of three equations for A, n and α, which are obtained by taking three points in the histogram, or the least squares method can be used.

Residual defectiveness N_{res} is defined as the difference between N_{in} and N_{det}:

$$N_{res}(\chi) = N_{in}(\chi) - N_{det}(\chi).$$

Moreover, the number of defects remaining in the product after NDT and repair is defined in three ranges.

The residual defectiveness $N_{res.cr}^{\Sigma}$ in the range of the defects that are important for safety is determined as the number of defects in the product whose size is equal to or greater than the critical size χ_{cr} in the operating mode of the product:

$$N^{\Sigma}_{res.cr} = \frac{1}{m} \int\limits_{\div..}^{\chi_{bef}} N_{res}(\chi)d\chi.$$

When $N^{\Sigma}_{res.cr} < 1$ the safety S of the product (reliability criterion based on fracture resistance) is estimated by the expression:

$$S = 1 - N^{\Sigma}_{res.cr}.$$

If $N^{\Sigma}_{res.cr} \geq 1$ the produce does not have any safety margin and can not be allowed to operate.

Residual defectiveness $N^{\Sigma}_{res.d.s}$ in the range defects which are important for reliability is determined as the number of defects that are larger than the maximum size of defects $[\chi]_{d.s}$ in service

$$N^{\Sigma}_{res.d.s} = \frac{1}{m} \int\limits_{\div_{d.s.}}^{\chi_{bef}} N_{res}(\chi)d\chi \text{ at } N^{\Sigma}_{res.d.s.} < 1.$$

The reliability of the product R is defined by the equation

$$R = 1 - N^{\Sigma}_{res.d.s}.$$

If $N^{\Sigma}_{res.d.s} \geq 1$, the product does not have any safety margin. The third range describes accurately the quality of manufacture.

In constructing the histograms, the horizontal axis χ should include the critical defect size, even if all identified defects did not reach critical sizes in inspection.

In inspection of several similar products all the results are added up and presented in the form of a histogram. The greater the number of inspected items the higher is the reliability of the final result.

In the *second variant* a test sample is produced to determine the dependencies P_{dd}, $N_{in} = f(\chi)$ and $N_{res} = \varphi(\chi)$ (the residual defects) are made the test sample.

Taking into account the actual operating loads and service conditions for the product (e.g. pipeline in Fig. 5.42) the fracture mechanics methods (with safety factors taken into account) are used to determine the set of defects (discontinuities) of the critical size allowed in service as well as the allowable size of discontinuities in manufacture (Fig. 5.43).

The test sample is made in the form the product and on the scale of approximately 1:1 to the product or its most critical part. The most critical part of the product is the part of the product in which defects are most likely to form (welds, areas of maximum service effects, etc.) or destruction of which poses a risk. The test sample is produced from the same material

and using the same technology as the product. Artificial defects of three types are made in the test sample:
- defects of the critical size or close to it;
- defects whose size is in the range from the size permissible in operation to the critical size;
- defects whose size is in the range from the size permissible in manufacture to the size acceptable for service.

The defects (discontinuities) should simulate the defects of operational nature, i.e., defects that may develop from technological defects or nucleate and evolve under the influence of service loads (fatigue cracks, stress corrosion cracking, etc.). The embedded defects can also simulate the technological defects (if necessary).

All embedded defects should be hidden from NDT operators, i.e., should be internal (subsurface), or, if it is a surface defect, it should be located in the area where it cannot be detected by visual examination (or have dimensions that cannot be detected visually).

It should be noted that the defects are distributed randomly in the sample, for example, using tables of random numbers.

The minimum allowable distance between the defects is determined based on the conditions for the existence of single defects (if single defects are produced) or less – for a group of cracks (the conditions of mutual influence are known, for example, Ref. 92).

The number of defects of each type should be sufficient for statistical processing of the results, for example, at least nine defects (if the number of defects is smaller the results are less reliable).

Any defect can be modelled conservatively by a crack, and any crack can be described by an ellipse with the minor a and major c semiaxes.

There are various options for distributing defects in the test sample:
- in the form of ellipses with the ratio of the axes taken from the condition of maximum growth rate of the defect in the service stress field;
- in the form of ellipses with an arbitrary ratio of the axes, and the quantity characterising the size of the defect is the area of the planar defect or the area of projection of the volume defect on the plane of the probable development of the defect;
- in the form of ellipses, with the number of defects and the ratio of the axes chosen using mathematical methods of experiment planning, based on the condition of minimising the number of embedded defects.[93]

If the defects produced int the test sample do not have the shape of an ellipse, then they are schematised by ellipses.

After making a test sample, it is inspected using the same inspection means and methods and the operators with the same skills which will then be applied in inspection of products, and the results are compared with the actual defects embedded in the test sample.

Reliability is determined as a function of the probability of detecting defects $P_{dd}(\chi)$ for each of the characteristic defect size

$$P_{dd}(\chi) = N_{det.ts}(\chi) \, / \, N_{emb.ts}\,(\chi),$$

where $N_{det.ts}$ is the number of defects found in inspection of the test sample; $N_{emb.ts}$ is the number of defects embedded in preparation of the sample.

The inspection results are used to plot the curves of probability of detection of defects for the given product by the given NDT method, depending on the characteristic size of defects. The curve of the probability of detecting defects on the size of defects a and c (any defect in the material can be conservatively described by an ellipse with semiaxes a and c) can be approximated by the equation which describes most most accurately the experimental results, for example

$$P_{dd} = 1 - (1-\eta)\exp\left[-\alpha_{NDT}(a-a_0)(c-c_0)\right] - \eta;$$

$$P_{dd} = 1 - (1-\eta)\exp\left[-\alpha_{NDT}(a-a_0)\frac{a}{c}\right] - \eta;$$

$$P_{dd} = 1 - (1-\eta)\exp\left[-\alpha_{NDT}(\chi-\chi_0)\right] - \eta,$$

where α_{NDT} is the NDT reliability coefficient which characterises the increase in the detectability of defects, depending on its size; η is a constant characterising the detection limit of inspection by this method for an arbitrary large size of the defect; if the product is small, then this value can be ignored by adjusting the value α_{NDT}; χ is the characteristic size of the defect, for example, its area; χ_0 is the minimum characteristic size of the defect; a_0, c_0 are the minimum sizes of defects available for detection by NDT.

This is followed by inspection of the product, and the test results are presented in the form of a histogram in the 'characteristic size of the defect χ – the number of identified defects of the given size $N_{det.prod.}$' coordinates.

Initial defectiveness N_{in} is defined as the ratio $N_{det.prod}/P_{dd}(\chi)$; the resulting histograms are approximated by the equation of the type $N_{in} = A_\chi \exp(-n_\chi\chi)$ or $N_{in} = A_\chi \chi^{-n\chi}$ or $N_{in} = A_a \chi^{-na}$, or $N_{in} = A_{a,c}(a^2,c)^{-na,c}$ or $N_{in} = A_F F^{-n_F}$ or $N_{in} = A_a^{-na}\rho(c)$ or

$$N_{in} = A_{a,c}\, a^{-n}\,\frac{1}{\sqrt{2\pi D}}\exp\left[(c-\bar{c})^2 / 2D^2(c)\right],$$

where a, c are the linear dimensions of the defect, ρ_c is the distribution function of c, for example, a normal distribution, F is the area of the defect, n, A, D, $\bar{c}$ are the coefficients, chosen from the condition of maximal pproximation of the analytical curve to the experimental data, in this case $\bar{c}$ is the average value of c and D is dispersion.

The characteristic size can be represented by the minor axis a of the ellipse which schematizes the defect, and the ratio a/c is taken to be constant for all a, starting from the conditions of maximum growth rate of the defect in service; here

$$N_{\text{in}} = \int_0^{c_{\text{max}}} \varphi(a,c)\,dc = f(a),$$

for example, in the case of a uniform stress field $a/c = 2$ and the normal law for the distribution c with the mean value $c = 2a$ and the dispersion $D = a/2$ one obtains

$$N_{\text{in}} = \int_0^{c_{\text{max}}} Aa^{-n} \frac{1}{\sqrt{2\pi D}} \exp\left[\frac{-(c-\bar{c})}{2D^2(c)}\right] dc =$$

$$= Aa^{-n} \int_0^{c_{\text{max}}} \frac{\sqrt{2}}{a\sqrt{\pi}} \exp\left[-(c-2a)^2 / 0.5a^2\right] dc = Aa^{-n}.$$

Residual defectiveness is the difference between N_{in} and $N_{\text{det.prod}}$. In this case $N_{\text{det.prod}}$ is determined from the analytical expression $N_{\text{in}}P_{\text{dd}}(\chi)$, i.e. residual defectiveness N_{res} can be expressed by the equation

$$N_{\text{res}} = N_{\text{in}}(1 - P_{\text{dd}}).$$

The product safety S is defined as the probability of absence in the product of defects larger than or equal to χ_{cr}, where the residual defectiveness $N^\Sigma_{\text{res.safe}}$ in the range of defects that are important for safety, are determined as the probability of absence in the product of defects whose size is equal to or greater than the critical size χ_{cr} in the service mode of the product:

$$S = 1 - N^\Sigma_{\text{res.safe}} = 1 - \int_{\chi_{\text{cr}}}^{\chi_{\text{bef}}} N_{\text{res}}(\chi)\,d\chi;$$

Reliability R is determined as the residual defectiveness $N^\Sigma_{\text{res.rel}}$ in the range of defects and is defined as the probability of absence of defects, whose size exceeds the maximum size of the defects $\chi_{\text{d.ser}}$ allowable is service:

$$R = 1 - N^\Sigma_{\text{res.rel}} = 1 - \int_{\chi_{\text{d.ser}}}^{\chi_{\text{bef}}} N_{\text{res}}(\chi)\,d\chi;$$

Further, the residual defectiveness is divided into the reliable part in which defects with sizes $\chi \leq \chi_{\text{d}}$ are reliably detected, and the probabilistic part in which defects with sizes $\chi > \chi_{\text{d}}$ may or may not be.

The boundary between the reliable and probabilistic parts of residual defectiveness χ_{d} is determine the condition:

$$\int_{\chi_d}^{\chi_{max}} \varphi(\chi)\, d\chi = 1,$$

where χ_{max} is the maximum possible size of the defects in this product.

The probabilistic part of residual defectiveness is used to determine the probability of the existence of defects that are larger than χ_{cr} and defects that are larger than $[\chi]_{d.ser}$.

The safety of the product is defined as the probability of existence of defects that are larger than χ_{cr} and the reliability of the product is defined as the probability of finding defects that are larger than $[\chi]_{d.ser}$.

The reliability determined using the criterion of leakage of pressure vessels and piping (for 'leak before break' state) described by the equation:

$$N_{leak} = 1 - P_{def}(a^* \geq s,\ t),$$

where a is the size of the defect in the direction of the wall thickness, s is the wall thickness of the vessel or pressure pipeline.

The method is illustrated by the following examples.

Example 1

It is necessary to ensure the quality of two pipelines with the inside diameter $D = 800$mm and the wall thickness $s = 34$mm made of a pearlitic steel. The critical size of the defects in transverse welds are shown in Fig. 5.43 (curve 3). The defects permissible in service were identified by the fracture mechanics equations and the safety factor (curve 2 in Fig. 5.43). Norms of defects in manufacture are shown by curve 3.

The standard NDT method applied before operation (after installation) identified 60 discontinuities.

All the detected discontinuities (defects) are presented as histograms in Fig. 5.44.

The characteristic defect size is the width of the defect in the wall thickness direction or, more accurately, the minor axis of the ellipse which was used to schematize all known defects.

At the ratio $a/c \approx 0.5$ the critical size of the defect corresponds to $a = 15$ mm, $[a]_{d.ser} = 6$mm. For the maximum size in manufacture $[a]_{man} = 1.15$ mm (Fig. 5.43).

Despite the fact that the maximum size of the defect was identified $a_{max} = 13$mm, the horizontal axis shows the critical size of $a_{cr} = 15$ mm.

The equation describing the number of detected defects N_{det} depending on the size a is:

$$N_{det} = Aa^{-n}\,[1-\exp\,[-\alpha\,(a-a_0)].$$

According to the inspection results, the minimum size of the detected defect was $a = 0.6$ mm.

The constants A, n, α are determined by solving a system of three equations for these constants:

The first equation is derived for a point with the coordinates (a = 1 mm, N_{det} = 20) in Fig. 5.43:

$$20 = A \cdot 1^{-n} \{1 - \exp [-\alpha (1-0.6)]\};$$

The second equation is obtained for a point with the coordinates (a = 5mm, N_{det} = 4) in Fig. 5.43:

$$4 = A \cdot 5^{-n} \{1-\exp [-\alpha(5-0.6)]\};$$

The third equation is derived for a point with the coordinates (a = 13mm, N_{det} = 0.66) in Fig. 5.43:

$$0.66 = A \cdot 13^{-n} \{1- \exp [-\alpha (13-0.6)]\}.$$

For the third equation the value N_{det} = 0.66 was obtained by averaging the number of identified defects in the range from 11 to 13 mm, which was equal to 2/3, where 2 is the number of identified defects, 3 is the number of intervals.

Finally, the system of equations has the form:

$$20 = A \cdot [1- \exp (-0.4\ \alpha)]$$
$$4 = A \cdot 5^{-n} [1- \exp (-4.4\ \alpha)]$$
$$0.66 = A \cdot 13^{-n} [1-\exp (-12.4\ \alpha)].$$

The system of equations with respect to A, n, α produced the following results:

$$A = 1000 \text{ mm}, \ n = 2.56, \ \alpha = 0.05 \text{ mm}^{-1}.$$

Substituting the constants A, n, α in the corresponding equations gives:
- equation for the initial defectiveness:

$$N_{in} = 1000a^{-2.56} \text{ (curve 2 in Fig. 5.44)};$$

- equation for the probability of detecting a defect:

$$P_{dd} = 1 - \exp [-0.05\ (a-0.6)];$$

- equation for residual defectiveness (curve 3 in Fig. 5.44):

$$N_{res} (\chi) = N_{in} (\chi) - N_{det} (\chi).$$

The following equations are solved:

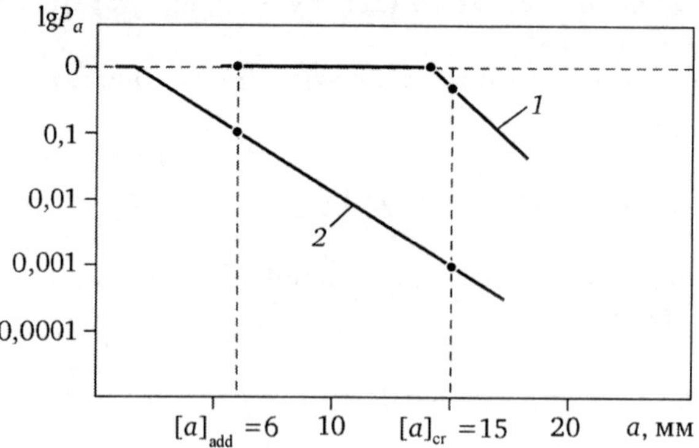

5.45 Curves of residual defectiveness after inspection nd repair of detected defects 1 and after more detailed inspection and repair of defected defects 2.

$$\int_{a_d}^{a_{max}} Aa^{-n}\exp\left[-\alpha(a-a_0)\right]da = 1; \quad N^{\Sigma}_{res.rel} = \int_{[a]_{d.ser}}^{a_{max}} Aa^{-n}\exp\left[-\alpha(a-a_0)\right]da;$$

$$N^{\Sigma}_{res.safe} = \int_{a_{cr}}^{a_{max}} Aa^{-n}\exp\left[-\alpha(a-a_0)\right]da.$$

In this case $a_{mas} = s$, where s is the wall thickness of the pipe. The results of the solutions are presented in Fig. 5.45 as curve 1. The defects permitted in service operate and critical size defects marked accordingly as $[a]$ and a_{cr}. Finally,

$$a_d = 14 \text{ mm}, R = 1{-}1 = 0, S = 1{-}0.55 = 0.45.$$

Since the values of reliability and safety for the characteristics of the residual defectiveness in accordance with curve 1 are unacceptably low, the product was rejected and sent back for further processing. The product was the subjected to repeated inspection and repair and the results are described by curve 2. In this case, reliability was

$$R = 1{-}N^{\Sigma}_{res.rel} = 1 - 0.1 = 0.9$$

and safety

$$S = 1 - N^{\Sigma}_{res.safe} = 1 - 0.001 = 0.999.$$

On the basis of the results the product was then admitted into service.

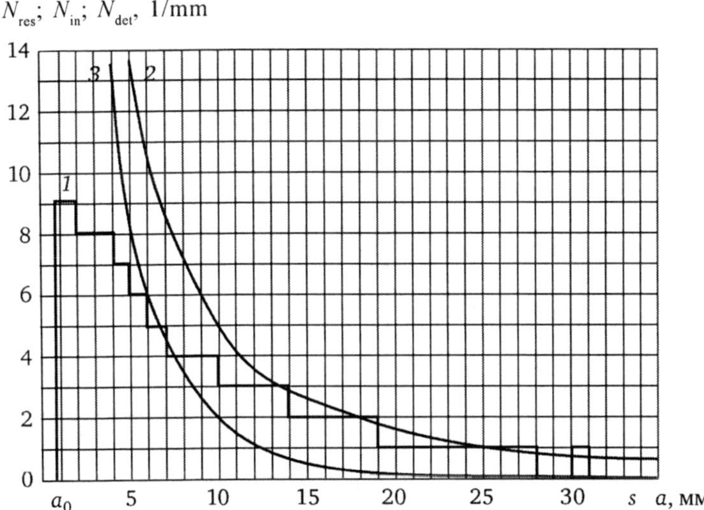

5.46 Histogram of defects detected in NDT and the envelope of the defects 1; the curve of initial detectiveness 2; curve of residual defectiveness 3.

Example 2

It is required to determine the probability of safe operation of pipelines which can be in the 'leak before fracture' state.

There are 100 pipelines of the same type produced from seamless steel pipes. In the welded joints distributed across the pipeline axis, ultrasonic inspection prior to the start of service revealed technological defects oriented in the transverse plane situated normal to the pipeline axis. The defects were schematised by the same procedure as in the previous example (Fig. 5.42).

The service life of the pipelines was 20 years. During this period, each pipeline was subjected to 1500 internal pressure cycles.

Non-destructive inspection of all pipelines revealed a specific number of crack-like defects in each pipeline. All the detected defects were schematised by ellipses, and the minor semi-axis $a = F/\pi c$, where F is the area of the defect, was determined for each ellipse.

The number of cracks of each standard size, detected in the 100 pipelines, was summed up and the mean value was determined:

$$N_{det}(\chi) = N_{det.mean}(\chi) = [N_{det1}(\chi) + N_{det2}(\chi) + N_{det3}(\chi) + ... N_{det100}(\chi)]/100.$$

The average values of the number of the detected defects were used to construct a histogram (Fig. 5.46, curve 1).

A test specimens with planar hidden defects was produced. The results of inspection of the specimens were used to determine the reliability

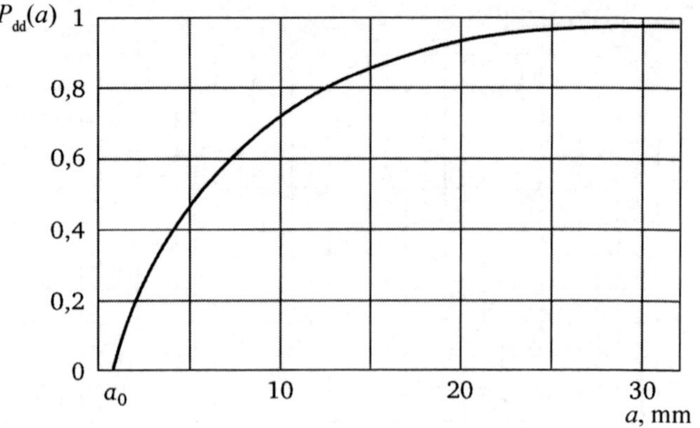

5.47   Probability of detection of defects with size a.

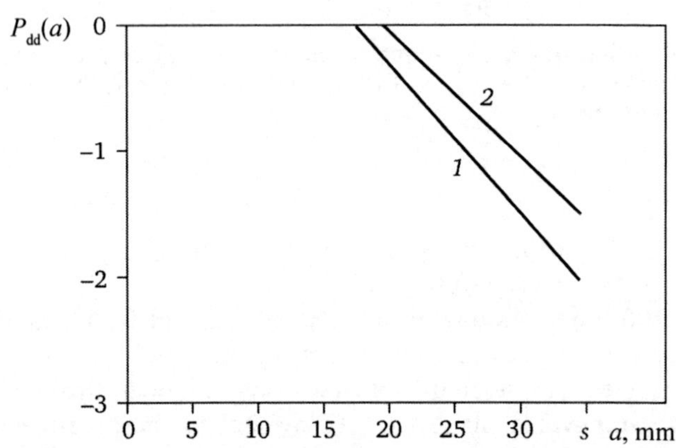

5.48   Probabilistic part of residual defectivness at the beginning of service 1 and at the end of service of the component 2.

characteristics of non-destructive inspection in the form of the curve of detectability of defects (Fig. 5.47).

The initial defectiveness of the pipeline was determined for each standard defect size

$$N_{in}(a) = N_{det}(a)/P_{dd}(a)$$

(curve 2 in Fig. 5.46) and the residual defectiveness of the pipeline after inspection and removal (repair) of the detected defects (curve 3 in Fig. 5.46) was determined from the equation:

$$N_{res}(a) = N_{in}(a)/N_{det}(a).$$

The residual defectiveness characteristics $N_{res}(a)$ were used to determine the probabilistic part of the residual defectiveness of the pipeline. The results of the calculations carried out using the equation:

$$P_{\text{def}}(a^* \geq a) = \int_a^{a_{\text{max}}} N_{\text{res}}(a)\,da.$$

in the graphical form are presented in Fig. 5.48 in the form of curve 1:

$$P_{\text{def}}(a^* \geq a) = \int_a^{a_{\text{max}}} N_{\text{res}}(a)\,da.$$

The growth of the cracks with the size in the range 17 mm $< a <$ 34 mm in 1500 load cycles was determined by the standard procedure M-02-91 using the equation:

$$da/dN_{\text{c}} = 5.80 \cdot 10^{-11}\, \Delta k^{2.6}\,(1-R)^{-1.3},$$

Here Δk is the range of the stress intensity factor; R is the stress ratio of loading of the component.

The probabilistic part of residual defectiveness at the end of service, determined taking into account the growth of the crack, is shown by curve 2 in Fig. 5.48.

Analysis of the position of the curves 1 and 2 in Fig. 5.48 shows that the probability of existence of a continuous defect in the pipeline prior to the start of the service is 0.01 and at the end of the service it is 0.03. Consequently, the statistical estimate of the probability of failure-free operation on the basis of the leak criterion for a period of 20 years in service is $P(t) - M_{\text{p}}/M = 1 - 0.03 - 0.97$. This means that out of 100 pipelines in service three pipelines will fail (leak). One of the pipelines shows leaks at the very start of service.

5.3.3 Quantitative relationship of the dependability indicators, determined by the criteria of fracture, leakage or defect detection in service with the NDT results

The method for determining the basic characteristics of reliability is described below:

- reliability;
- failure rate $\lambda(t)$;
- the number of failed structural elements according to the criteria of fracture, leaks or defects;
- density function $f(t)$ of time to failure during service time t.

The reliability of the product is one of the characteristics of its reliability. The probability of failure-free service in the probability that the products will not fail in the given operating life (GOST 27.002-89).

Statistical evaluation of the probability of failure-free operation for time t is determined from the relation:

$$P(t) = M_p/M, \qquad [5.27]$$

where M_p is the number of serviceable items at the end of time t; M is the number of items set for inspection or service.

As seen from the above expression, to estimate the probability of failurer-free operation it is necessay to collect have a sufficient number of products and test them or put into operation.[95]

This approach has drawbacks because it requires a certain time and material resources for testing or service with possible losses from unreliable operation of the product. In addition, this approach is not applicable in the case of unique expensive and/or highly dangerous products, for example, the reactor vessel of modern nuclear power plant,.

In this regard, a special method of predicting reliability characteristics was developed.[94]

The method for determination of the reliability products includes the definition of the probability of faultless operation of the product $P(t)$ at time t. To determine $P(t)$, the reliability criteria for defects, leaks or failure are determined before operation without prior testing or without prior use of the product based on the results of non-destructive testing, the reliability characteristics of the NDT method used and the growth rate of probable defects during time t. To do this, the product is subjected to NDT and the detected defects are repaired, and a histogram of the detected defects in the coordinates 'characteristic defect size χ – number N_{det} of detected defects of the given size χ' is constructed. The quantitative characteristic of the reliability of the NDT method used in the form of the probability of detection of defects $P_{dd}(\chi)$ is determined by the initial defectivness of the product $N_{in}(\chi)$ is calculated from the equation:

$$N_{in}(\chi) = N_{det}(\chi) / P_{dd}(\chi). \qquad [5.28]$$

The initial defectiveness of the product is a combination of all the defects in it before NDT; the initial defectiveness of a product can be defined in different ways described in the invention[97].

The residual defectiveness of the product $N_{res}(\chi)$:

$$N_{res}(\chi) = N_{in}(\chi) - N_{det}(\chi). \qquad [5.29]$$

Residual defectiveness is the total set of defects undetected because of imperfenctions of NDT and of the defects missed in NDT and sent to service and remaining in the product after NDT and repair (removal) of identified defects; more details can be found in Ref. 97..

Residual defectiveness is divided into reliable and probabilistic parts. The boundary between the reliable and probabilistics parts of the residual defectiveness is denoted by χ_d; in the reliability part, the probability of existence of a defect with the size $\chi \leq \chi_d$ is equal to 1:

In the probabilistic part of residual defectiveness the probability of existence of the defect with the size $\chi^* > \chi_d$ is less than 1:

$$P_{def}\left(\chi^* \geq \chi\right) = \int_{\chi}^{\chi_{max}} N_{res}(\chi)\,d\chi, \ \chi > \chi_d. \qquad [5.30]$$

Here χ_{max} is the maximum possible size of the defect in this product, and χ χ^* varies from χ_{max} to $\chi > \chi_d$.

Defect sizes χ_d at the border between the reliable and probabilistic parts of the residual defectiveness in the product is determined by the condition:

$$\int_{\chi_d}^{\chi_{max}} N_{res}(\chi)\,d\chi = 1. \qquad [5.31]$$

(more on the definition of the residual defectiveness of the product in the reliable and probabilistic parts can be found in the patent RF No. 2243586 'Method for determining the quality of products on the basis of the reliable and probabilistic parts of residual defectiveness').

The characteristics of the probabilistic part of residual defectiveness $P_{def}(\chi^* \geq \chi; t)$ at the end of life of the product is determined. To do this, find the extent of growth $\Delta\chi_t$ of defects during operating time t is ascertained. The extent of growth is determined for the defects with the dimensions $[\chi]_{serv} \leq \chi < \chi_{cr}$.

The characteristics of the probabilistic part of residual defectiveness:

$$P_{def}\left(\chi^* \geq \chi; t\right) = P_{def}\left[\chi^* \geq (\chi - \Delta\chi_t); t = 0\right] = \int_{\chi - \Delta\chi_t}^{\chi_{max}} N_{res}(\chi)\,d\chi. \qquad [5.32]$$

The extent of growth of the defect $\Delta\chi_t$ during operating time t is determined by the formula $\Delta\chi_t = \chi_e - \chi_b$ in which χ_b is the size of the defect the beginning of operation and χ_e is the size of the same defect at the end of operation.

The value χ_e determined by the equation of type:

$$\frac{d\chi}{dt} = c_i k^{m_i} \quad \text{in the case of stress corrosion cracking}$$

$$\frac{d\chi}{dN_c} = c_N \Delta k^{m_N} \frac{1}{\sqrt{1-R}} \quad \text{in fatigue,}$$

where c and m are constants; k, Δk are the stress intensity factor and the range of the stress intensity factor; R is the stress ratio of loading force of the product; N_c is the number of load cycles; the nature of crack growth (stress corrosion cracking or fatigue) is determined from corrosion and other conditions that are extensively described in scientific literature.

To determine the reliability indices of group M of identical (similar) products (i.e. products that have the same geometrical dimensions, are made of the same material and by the same technology and operate in the same conditions) the probability of failure-free operation of the product $P(t)$ during operating time t is defined for a single product as described above. The result can be attributed to the entire batch of products, provided that all the M products were subjected to NDT and repaired in accordance with the NDT results.

More accurate statistical estimates can be obtained on the basis of statistical estimates of failure-free operation for each of the M products: $P_1(t)$; $P_1(t)$; ...; $P_M(t)$

The statistical estimate of the probability of failure-free operation is determined by the formula:

$$P(t) = \frac{1}{M}\left[P_1(t) + P_2(2) + ... + P_M(t)\right]. \qquad [5.33]$$

Other options for averaging the results can also be used. For example, the NDT results can be averaged-out using the:

$$N_{det}(\chi) = N_{det.mean}(\chi) = [N_{det1}(\chi) + N_{det2}(\chi) + N_{det3}(\chi)+...$$
$$+N_{detM}(\chi)] / M \qquad [5.34]$$

and then continue with the value $N_{det.mean}(\chi)$ as well as for a single product.

Using the determined values of $P(t)$ for a batch of M products other indicators of reliability, postulated by GOST 27.002-89, can be obtained.

The number of failed products $m_{cr}(t)$ according to the criterion of full or partial failure is service is calculated from the formula

$$m_{cr}(t) = M[1-P(t)] = M[1-P_{def}(\chi \geq \chi_{cr}; t)]. \qquad [5.35]$$

The number of products $m_{def}(t)$, failed in service according to the criterion of defectiveness, is:

$$m_{def}(t) = M[1-P(t)] = M[1-P_{def}(\chi \geq [\chi]_{serv}; t)]. \qquad [5.36]$$

The number of vessels or pressure piping showing leaks during operating time:

$$m_{leak}(t) = M[1-P(t)] = M[1-P_{def}(a \geq s; t)]. \qquad [5.37]$$

The total number of failed product $m(t) = m_{def}(t)$, of which $m_{leak}(t)$ products leaked (if they are vessels or pressure piping) and $m_{cr}(t)$ products fractured.

Function of density distribution $f(t)$ of the time to failure has the form:

$$f(t) = \frac{\Delta m(t)}{M\,\Delta t} = \frac{dP_{def}\left(\chi^* \succ \chi; t\right)}{dt}, \qquad [5.38]$$

where $\Delta m(t)$ is the increase in the number of failures over time Δt.

The failure rate $\lambda(t)$

$$\lambda(t) = \frac{\Delta m(t)}{M_p\,\Delta t} \quad \text{or} \quad \lambda(t) = \frac{f(t)}{P(t)} = \frac{1}{P(t)} P_{def}\left(\chi^* \succ \chi; t\right). \qquad [5.39]$$

Hence, the probability of failure-free operation can be determined by the formula which is one of the basic equations of the mathematical theory of reliability:

$$P(t) = \exp\left[-\int_0^t \lambda(t)\,dt\right], \qquad [5.40]$$

$$P(t) = \exp\left\{-\int_0^t\left[\frac{1}{1 - P_{def}\left(\chi^* \geq \chi; t\right)}\frac{d}{dt}P_{def}\left(\chi^* \geq \chi; t\right)\right]dt\right\}. \qquad [5.41]$$

The technical result of the method is that the indicators of failure-free operation of the product according to the criteria for defects, leaks or failure are determined before the start operation without preliminary tests with or without prior service of the product based on the results of non-destructive testing of products, the reliability characteristics of the NDT method used and the growth rate of possible defects during operating time t.

5.3.4 Methods of determining gamma-percentile life

The duration of safe operation in the reliability aspect can be defined as the time during which the probability of the product maintaining its strength is

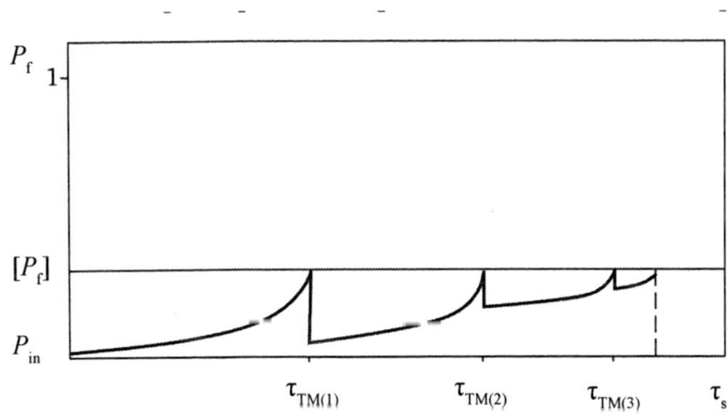

5.49   Determination of operating life in the probabilistic aspect.

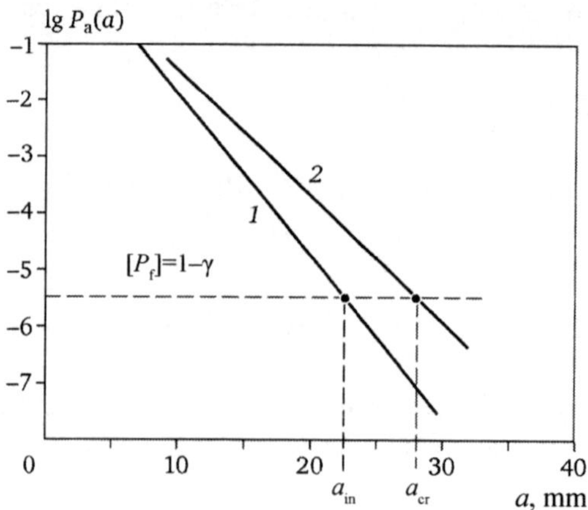

5.50 To evaluation of gamma-percentile life of the product: 1) initial curve of residual defectiveness; 2) curve of residual defectiveness at the end of gamma-percentile life.

on the level not lower than a certain value which is denoted as the allowable level of reliability equal to

$$[H] = 1 - [P_f] = \gamma,$$

where $[P_f]$ is the permissible (standard) probability of failure of the product.

Figure 5.49 schematically shows the definition of service life in the probabilistic aspect.

If the initial level of probability is equal to P_{in}, then during operation due to the development of various processes of damage in the metal structure under consideration the probability of failure will increase. When the probbility P_f reaches the allowable value of $[P_f]$ the structure should be either be decommissioned or technical mesures reducing the likelihood of its destruction should be taken. Such activities include technical maintenance (TM) of the structure (e.g. NDIS + repair of detected defects).

The gamma-percentile life of the product, i.e. life that is provided with reliability equal to the value of γ, can be defined using the probabilistic curve of residual defectiveness.

Evaluation of the residual defects and divides it into accurate and the probability of as described in Section 5.3.2.

The characteristics of the probabilistic part of residual defectiveness P_{def} $(\chi^* \geq \chi;\ t)$ at the end of the gamma-percentile life of the product are determined. In this case, the service life determines the time of growth of the defect from the size a_{in} to the size a_{cr} in Fig. 9.50 where the characteristic size of the defect is the size of the minor semiaxis of the ellipse which schematises the defect, i.e. the growth time of the defect a_{in} (the probability of existence of this defect in the product before operation is equal to $(1-\gamma)$) to the critical size a_{cr} is the gamma-percentile life of the product.

The magnitude and time of growth of the defect is determined by the equations presented for the cases of stress corrosion cracking and fatigue.

5.3.5 Influence on the probability of failure of statistical nature of strength properties and loading (generalised method)

The generalised method takes into account the distribution laws of the mechanical properties, loading, defects, and also the effects of ageing, cyclic loadin, and the influence of corrosive environment and destruction mechanisms.

The basic equation for calculating the probability of failure P_f is as follows:

$$P_f = \iiint_{\omega} \rho(x_1)\rho(x_2)\rho(a)da\,dx_1\,dx_2, \tag{5.42}$$

where $\rho(x_1)$ is the differential distribution function of the strength characteristics of the specific form of which depends on the criterion of strength or fracture criterion; $\rho(x_2)$ is the differential distribution function of the characteristics of the load the concrete form of which depends on the criterion of strength or fracture criterion; a is the characteristic size of the defect; ω is the integration domain which depends on the strength or failure criteria used.

The generalised equation has its own form for each failure mechanism (brittle, ductile and quasi-brittle). For example, the equation for calculating the probability of fracture of the components in the brittle state is as follows:

$$P_f = \int_{K_{1c\,min}}^{K_{1c\,max}} \rho_{K_{1c}}(K_{1c}) \int_{\sigma_{c\,min}}^{\sigma_{max}} \rho_\sigma(\sigma)P_a(a \geq a_{cr})da\,dK_{1c}, \tag{5.43}$$

where P_f is the probability of failure; ρ_{K1c} is the differential distribution function of the brittle fracture toughness coefficient of fracture toughness K_{1c}; ρ_σ is the differential distribution function of stress σ; P_a is an integral function of the size distribution of defects;

From equation [5.43] it is easy to get partial cases:
* in the absence of the scatter of load:

$$P_f = \int_{K_{1c\,min}}^{K_{1c\,max}} \rho_{K1c}(K_{1c})P_a(a \geq a_{cr})dK_{1c}da; \tag{5.44}$$

* in the absence of scatter of the fracture toughness coefficient:

$$P_f - \int_{\sigma_{min}}^{\sigma_{max}} \rho_\sigma(\sigma)P_a(a \geq a_{cr})d\sigma\,da; \tag{5.45}$$

* in the absence of scatter of the fracture toughness factor and load:

$$P_f = P_a(a \geq a_{cr}).$$ [5.46]

It can be seen that this equation is the case described in sections 3.3.1 and 3.3.3.

If there are no defects in the structure, then the term associated with the size of the defect is removed from equation [5.43]. For example, the following equation is obtained for the probability of failure of a defect-free strcucture with the strength criterion of the material in the form of yield stress σ_T and random static load, characterised by stress σ:

$$P_d = \int_{\sigma_{Tmin}}^{\sigma_{Tmax}} \rho_{\sigma_T}(\sigma_T) \int_{\sigma_{min}}^{\sigma_T} \rho_\sigma(\sigma)d\sigma\,d\sigma_T,$$ [5.47]

It can be seen that in this case the model proposed by A.R. Rzhanitsyn (section 3.1) is described.

5.3.6 Comparison of results of calculations using procedures described in sections 5.3.5 and 3.1

Software PN1.1[96] (a brief description is given in the appendix) was developed for calculations using the generalised method (section 5.3.5). The software was used for calculated on the basis of an example for the calculation of steel structures using the method proposed by Rzhanitsyn (see section 3.1). The calculations yielded the probabilities of fracture for the input data from the example given by Rzhanitsyn (Fig. 5.51):

Calculation No. 1:

$\overline{\sigma}_T = 2663$ kg/cm^2 = 266.3 MPa – Average expected yield stress;

$\sqrt{D_{\sigma_T}} = 284$ kg/cm^2 = 28.4 MPa – Standard deviation of yield stress;

$\overline{\sigma} = 1400$ kg/cm^2 = 140 MPa – Average value of stress;

$\sqrt{D_\sigma} = 140$ kg/cm^2 = 14 MPa – Standard deviation of stress;

Calculation No. 2:

$\overline{\sigma}_T = 2663$ kg/cm^2 = 266.3 MPa – Average expected yield stress;

$\sqrt{D_{\sigma_T}} = 284$ kg/cm^2 = 28.4 MPa – Standard deviation of the yield stress;

$\overline{\sigma} = 1600$ kg/cm^2 = 160 MPa – Average value of stress;

$\sqrt{D_\sigma} = 160$ kg/cm^2 = 16 MPa – Standard deviation of stress.

The results of the calculations show that at the allowable stress in the structure of $[\sigma] = 1400$ the failure probability is $P_f = 3.3 \cdot 10^{-5}$, and at $[\sigma] = 1600$ the probability of failure $P_f = 5.5 \cdot 10^{-4}$ and $4.7 \cdot 10^{-4}$, close to the probability values, obtained by the Rzhanitsyn' method ($3.2 \cdot 10^{-5}$ and $4.7 \cdot 10^{-4}$, respectively).

It is also necessary to mention another result which follows from the numerical solution of the Rzhanitsyn's problem. The range of the stress

Calculation No. 1

	$2\sqrt{D_\sigma}$	$3\sqrt{D_\sigma}$	$4\sqrt{D_\sigma}$	$5\sqrt{D_\sigma}$	$6\sqrt{D_\sigma}$
$2\sqrt{D_{\sigma_T}}$	–	–	–	$0.2 \cdot 10^{-10}$	$2.0 \cdot 10^{-9}$
$3\sqrt{D_{\sigma_T}}$	–	$2.2 \cdot 10^{-8}$	$1.4 \cdot 10^{-6}$	$1.6 \cdot 10^{-6}$	$1.6 \cdot 10^{-6}$
$4\sqrt{D_{\sigma_T}}$	$8.7\ 10^{-6}$	$1.9 \cdot 10^{-5}$	$2.2 \cdot 10^{-5}$	$2.2 \cdot 10^{-5}$	$2.2 \cdot 10^{-5}$
$5\sqrt{D_{\sigma_T}}$	$2.0 \cdot 10^{-5}$	$3.0 \cdot 10^{-5}$	$3.2 \cdot 10^{-5}$	$3.2 \cdot 10^{-5}$	$3.2 \cdot 10^{-5}$
$6\sqrt{D_{\sigma_T}}$	$2.0 \cdot 10^{-5}$	$3.0 \cdot 10^{-5}$	$3.2 \cdot 10^{-5}$	$3.3 \cdot 10^{-5}$	$3.3 \cdot 10^{-5}$

Calculation No. 2

	$2\sqrt{D_\sigma}$	$3\sqrt{D_\sigma}$	$4\sqrt{D_\sigma}$	$5\sqrt{D_\sigma}$	$6\sqrt{D_\sigma}$
$2\sqrt{D_{\sigma_T}}$	–	–	$9.2 \cdot 10^{-6}$	$1.1 \cdot 10^{-5}$	$1.1 \cdot 10^{-5}$
$3\sqrt{D_{\sigma_T}}$	$1.0 \cdot 10^{-4}$	$2.1 \cdot 10^{-4}$	$2.4 \cdot 10^{-4}$	$2.5 \cdot 10^{-4}$	$2.5 \cdot 10^{-4}$
$4\sqrt{D_{\sigma_T}}$	$3.2 \cdot 10^{-4}$	$4.8 \cdot 10^{-4}$	$5.2 \cdot 10^{-4}$	$5.2 \cdot 10^{-4}$	$5.3 \cdot 10^{-4}$
$5\sqrt{D_{\sigma_T}}$	$3.4 \cdot 10^{-4}$	$5.1 \cdot 10^{-4}$	$5.5 \cdot 10^{-4}$	$5.5 \cdot 10^{-4}$	$5.5 \cdot 10^{-4}$
$6\sqrt{D_{\sigma_T}}$	$3.4 \cdot 10^{-4}$	$5.1 \cdot 10^{-4}$	$5.5 \cdot 10^{-4}$	$5.5 \cdot 10^{-4}$	$5.5 \cdot 10^{-4}$

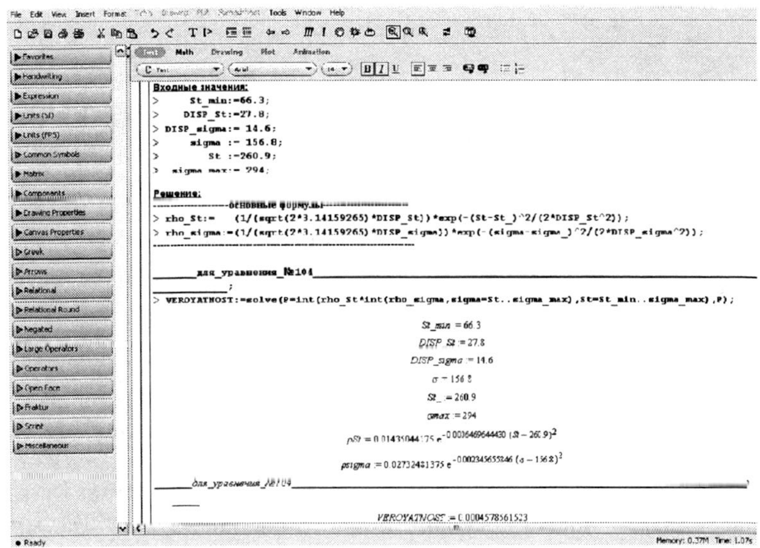

5.51   Calculation module window.

and yield stress has a significant impact on the final results. Only after 4–5 ranges counted in the calculation, can the results be regarded as final and differing only slightly. This is especially important when the so-called guaranteed values of mechanical properties, listed in the regulations, are used. For the yield stress this usually corresponds to $\left(\bar{\sigma}_T - \sqrt{\sigma_T}\right)$.

Example. It is required to determine the probability of brittle fracture of a structure which contains discontinuities, and its dependence on various characteristics[96] and section 5.3.5.

Input data: $K_{1cmin} = 259.9$; $\bar{K}_{1c} = 356.5$; $D^{0.5}_{K1c} = 48.3$ – the crack characteristics; $\sigma_{max} = 26.9$; $D^{\sigma}_{0.5} = 0.05$; $\bar{\sigma} = 26.8$ – the stress characteristics; $a_{max} = 48.12$; $a_0 = 3.19$; $\gamma = 0.307$ – the defect characteristics.

Figures 5.52–5.56 show the dependences of the probability of failure on different characteristics.

For the input data $\bar{K}_{1c} = 356.5$; $D^{0.5}_{K1c} = 48.6$; $\bar{\sigma} = 26.8$; $D^{0.5}_{\sigma} = 2$; $a_{max} = 48.12$; $f = 1$; $a_0 = 3.19$; $\gamma = 0.307$ calculations yielded the graphical dependences shown in Fig. 5.57–5.59.

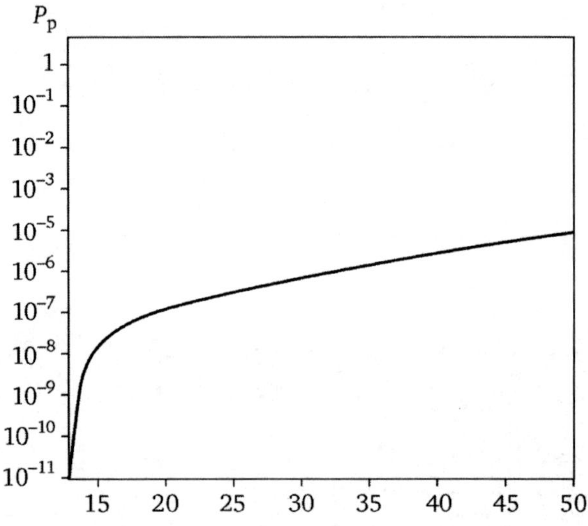

5.52   Dependence of probability of failure on $D^{0.5}_{K1c}$.

5.3.7 Changes in strength properties of steels in operation due to ageing

The generalised model delas with different types of ageing. Ageing in the form of growth of defects during operation is taken into account by the methods described in sections 4.1, 5.3.3 and 5.3.4. Ageing as thinning of the walls of equipment by corrosion is taken into account by substituting the stress, taking into account the new size and shape of the structure.

Ageing as a change in the mechanical properties during operation was studied by experimental studies of the properties of metal cut from the

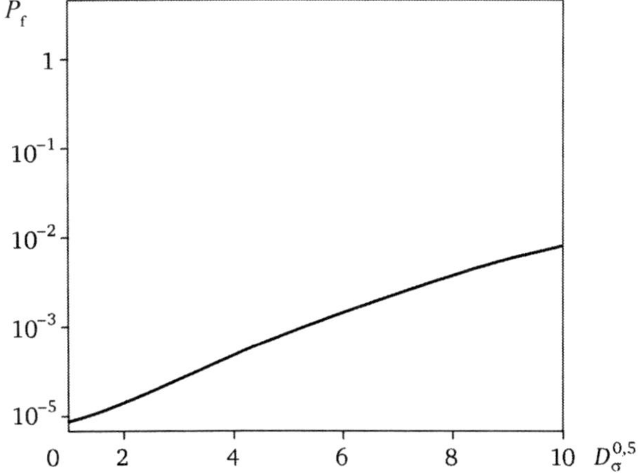

5.53   Dependence of the probability of failure on $D_\sigma^{0.5}$.

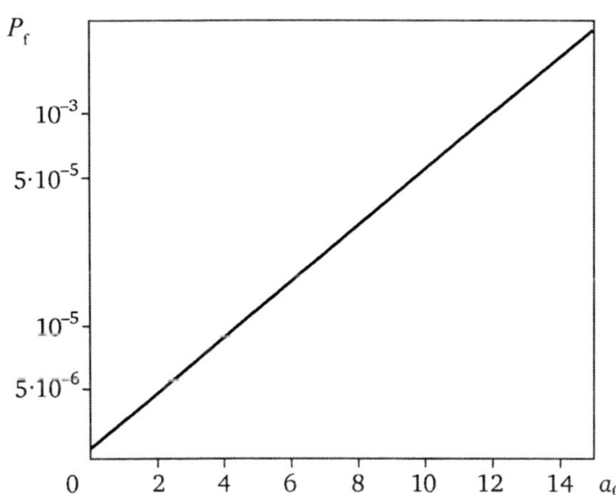

5.54   Dependence of the probability of failure on a_0.

elements of nuclear power plant. This kind of ageing is further explained below.

At Russian nuclear power plants special attention is paid to experimental studies of the ageing of structural steels during their operation. These studies are performed on a regular basis every 100 000 hours of service, including after 150 000 and 200 000 hours of operation, and in some cases with a defect in the metal.

A list of major steel plants with VVER-440, VVER-1000 and RBMK-1000 reactors and the extent of studies of these during operation are shown in Table 5.11.

The results of investigation of the mechanical properties of the steel of the main pipeline of the VVER-440 reactor – 0Kh18N12T – are given as an example.

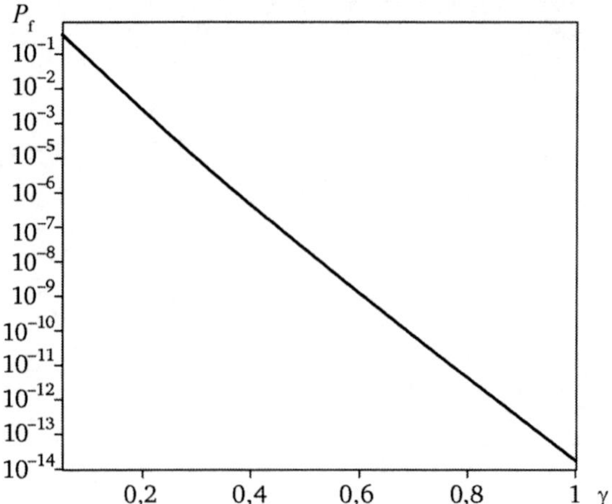

5.55   Dependence of the probability of failure on γ.

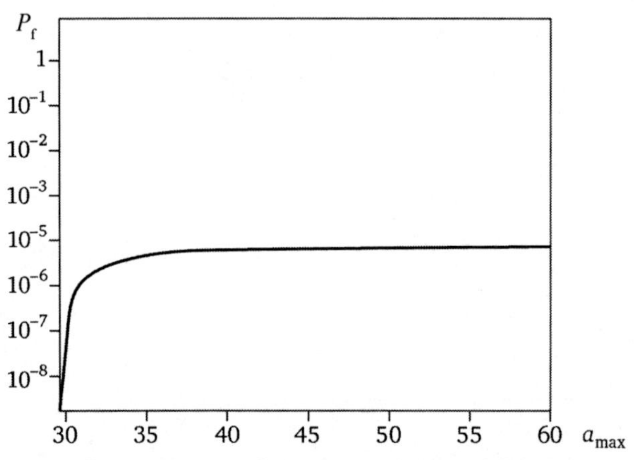

5.56   Dependence of the probability of failure on a_{max}.

The mechanical properties of steels 08Kh18N12T and of welds in these steels after prolonged operation in the pipeline of the reactors VVER-440 were determined in Ref. 98–103, 9, etc.

The test results obtained for samples cut from metal of MCO DN-500 after 100 000 hours of operation are presented in Table 5.12 and Fig. 5.60. The results of tests of the same metal after austenitising and tempering as well as the certificate data are also included. The certificate data and test results after long-term service as well as after long-term servive and heat treatment are presented. The comparison showed that the differences of values of mechanical properties (R_m, $R_{p0.2}$, A_s, Z) are within the natural variability of the mechanical properties and the effect of errors in the tests.

Further studies of the effect of operational impacts (150 000 hours of operation) showed no significant changes in the mechanical properties of

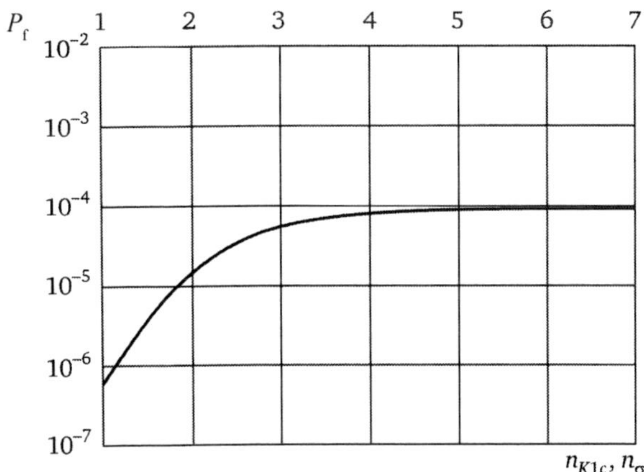

5.57   Variation of n_σ and n_{K1c}.

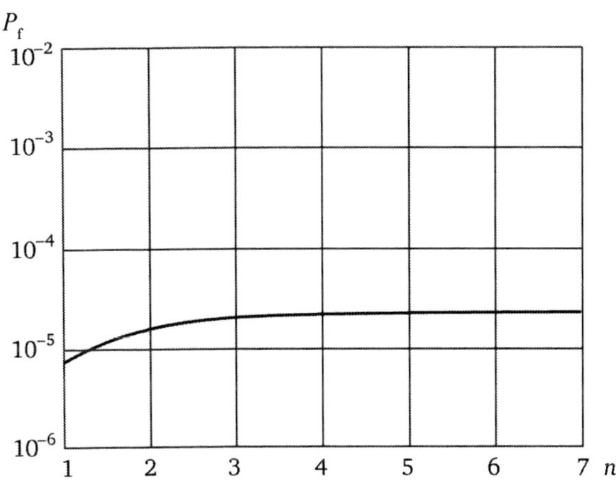

5.58   Variation of n_σ at $n_{K1c} = 2$.

08Cr18Ni10Ti steel (Fig. 5.61 and 5.62).

Additional data on changes in mechanical properties of steels of the type 0Cr18Ni12Ti and its welded joints, obtained by testing specimens cut from pipes of RBMK-1000 and thermal power plants are presented in Fig. 5.62[98–101].

These results clearly show tha there has been no degradation of mechanical properties of type 08Cr18Ni12Ti steels and of welded joints produced by manual arc welding with EA-400/10U electrodes.

To address the issues of reliability and durability of pipelines with defects it is important to determine cracking resistance characteristics. Figure 5.63 and 5.64 show the results of cyclic crack resistance tests of samples cut from the MCP of the units 1 and 3 of the Novovoronezh nuclear plant after prolonged use.[100,9] The experimental data were compared with

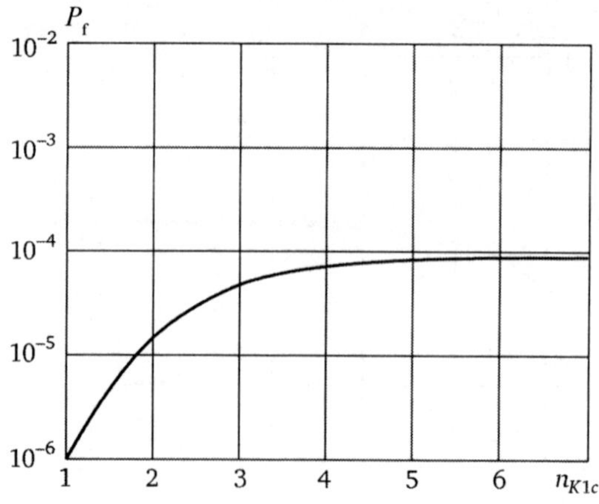

5.59 Variation of n_{K1c} at $n_\sigma = 2$.

Table 5.11 Variation of the structural properties of steels (ageing) during operation of nuclear power plants

Steel	Component of nuclear plant	Mechanical properties	Crack resistance	Microstructure	Chemical compound
St20, 22K	Main pipeline of RBMK reactor	+	+	+	+
10GN2 10E	Steam line and casing of VVER-1000 reactor	+	+	+	+
0Kh18N 10E	Main pipeline of VVER-40 reactor	+	+	+	+
48TS 15Kh2N MFA	High-pressure casing of VVER-440 and VVER-1000 reactors	+	+	+	+

the normative curve for cyclic crack resistance specified in the document in Ref. 104.

The experimental results lie below the normative values of cyclic crack resistance which indicates that long-term operation has no significant effect on the cyclic crack resistance of the pipes.

This conclusion can also be fully attributed to the static crack resistance of austenitic steel pipelines. As shown in Ref. 9, the limiting condition of these pipes with cracks determined by the 'plastic hinge' mechanism which is controlled by yield strength. It was already shown that long-term operation has no significant effect on the mechanical properties, therefore, the static fracture toughness of pipelines made of 08Cr18Ni12Ti steel

Fatigue of specimens of MCP of block 1 of the Novovoronezh NPP after prolonged use was invsetigated in Ref. 100. The fatigue characteristics were higher than the minimum values established by the norms.[4] Thus, the endurance limit according to the test results is equal to 246 MPa[11], which is higher than the normative values for steel 08Cr18Ni12Ti equal to 220 MPa.[4]

Table 5.12 Change of the mechanical properties of base metal of pipeline DN 500 of the main circulation pipeline during operation

Melt	$T_{test} = 20°C$				$T_{test} = 300°C$				Grain size
	TS, MPa	YS, MPa	EL, %	RA, %	TS, MPa	YS, MPa	EL, %	RA, %	
Initial state									
161 944	534	240	53.5	52.0	-	215	-	-	5
	534	240	54.0	54.5		211			
160 666	519	284	51.5	52.0	-	235	-	-	2
	524	284	56.0	56.5		240			
160 679	519	269	52.0	60.0	-	235	-	-	5
			54.0	63.5		240			
100 000 hours of operation									
161 944	529	253	54.4	65.5	266	212	29.0	65.0	-
	527	238	56.8	65.2	385	196	30.5	64.0	
					370	187	22.0	62.5	
160 666	568	284	58.0	62.0	-	-	-	-	-
	549	274	61.0	60.0					
	509	250	57.0	62.0					
160 679	560	358	52.5	63.0	-	-	-	-	-
	545	304	58.5	63.0					
	550	348	57.0	63.0					
100 000 of operation + austenitizing and tempering									
160 666	478	214	45.3	61.2	-	-	-	-	-
	490	218	57.7	67.1					

5.3.8 Initial data for calculating the probability of destruction of equipment and pipelines of NPP

All information necessary to determine the probability of failure can be found at the NPP and the organisations that support operation, construction and design of NPPs.

For example, the power operating conditions are defined by regulations for operation and may be made more accurate by NPP documentation. In recent years, nuclear power plants in Russia and abroad have been using automated inspection systems for the residual life of the type SAKOR in Russia or Famos in Germany, in which the entire history of thermal force loading of equipment and pipelines of NPPs is inspected.

The characteristics of the scatter of strength properties can be determined according to the certificate data or appraisal reports. In the first approximation, these regulations can be used, taking the variance to be equal to zero. In this case, the estimates will be the greatest error in the margin of safety.

The baseline data for evaluation of residual defects discussed above are in sections 3.2.1 and 3.2.3. These data can be obtained from the results of

R_m; $R_{p0.2}$; HB, MPa

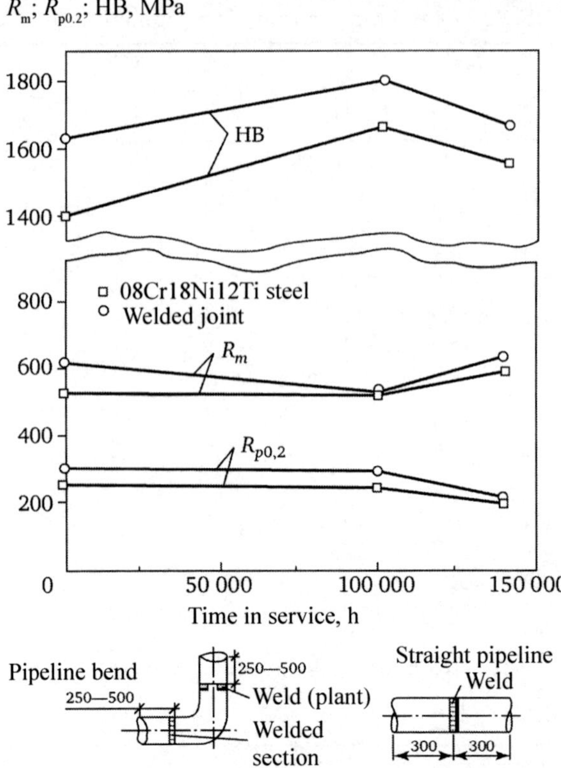

5.60   Variation of the strength properties and hardness of base metal and welded joint of DN500 pipelines of power unit 3 of the Novovoronezh nuclear power plant in long-term service.

non-destructive testing in nuclear power plants and/or the manufacturer of equipment.

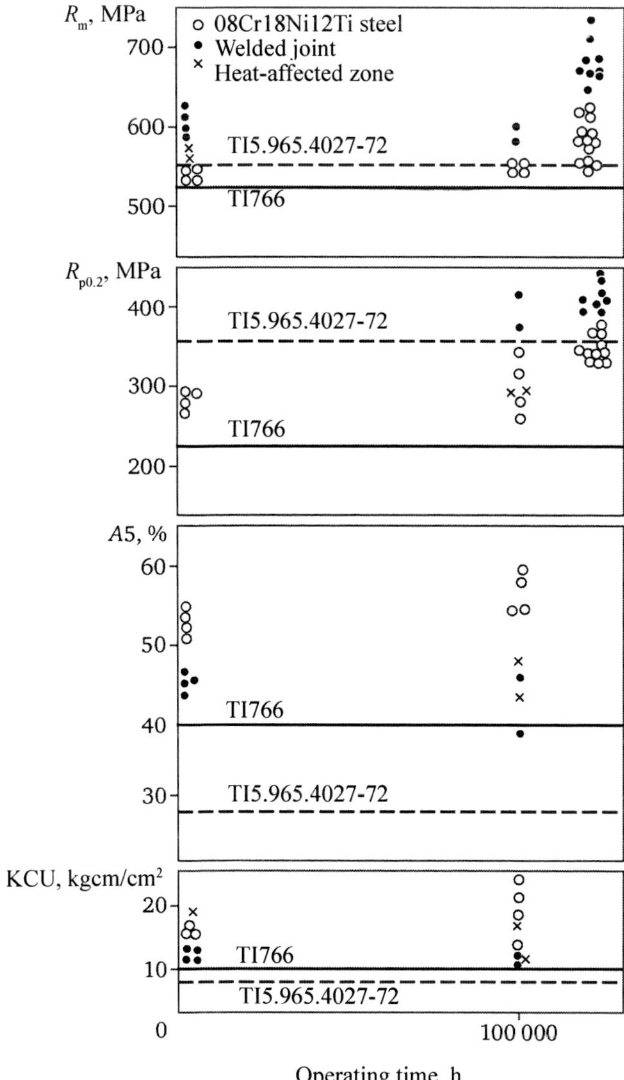

5.61 Variation of the mechanical properties of welded joints in DN-500 pipeline of power unit 1 of the Kola nuclear power plant in long-term service.

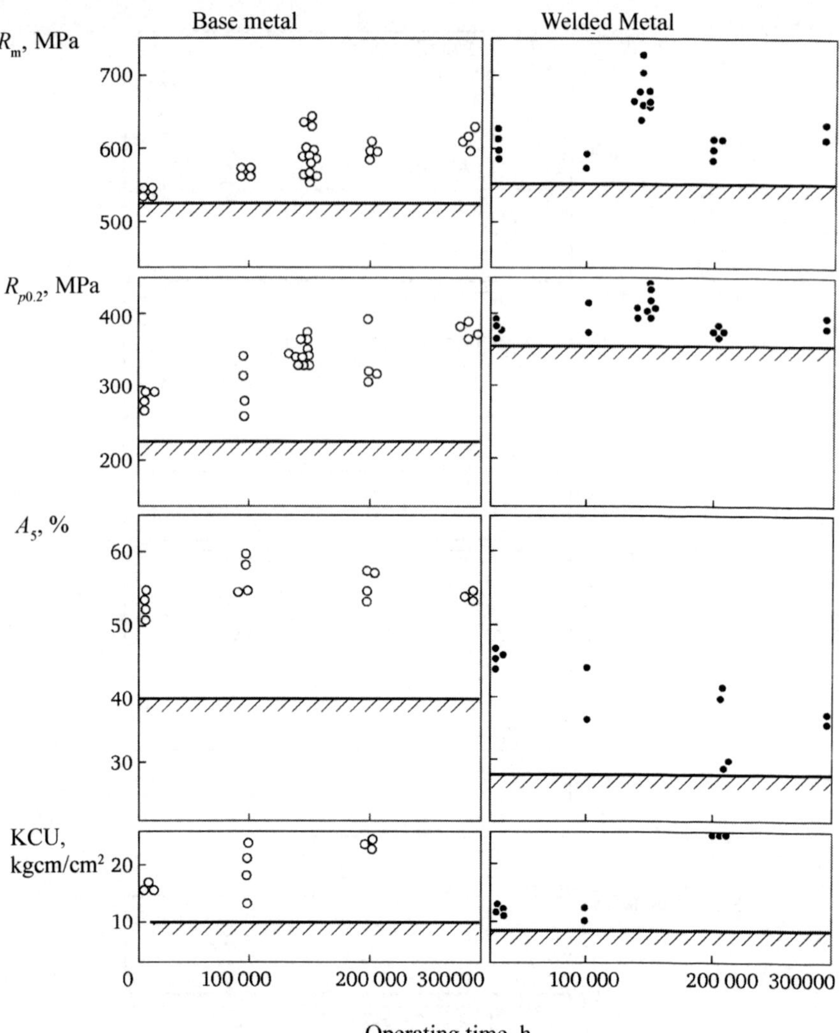

5.62 Variation of the mechanical properties of 08Cr18Ni10Ti and 08Cr18Ni12Ti steels and of welded joints in these steels during service.

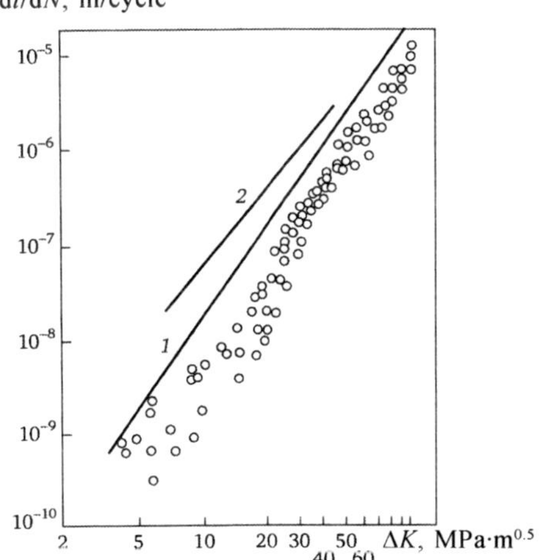

5.63 The results of tests of specimens of 08Cr18Ni12Ti steel taken in the longitudinal and transverse directions from the pipeline after service in the conditions of the power unit 1 of the Novovoronezh nuclear power plant in the 'growth rate of fatigue cracks – the range of the stress intensity factor' coordinates: 1) the upper envelope of the experimental results; 2) the normative curve.

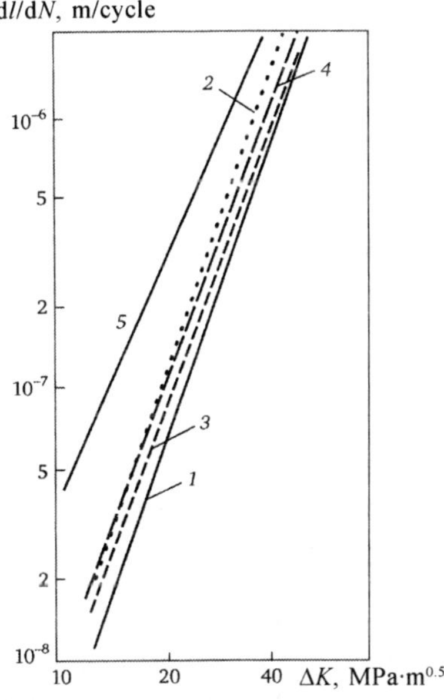

5.64 Generalised results of testing of specimens taken from the pipelines of the power unit 3 of the Novovoronezh power plant after 100 000 hours in service in the axial and circumferential directions: 1) axial direction, temperature 20°C; 2) axial direction, the temperature 300°C; 3) circumferential direction, temperature 20°C; 4) circumferential direction, temperature 300; 5) normative curve.

6

Probabilistic analysis of safety: Increasing the reliability and safety of nuclear power plant components

6.1 Probabilistic safety analysis model taking into account the initiating event 'a large break loss-of-coolant accident'

6.1.1 Understanding the model

For probabilistic safety analysis (PSA) it is important to have data on the reliability of passive safety components. In particular, in the analysis of the maximum design-basis accident the final result of PSA is decisively influenced by the probability of the initiating event – a large diameter pipeline rupture. The method of taking into account the probability of pipeline failure is studied below and described in Ref. 31.

The PSA model taking into account the initiating event (IE) 'large break LOCA' was investigated in the calculations of the risk of melting of the reactor core, taking into account the effects of aging of equipment. The IE large break LOCA means the breaking of pipelines with the equivalent leak diameter $D > 6$ inch (145 mm). This failure is characterised by a sharp drop in the pressure in the primary circuit with a simultaneous increase of the pressure in containment. The required safety functions – i.e. ensuring coolant inventory in the core and residual heat removal to the ultimate heat sink (UHS), are fulfilled by the following safety systems:

- a system of hydraulic accumulators,
- low pressure safety injection system
- a spray system.

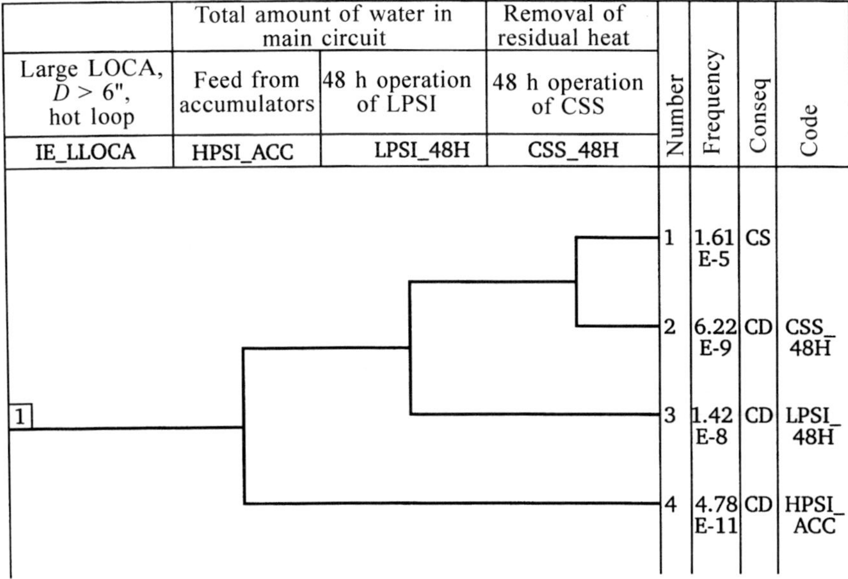

6.1 Example of the structure of the ET (large-break LOCA).

To calculate the core damage frequency, four event trees (ET) were developed with the following factors taken into account:
- the location of leaks: the hot (HL) or cold (CL) loop;
- the state of the reactor: under operation (PO) or hot shutdown (HS).

All four ETs have the same structure, Fig. 6.1. Thus, depending on the location of the leak and the state of the reactor, the successful course of the accident requires a different configuration of the system of hydraulic accumulators and low pressure safety injections. These differences are accounted for at the fault tree level using the initiating events and the appropriate specification of the boundary conditions for calculations.

6.1.2 The main results of core damage frequency analysis of fusion

The total frequency of core damage (CDF) as a result of IE a large-break LOCA is $7.18 \cdot 10^{-8}$ 1/(reactor·year). The frequency of IE, when taking account of the reactor states is $4.06 \cdot 10^{-5}$ 1/(reactor·year).

Table 6.1 and Fig. 6.2 show the distribution of risks depending on the location of the leak and the state of the reactor.

In all ETs the dominant accident sequence (AS) is a scenario with the failure of the low-pressure safety injection system AS 3, Table 6.2. This AS provides more than 75% contribution to the total rate of core melting in the case of large leaks. The contribution of accident sequences with the failure of the spray system is more than 33%.

Table 6.1 Distribution of risk in relation to the location of the leak and the state of the reactor

IE	IE code	IE frequency	CDF	Contribution to total risk of TRMS
Large leak from hot loops, power operation	LLOCA PO /HL	$1.61 \cdot 10^{-5}$	$2.05 \cdot 10^{-8}$	29
Large leak from cold loop power operation	LLOCA PO /CL	$2.41 \cdot 10^{-5}$	$5.03 \cdot 10^{-8}$	70
Large leak from the hot loop, hot shutdown	LLOCA HS /HL	$1.79 \cdot 10^{-7}$	$3.34 \cdot 10^{-10}$	> 0.5%
Large leak from cold loop and hot shutdown	LLOCA HS /CL	$2.68 \cdot 10^{-7}$	$7.21 \cdot 10^{-10}$	1
Total		$4.06 \cdot 10^{-5}$	$7.18 \cdot 10^{-8}$	

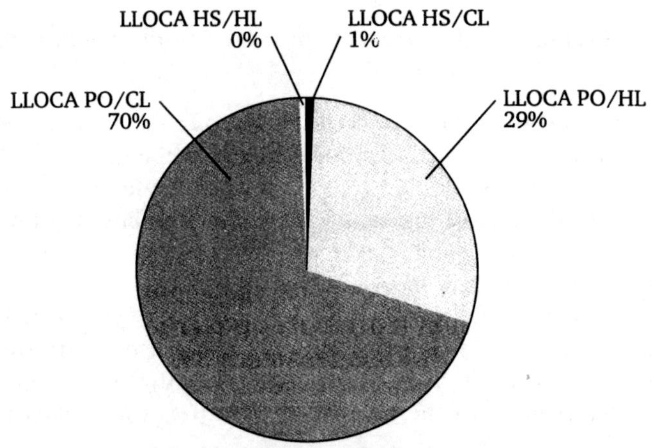

6.2 Main factors causing core melt under large-break LOCA.

Analysis of minimal cross sections (see Table 6.3) determines the most significant failures of system components and erroneous personnel actions, leading to reactor core melting. Large-break LOCA accidents include:

- in the AS 3 ET LLOCA PO/CL (leak from the cold loop, reactor at power),

– failure to open the check valves on the injection lines into the cold loop (21% of the total risk of CDF);

– common cause failures of emergency pumps make-up pumps (7% of the total risk);

– failure of the control system for automatic switching to recirculation through the sump due to human error when setting four relays (5% of the total risk of CDF);

- in the AS 3 ET LLOCA PO/HL (leak from the hot loop, reactor at power)

– common cause failure of emergency make-up pumps (5% of the total risk);

– failure of the control system for automatic switching to recirculation through the sump due to human error when configuring four relays (3% of the total risk of CDF);

- in the AS 2 ET LLOCA PO/GL (leak from the cold loop, reactor under power) ;

– common cause failures of RWST level sensors (8% of the total risk);

– common cause failures of spray pumps function and the failure to open motor-driven valves V13 and V14 contribute less than 3% to the total CDF.

It should be noted that the common cause failures of the RWST level sensors and control system failure in the automatic switching to recirculation through the sump due to human error when configuring the four relays, have already been observed when analyzing the reliability of spray systems in various studies as major contributors to the unavailability of the system.

In this case, in the framework of analysis at the level of event trees, these minimal cross sections (MC) correspond to accident sequences associated

Table 6.2 Distribution of risk in relation to the location of the leak and the state of the reactor

IE code	IE frequency	AS number	CDF	Contribution to ET, %	Contribution to total risk of CDF
LLOCA PO/HL	$1.61 \cdot 10^{-3}$	2	$6.22 \cdot 10^{9}$	30	
		3	$1.42 \cdot 10^{-8}$	69	20%
		4	$4.78 \cdot 10^{-11}$	0	
		Total ET	$2.05 \cdot 10^{-8}$		
LLOCA PO / CL	$2.41 \cdot 10^{-5}$	2	$9.32 \cdot 10^{-9}$	19	13
		3	$3.97 \cdot 10^{-8}$	79	55
		4	$1.27 \cdot 10^{-9}$	3	
		Total ET	$5.03 \cdot 10^{-8}$		
LLOCA HS / HL	$1.79 \cdot 10^{-7}$	2	$6.89 \cdot 10^{-11}$	21	
		3	$2.65 \cdot 10^{-10}$	79	
		4	$5.33 \cdot 10^{-13}$	0	
		Total ET	$3.34 \cdot 10^{-10}$		
LLOCA HS / CL	$2.68 \cdot 10^{-7}$	2	$1.03 \cdot 10^{-10}$	14	
		3	$6.04 \cdot 10^{-10}$	84	
		4	$1.42 \cdot 10^{-11}$	F2	
		Total EW	$7.21 \cdot 10^{-10}$		

Table 6.3   Analysis of the minimum cross sections

Code	CDF	Contri-bution, %	IS	AE
LLOCA PO/CL/AS3	$7.53 \cdot 10^{-9}$	10.45	LLOCA_PO_CL	RCP222VP_RO
	$7.53 \cdot 10^{-9}$	10.45	LLOCA_PO_CL	RCP122VP_RO
	$5.14 \cdot 10^{-9}$	7.13	LLOCA_PO_CL	RIS2PO # 1DF_48H-ALL
	$3.62 \cdot 10^{-9}$	5.02	LLOCA_PO_CL	HE_I & C
LLOCA PO/HL/AS3	$3.43 \cdot 10^{-9}$	4.76	LLOCA_PO_HL	RIS2PO # 1DF_48H-ALL
	$2.41 \cdot 10^{-9}$	3.35	LLOCA_PO_HL	HE_I & C
LLOCA PO/CL/AS2	$1.11 \cdot 10^{-9}$	1.54	LLOCA_PO_CL	CCFRWSTLS1-123
	$1.11 \cdot 10^{-9}$	1.54	LLOCA_PO_CL	CCFRWSTLS1-124
	$1.11 \cdot 10^{-9}$	1.54	LLOCA_PO_CL	CCFRWSTLS1-ALL
	$1.11 \cdot 10^{-9}$	1.54	LLOCA_PO_CL	CCFRWSTLS1-234
	$1.11 \cdot 10^{-9}$	1.54	LLOCA_PO_CL	CCFRWSTLS1-134
	$1.03 \cdot 10^{-9}$	1.43	LLOCA_PO_CL	CSS2PO # 1DF_48H-ALL
	$9.85 \cdot 10^{-10}$	1.37	LLOCA_PO_CL	CSS2VBE2RO-ALL

with the failure of the safety injection (AS3) and not of the spray system (AS2). This is explained by the fact that both systems (safety injection and spray) are common elements in a control system and this is taken into account by the logic of simulation ET/FT (failure tree).

The results of the calculations are presented in Tables 6.1–6.3

Thus, the probability of core melt is significantly affected by probability of the initiating event – pipeline rupture. In particular, this relationship is clearly visible in the data in Table 6.2

6.2   Taking into account in PSA models the first level of ageing effects of systems and equipment in nuclear power plant

6.2.1 The set of the input data

The set of the input data, including the list of components and systems sensitive to the ageing effects and point values of the reliability parameters, calculated for the age of the unit of 10, 20, 30 and 40 years, was prepared in accordance with the procedure described in Ref. 151. The input data refer to mechanical and electrical equipment of the high pressure safety injection system and the spray system, and the components of the control and protection system (Table 6.4).

For the majority of the components, the relative increase of the reliability parameters with time is not very large (Fig. 6.3). The relative increase of the

Table 6.4 Initial date for reliability with ageing effects taken into account

Equipment	Rate of failure (initial value)	Selected reliability model	Data correction	$\lambda(10)$	$\lambda(20)$	$\lambda(30)$	$\lambda(40)$
Electric batteries	$2.80 \cdot 10^{-6}$	Linear	Running-in failures excluded	$2.72 \cdot 10^{-6}$	$4.30 \cdot 10^{-6}$	$5.88 \cdot 10^{-6}$	$7.46 \cdot 10^{-6}$
Flow rate sensors	$2.40 \cdot 10^{-6}$	Weibull	Running-in failures excluded	$2.23 \cdot 10^{-6}$	$4.92 \cdot 10^{-6}$	$7.83 \cdot 10^{-6}$	$1.09 \cdot 10^{-6}$
Level sensors	$2.10 \cdot 10^{-6}$	Log-linear	-	$2.10 \cdot 10^{-6}$	$3.62 \cdot 10^{-6}$	$6.26 \cdot 10^{-6}$	$1.08 \cdot 10^{-5}$
Pressure sensors	$1.40 \cdot 10^{-6}$	Weibull	-	$1.33 \cdot 10^{-6}$	$1.64 \cdot 10^{-6}$	$1.86 \cdot 10^{-6}$	$2.03 \cdot 10^{-6}$
380 V switches (pumps) FF	$3.10 \cdot 10^{-6}$	Weibull	-	$3.03 \cdot 10^{-7}$	$4.18 \cdot 10^{-7}$	$5.05 \cdot 10^{-7}$	$5.78 \cdot 10^{-7}$
380 V switches (pumps) FD	$8.40 \cdot 10^{-6}$	Log-linear	Running-in failures excluded	$6.45 \cdot 10^{-6}$	$2.32 \cdot 10^{-5}$	$8.35 \cdot 10^{-5}$	$3.0 \cdot 10^{-6}$
6.6 kV switches FD	$4.80 \cdot 10^{-6}$	Weibull	-	$5.54 \cdot 10^{-5}$	$6.48 \cdot 10^{-5}$	$7.10 \cdot 10^{-5}$	$7.57 \cdot 10^{-5}$
Pump electric motors 6.6 kV FR	$1.90 \cdot 10^{-6}$	Weibull	Running-in failures excluded	$2.04 \cdot 10^{-6}$	$2.87 \cdot 10^{-6}$	$3.50 \cdot 10^{-6}$	$4.04 \cdot 10^{-6}$
Pump electric motors 6.6 kV FS	$1.80 \cdot 10^{-5}$ $1.80 \cdot 10^{-5}$	Log-linear Weibull	-	$1.57 \cdot 10^{-5}$ $1.89 \cdot 10^{-5}$	$2.14 \cdot 10^{-4}$ $8.20 \cdot 10^{-5}$	$2.94 \cdot 10^{-3}$ $1.94 \cdot 10^{-4}$	$4.02 \cdot 10^{-2}$ $3.56 \cdot 10^{-4}$
LPSI and CSS FR pumps	$8.90 \cdot 10^{-5}$	Weibull	-	$5.91 \cdot 10^{-5}$	$1.38 \cdot 10^{-4}$	$2.26 \cdot 10^{-4}$	$3.21 \cdot 10^{-4}$
CCS FD pumps	$3.60 \cdot 10^{-5}$	Log-linear		$3.85 \cdot 10^{-5}$	$8.25 \cdot 10^{-5}$	$1.77 \cdot 10^{-4}$	$3.78 \cdot 10^{-4}$
CSS FO valves	$6.80 \cdot 10^{-4}$	Weibull	Running-in failures excluded	$7.59 \cdot 10^{-4}$	$1.49 \cdot 10^{-3}$	$2.21 \cdot 10^{-3}$	$2.92 \cdot 10^{-3}$
CSS RF reverse valves	$2.30 \cdot 10^{-6}$	Log-linear	-	$1.74 \cdot 10^{-6}$	$3.93 \cdot 10^{-6}$	$8.89 \cdot 10^{-6}$	$2.01 \cdot 10^{-5}$
LPSI FD valves	$1.10 \cdot 10^{-4}$	Log-linear	-	$7.59 \cdot 10^{-4}$	$1.49 \cdot 10^{-3}$	$2.21 \cdot 10^{-3}$	$2.02 \cdot 10^{-3}$

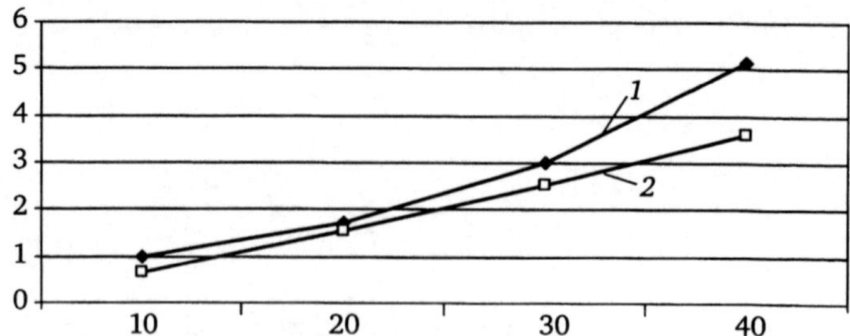

6.3 Relative increase of the rate of failure with time for different groups of equipment: 1) RWST level sensors; 2) pumps of the spray system.

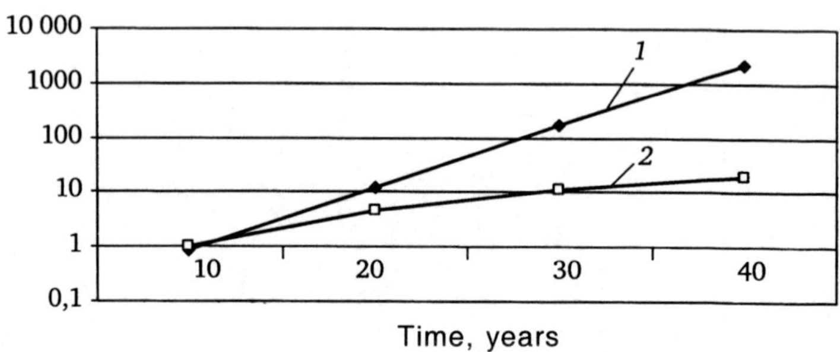

6.4 Relative increase of the probability of failure on demand for electric drives of centrifugal pumps: 1) log-linear model; 2) Weibull model.

reliability parameters in 40 years of operation varies in the range 2–10. An exception is the group of equipment 'electric drives of centrifugal pumps' for which the reliability models indicate a large increase of the probability of failure demand with time (Fig. 6.4). For this group, the best extrapolation of the primary data is obtained by the log-linear model ($P = 0.98$). The relative increase of the probability of failure at 40 years in operation is more than three orders of magnitude. At the same time, the Weibull model, which is also characterised by good extrapolation of the primary data with the significance level of 0.96, shows the relative increase of the probability of failure at 40 years in operation by approximately 20 times.

For all the components for which the ageing effects were taken into account, it was assumed that technical maintenance is carried out without reconditioning.

Calculations were performed for units operated for 10, 20, 30 and 40 years. The results were compared with the basic data (section 6.1).

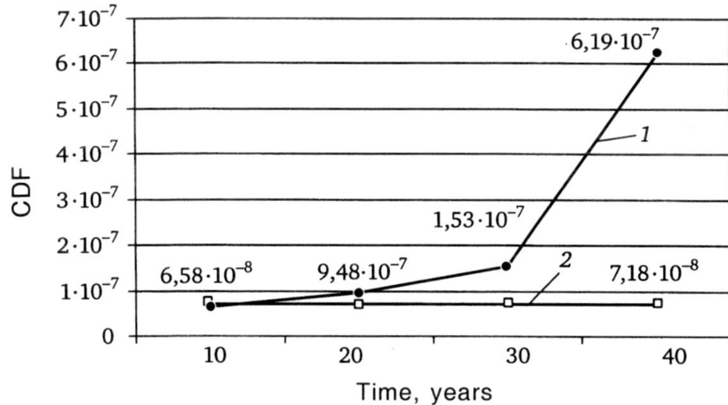

6.5 Time dependence core damage frequency for the case of 'large-break LOCA. 1) CDF as a function of time; 2) CDF (basic variant).

6.2.2 The effect of ageing on core damage frequency (CDF)

Figure 6.5 shows the effect of ageing of the components and the core damage frequency. Calculations, carried out using the reliability models, dependent on time, show the monotonic increase of the core damage frequency from the value of 6.58 · 10^{-8} 1/(reactor · year) in the initial operation period (10 years) up to 6.19·10^{-7} 1/(reactor · year) to the end of the operating life (40 years). In comparison with the basic calculation variant (7.18 · 10^{-8}), the cored damage frequency 8.6 times up to 40 years of operating life.

Table 6.5 shows the results of calculation of the failure consequences and the distribution of risk in relation to the position of the leak and the state of the reactor. As in the basic calculation variant (section 6.1), the dominant failure sequence in all ESs is the scenario with the failure of the systems for low-pressure safety injection ES3. However, its the contribution to the total core damage frequency in the presence of large leaks decreases with time. For example, after operation for 40 years, the given AS contributes approximately 56% of the total core damage frequency, whereas in the basic calculation variant it is more than 75%. The contribution of the failure consequences with the failure of the spray system (ES2) at the end of the operating life increases to 60% (33% in the basic variant).

Since the components characterised by the highest sensitivity to ageing (the electric drives of the pumps and level sensors) are included in the minimum sections of the dominant failure sequences (section 6.1), the distribution of risk in relation to the state of the reactor and the position of the leak changes only slightly with time.

The analysis of the minimum sections shows the time dependence of the significance of the failure of components of the systems and erroneous

Table 6.5 Results of ca calculations of failure sequences and risk distribution

IE code	IE frequency	AS number	CDF (basic value)	*CDF as a function of time				Contribution to ET, %	Contribution to total CDF for LLOCA, %
				10 years	20 years	30 years	40 years		
LLOCA PO/ HL	$1.61 \cdot 10^{-5}$	2	$6.22 \cdot 10^{-9}$	$5.65 \cdot 10^{-9}$	$9.20 \cdot 10^{-9}$	$1.74 \cdot 10^{-8}$	$1.07 \cdot 10^{-7}$	45	17
		3	$1.652 \cdot 10^{-8}$	$1.24 \cdot 10^{-8}$	$2.02 \cdot 10^{-8}$	$3.49 \cdot 10^{-8}$	$1.29 \cdot 10^{-7}$	55	21
		4	$4.78 \cdot 10^{-11}$	$4.78 \cdot 10^{-11}$	$4.78 \cdot 10^{-11}$	$4.78 \cdot 10^{-11}$	$4.78 \cdot 10^{-11}$	0	-
		Total ET	$2.05 \cdot 10^{-8}$	$1.81 \cdot 10^{-8}$	$2.94 \cdot 10^{-8}$	$5.23 \cdot 10^{-8}$	$2.36 \cdot 10^{-7}$	-	-
LLOCA PO/ CL	$2.41 \cdot 10^{-5}$	2	$9.32 \cdot 10^{-9}$	$8.47 \cdot 10^{-9}$	$1.38 \cdot 10^{-8}$	$2.61 \cdot 10^{-8}$	$1.60 \cdot 10^{-7}$	43	26
		3	$3.97 \cdot 10^{-8}$	$3.70 \cdot 10^{-8}$	$4.89 \cdot 10^{-8}$	$7.13 \cdot 10^{-8}$	$2.14 \cdot 10^{-7}$	57	35
		4	$1.27 \cdot 10^{-9}$	$1.27 \cdot 10^{-9}$	$1.27 \cdot 10^{-9}$	$1.27 \cdot 10^{-9}$	$1.27 \cdot 10^{-9}$	0	-
		Total ET	$5.03 \cdot 10^{-8}$	$4.67 \cdot 10^{-8}$	$6.40 \cdot 10^{-8}$	$9.87 \cdot 10^{-8}$	$3.75 \cdot 10^{-7}$	-	-
LLOCA HS/ HL	$1.79 \cdot 10^{-7}$	2	$6.80 \cdot 10^{-11}$	$6.27 \cdot 10^{-11}$	$1.02 \cdot 10^{-10}$	$1.93 \cdot 10^{-10}$	$1.19 \cdot 10^{-9}$	43	-
		3	$2.65 \cdot 10^{-10}$	$2.42 \cdot 10^{-10}$	$3.46 \cdot 10^{-10}$	$5.23 \cdot 10^{-10}$	$1.59 \cdot 10^{-9}$	57	-
		4	$5.33 \cdot 10^{-10}$	$5.33 \cdot 10^{-15}$	$5.33 \cdot 10^{-13}$	$5.33 \cdot 10^{-10}$	$5.33 \cdot 10^{-13}$	0	-
		Total ET	$3.34 \cdot 10^{-10}$	$3.05 \cdot 10^{-10}$	$4.49 \cdot 10^{-10}$	$7.17 \cdot 10^{-10}$	$2.78 \cdot 10^{-9}$	-	-
LLOCA HS/ CL	$2.68 \cdot 10^{-5}$	2	$1.03 \cdot 10^{-10}$	$9.39 \cdot 10^{-11}$	$1.53 \cdot 10^{-10}$	$2.89 \cdot 10^{-10}$	$1.78 \cdot 10^{-9}$	40	-
		3	$6.04 \cdot 10^{-10}$	$5.68 \cdot 10^{-10}$	$7.27 \cdot 10^{-10}$	$9.96 \cdot 10^{-10}$	$2.61 \cdot 10^{-9}$	50	-
		4	$1.42 \cdot 10^{-10}$	$1.43 \cdot 10^{-11}$	$1.42 \cdot 10^{-11}$	$1.42 \cdot 10^{-10}$	$1.45 \cdot 10^{-11}$	0	-
		Total ET	$7.21 \cdot 10^{-10}$	$6.76 \cdot 10^{-10}$	$8.94 \cdot 10^{-10}$	$1.30 \cdot 10^{-9}$	$4.40 \cdot 10^{-9}$	-	-
Total CDF for LLOCA				$7.18 \cdot 10^{-8}$	$6.58 \cdot 10^{-8}$	$9.48 \cdot 10^{-8}$	$1.53 \cdot 10^{-7}$	$6.19 \cdot 10^{-17}$	-	-

action of the personnel leading reactor core meltdown. For example, in ES3 ET LLOCA PO/CL (leak from a cold loop, reactor at power), the following factors were most significant in the basic calculation variant;

- failure to open the reverse valves in the injection lines into the cold loops (21% in the total risk of the core damage frequency);
- common cause failures of safety injection pumps (7% in the total risk);
- failure of the control system in automatic transition to recirculation through the sumps because of erroneous personal actions when setting four relays (5% in the total risk of the core damage frequency).

In the calculations taking into account ageing, these failures remain the dominant contributors to the frequency of melting of the reactor up to the operating life of the unit of 30 years. In this case, if the contribution of failure due to malfunction in opening of the reverse valves in the injection lines after the operating life of 10 years was 41%, common cause failure of the emergency injection pumps, 90%, and the failure of the system of controlling automatic transition to recirculation 10%, then after operation for 30 years the contribution of these failures to the total risk equals 21, 18 and 5%, respectively.

However, after operation for 40 years, the list of the main contributors greatly changes. The dominant contributors are the failures of the following elements:

- failure of the electric drives of safety injection pumps (39% in the total risk);
- failure of the electric drive combined with failure of the safety injection pumps (13%);
- common cause failures of safety injection pumps (9%).

The same change in the main contributors to the total risk of melting of the reactor core is also found for other dominant failure consequences.

The time dependence of the relative contribution of the failures of the components to the core damage frequency was analysed using the factors of significance of risk according to Fussell and Vesely (FV) and the factors of the reduction and increase of the risk. Figure 6.6 shows the variation of FV (fractional contribution).

According to the calculations, the most significant changes of FV took place in the last decade, i.e., between 30 and 40 years in operation. These changes are associated with the rapid increase of the probability of failure of electric drives of the pumps.

Comparison of the values of the relative contribution to the risk, obtained at the end of the operating life, with the values for the basic variant shows different dynamics of FV for the components sensitive to ageing, and the reduction of this parameter for the components not sensitive to ageing.

For example, for the RWST sensors (the components sensitive to ageing), the value of this parameter, calculated for the operating life of 40 years,

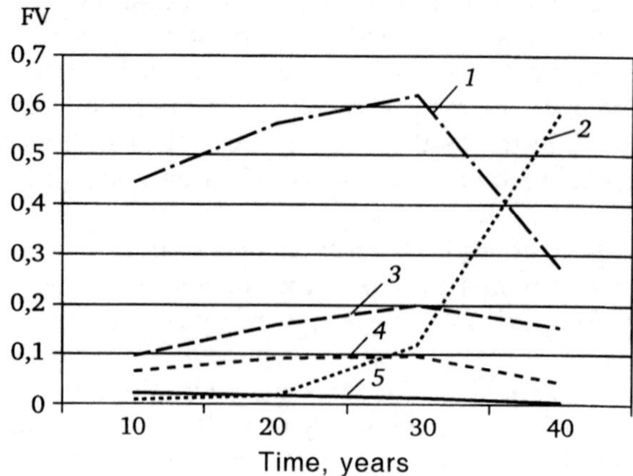

6.6 **Risk significance factor according to Fussell and Vesely as a function of time: 1) RSWT sensors; 2) pump electric drives; 3) pumps (failure to start); 4) V13-V14 slide valves; 5) I&C relay.**

is 0.272, whereas the basic value is 0.428. At the same time, the values of these parameters for the electric drives of the pumps (very sensitive to ageing) is $5.86 \cdot 10^{-1}$ and $7.53 \cdot 10^{-3}$, respectively, i.e., the differences more than two orders of magnitude). For the I&C relay (components not sensitive to ageing), the value of FV after 40 years in operation is equal to $4.36 \cdot 10^{-3}$ which is almost 5 times lower than the basic value of $2.07 \cdot 10^{-2}$.

Analysis of the variation of the risk increase factor with time (RIF) (Fig. 6.7) shows its monotonic reduction with time for the two dominant contributors to the core damage frequency: reverse valves LPSI (not sensitive to ageing) and electric drives of the pumps (with a very strong dependence of the probability of failure on time).

It should be mentioned here that in both cases the values of RIF at the end of the operating life are considerably lower than the values calculated for the basic case.

As shown by the analysis results, taking into account the ageing effects in the PSA model may result in significant changes of both the core damage frequency and of the main contributors to the risk of melting AS. The change of the factors of the significance of risk for different components may also be large.

Since the factors of the significance of risk are used widely as criteria for prioritisation and optimisation of the problems arising during operation, technical maintenance and design of nuclear power plant, ignoring the effect of ageing may result in errors when making decisions.

Analysis of results sensitivity to a model of component reliability

Figure 6.7 shows the results of analysis of the sensitivity of the core

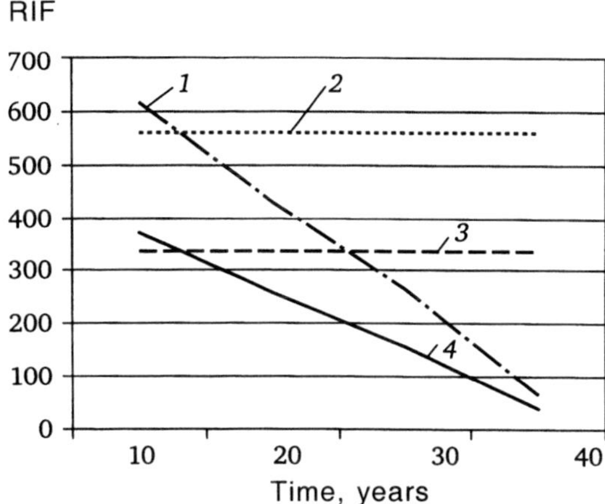

6.7   Risk increase factor as a function of time: 1) CCF of electric drives of pumps; 2) CCF of electric drives of pumps (basic variant); 3) reverse LPSI valves (basic variant); 4) LPSI reverse valves.

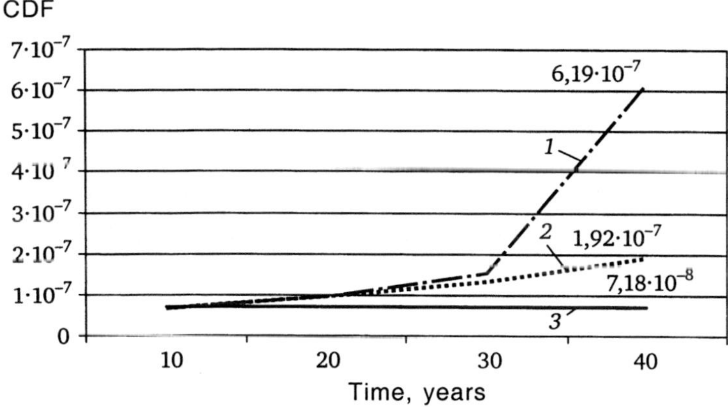

6.8   Time dependence of CDF using different reliability models: 1) CDF as a non-linear function of time; 2) CDF as a function of time (analysis of sensitivity); 3) CDF (basic variant)

damage frequency in relation to the selected reliability model for the most significant (from the viewpoint of ageing) components – the electric drives of the pumps. As mentioned previously for this group of components, the best interpolation of the data is obtained using the log-linear model. The Weibull model also describes efficiently the initial data. However, the extrapolated values of the probability of failure up to the end of the operating life of the unit, obtained using the two models, differ by more than three orders of magnitude. Taking this into account, two variants of the calculations were carried out:

1. Using the log-linear model

2. Using the Weibull model

As indicated by Fig. 6.8, the both models provide similar estimates of the core damage frequency up to 30 years of operation of the unit. However, at 40 years in operation, the core damage frequency, calculated using the log-linear model, differs by more than three times from the estimate obtained using the Weibull model. It should also be mentioned that in comparison with the basic calculation variant, the log-linear model provides the most conservative estimate of the core damage frequency (increased by almost an order of magnitude). Calculations, carried out using the Weibull model, show that the basic value of the core damage frequency increases 2.7 times. As shown in Ref. 152, the indeterminacy of extrapolation of the low-linear model is considerably higher than the indeterminacy of the extrapolations obtained using the Weibull model.

The conclusion from this analysis of sensitivity is that it is important to verify and substantiate the selection of a specific model for application in the PSA from a number of alternatives.

The results of the calculations show that both the absolute and relative characteristics of the risk can vary with time. Since the core damage frequency, the unavailability of the system and of the component, and the factors of risk significance are used widely as criteria for prioritisation and optimisation of the problems formed in operation, technical maintenance and design of nuclear power plant, ignoring the ageing effects may result in errors in taking decisions.

6.3 Method of bringing the product to the desired level of quality, reliability and safety security

The field of application of the method is manufacturing, installation and operation of products of modern engineering, including nuclear power.

It is well know that to bring the product to a specified level of quality, reliability and safety security, non-destructive testing of components is carried out either immediately after manufacture of the product (see the rules 'The equipment and piping of nuclear power plants. Welds and surfacing. Control Rules PN AE G-7-010-89, approved by Gosatomenergonadzor, USSR, May 11, 1989), or during operation, (see 'Rules for installation and safe operation of equipment and pipelines of nuclear power plants', No. AE G-7-008-89), and defects found in the product are then repaired. After the repair it is considered that the product has an acceptable level of quality, safety and reliability.

In fact, due to imperfection of NDT methods, not all defects are detected and a significant proportion of defects remain unnoticed.

The result, achieved using the technique described, is to improve the accuracy and objectivity of the evaluation of the actual state of the product so that its quality, reliability and safety can be purposefully improved.

Taking into account the actual operating loads and conditions determined for the products by fracture mechanics methods, the size of the following discontinuities is determined:

- critical,
- suitable for operation,
- suitable for manufacture (the norms of defects in the component) on the basis of the current valid standard documents and/or Technical Instructions for manufacture.

For the selected product, it is necessary to define the probability of the existence of defects, determining its quality, reliability and safety (the regulatory requirements for its quality, reliability, safety). At the same defects that determine the quality are the defects whose sizes range from the minimum sizes allowable for detection of defects (search) to the size of defects that are allowed in manufacturing and larger; defects determining the reliability – the defects whose sizes are in the range from rejection in manufacture to an acceptable level for operation and larger; defects determining the safety – ranging from allowable in operation to the critical size and larger.

This probability is given on the basis of socio-economic requirements for safety and reliability of products, taking into account possibilities of the current level of technology, and may be called the regulatory requirements for quality, reliability and product safety.

Further, the regulatory curve is plotted in the coordinates lg P–χ for the state of the product, which reflects the given probability requirements for

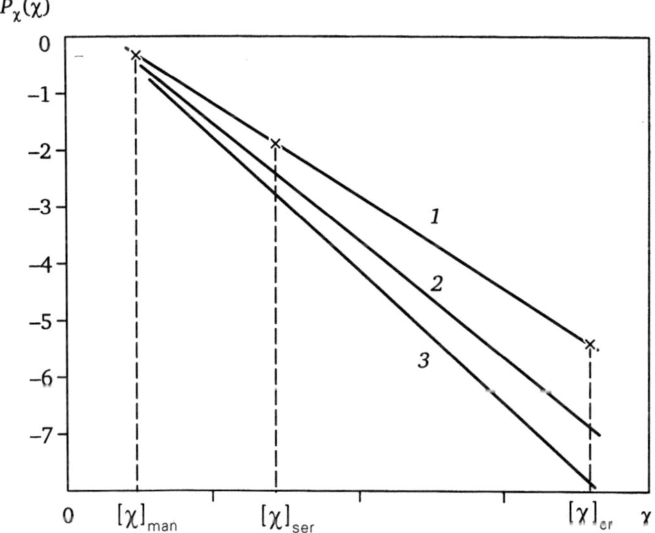

6.9 Comparison of the curves of residual defectiveness after manufacture 3, after inspection and repair of detected defects 2 with the regulatory curves of residual defectiveness 1.

quality, reliability and product safety (curve 1 in Fig. 6.9). The regulatory curve of the state of the product characterises the probability of the existence of defects in the range from allowable manufacturing defects $[\chi]_{man}$ to critical size defects χ_{cr}, including permissible defects in operation $[\chi]_{ser}$.

The curve is based on the assumption that in the logarithmic scale of the coordinates the relationship between the logarithm of probability and the characteristic size of the defect is almost linear. This assumption is confirmed by experiment[89]. The curve is plotted using the fixed points of this curve: the points that characterise the probability of a defect of the critical characteristic size and the point which characterises the probability of occurrence of the defect of the characteristic size allowed in operation.

The product is inspected by any selected NDT method. The results of inspection are used for repair (removal) of detected defects.

The inspection results are represented as a histogram in coordinates $N_{det} - \chi$, where N_{det} is the number of detected defects in inspection, χ is the characteristic defect size (usually, χ is represented either by the linear size of the defect or combination of the linear size of the defect, of the defect area, or the defect volume).

The resultant histogram is approximate by the equation

$$N_{det}(\chi) = A\chi^{-n}\left\{1 - (1-\eta)\exp\left[-\alpha(\chi - \chi_0)\right] - \eta\right\},$$

where A, n, α, η are the constants determined from the condition of maximum approximation of the equation $N_{obn}(\chi)$ to the inspection results presented in the form of a histogram; χ_0 is the minimum characteristic size of the defect that can be detected.

The initial defectiveness $N_{in} = f(\chi)$ is determined by the formula:

$$N_{in} = A\chi^{-n} \text{ or } N_{in} = A\exp(-n\chi).$$

The initial defectiveness is divided into reliable part $\chi \leq \chi_d$, and the probabilistic part $\chi > \chi_d$, where χ is the characteristic size of the defect, χ_d is the size of defects on the border between reliable and probabilistic parts.

The size of the defect χ_d is determined from the equation

$$\int_{\chi_d}^{\chi_{max}} f(\chi)\,d\chi = 1,$$

where χ_{max} is the maximum possible size of the defects in the given product.

Th probabilistic part of the curve of the initial defect characterises the probability of the existence of defects of a given size in the product.

The residual defectiveness $N_{res} = \varphi(\chi)$ is defined as the difference between the initial defectiveness N_{in} and the number of defects detected in inspection of the product $N_{det\ prod}$.

Residual defects are divided into the reliable part $\chi \leq \chi_d$ and the probabilistic part $\chi > \chi_d$.

χ_d is determined from the equation

$$\int_{\chi_d}^{\chi_{max}} \varphi(\chi)d\chi = 1,$$

where χ_{max} is the maximum possible size of the defects in this product.

The probabilistic parts of the initial and residual defects are plotted on the same graph which shows the regulatory curve of the status of the product.

The curve of probabilistic parts of the initial (hereinafter initial defectiveness) and residual defects (hereinafter - the curve of residual defectiveness) of the product is compared with the regulatory curves of the status of the product.

Analysis of the curve of the initial defectiveness indicates the state in which the product was before repair (removal) of the detected defects:

1) *The first variant of the position of the curves.* If the curves of the initial and residual defects are below the regulatory curve, then this is evidence of the satisfactory state of the product, and such a product (or products of this type) is not required to undergo inspection or the extent of inspection in the future can be minimised (Fig. 6.9);

2) *The second variant of position of the curves.* If the curve of initial defectiveness lies above the regulatory curve and the curve of residual defectiveness is below the regulatory curve, then the product after repair and removal of identified defects is in satisfactory condition and may be used. However, it must be periodically inspected and defects must be removed (Fig. 6.10). The position of the curves indicates satisfactory quality, reliability and safety of the product;

3) *The third variant of position of the curves* (Fig. 6.11). If the curves 2 and 3 of the residual and initial defectivness are positioned above the regulatory curve 1, this product requires additional inspection and repair defects.

Thus, if the curves are arranged as described in the first and second variants, they are restricted by the NDI and repairs which were already carried out, and if the third variant applies, the product should be again inspected and repairs should be carried out in accordance with the new inspection results.

This method can be applied to groups of similar products with approximately the same geometrical dimensions, material, manufacturing method and operating conditions.

The second variant may be described as follows. One defines the product or group of similar products of appropriate quality, reliability and safety of which shall be provided.

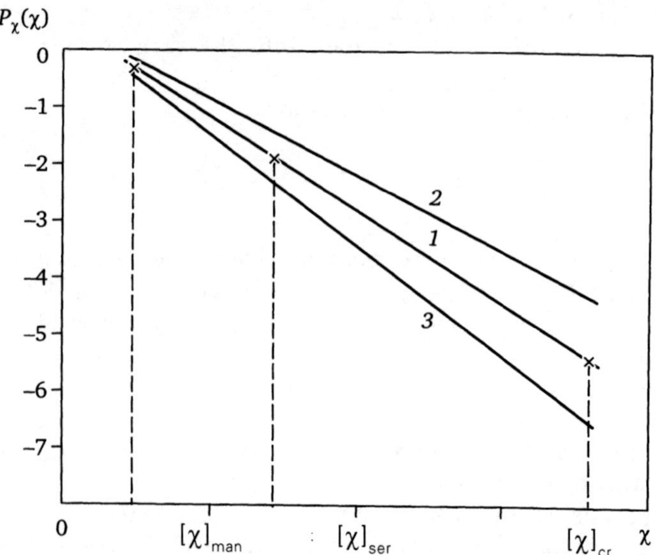

6.10   Comparison of the curves of initial residual defectiveness 2 and residual defectiveness after manufacture, NDT and repair of detected defects 3 with the regulatory curve of residual defectiveness 1.

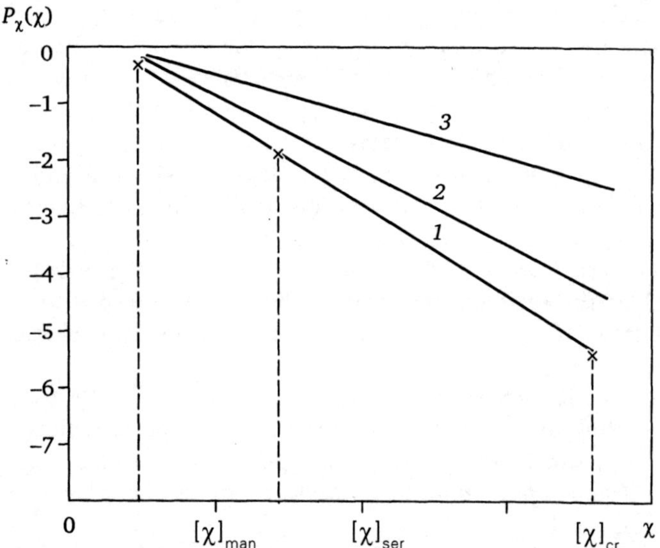

6.11   Determination of probabilistic characteristics.

The place of the most probable formation of technological and/ or operational defects is determined. Using the fracture mechanics methods, discontinuities rare determined taking into account the actual operating loads and conditions:

- Critical dimensions;
- Dimensions allowable for operation;
- Allowable dimensions of discontinuities in manufacture (the norms of defects in component) using the valid regulatory documents and/ or Technical Instructions for manufacture.

For the selected product, it is necessary to define the probability of the existence of defects which determine its quality, reliability and safety (regulatory requirements for its quality, reliability and safety).

Further, the regulatory curve is plotted in the coordinates lgP–χ fr the state of the product using the fixed points of this curve: the point that characterises the probability of formation of a defect of the critical size, and the point which characterises the probability of occurrence of the defect allowed in the operation of the characteristic size.

The product is inspected by any chosen NDT method. Repair is carried out in accordance with the inspection results.

.The test specimen is prepared for the given product, as described in section 5.3.2.

The selected method of non-destructive inspection is used to inspect the test specimen for the given product and the results of inspection are compared with the actual defects found in the test specimen. The ratio of 'the number of defects detected in the test specimen $N_{det.ts}$ (χ) divided by the number of the defects embedded in the test specimen $N_{emb.ts}$ (χ)' is determined for each characteristic defect size χ. This ratio characterises the probability of detection of the given characteristic size of the defect $P_{d.d}$ (χ):

$$P_{d.d}\ (\chi) = N_{det.ts}\ (\chi)/\ N_{emb.ts}\ (\chi)$$

The initial defectiveness of the component N_{in} (χ) = $f(\chi)$ is determined. This parameter is determined as the ratio of the number of defects of the given characteristic size $N_{det.comp}$ (χ), detected in inspection, to the probability of detection of defects by the given method of non-destructive inspection:

$$N_{in}(\chi) = N_{det.comp}\ (\chi)/\ P_{d.d}\ (\chi),$$

On the graph which shows the regulatory curve of the state of the component it is necessary to plot the probabilistic part of the curve of initial defectiveness P_{in} ($\chi \geq \chi^*$), which reflects the probability of existence of a defect with the characteristic size $\chi \geq \chi^*$, where χ^* is the characteristic size of the defect given as the lower boundary of the range of characteristic dimensions of the defects whose probability of existence resistance is being investigated.

The residual defectiveness N_{res} – $\varphi(\chi)$ is determined as the difference of the initial defectiveness N_{in} (χ) and other defects detected in inspection $N_{det.comp}$ (χ)

This is followed by the analysis of the position of the curves of the residual defectiveness in comparison with the regulatory curve of the residual defectiveness as in the first variant (Fig. 6.9-6.11).

6.4 Improving the safety of main circulation pipelines of nuclear power plant with first-generation VVER-440 reactors

In the preparation of a project for reconstructing the energy units of the first-generation with the incomplete safety systems, it was necessary to solve the problem of ensuring the reliability of main circulation pipelines (MCP) using the criterion of fracture resistance (beginning of the maximum design-basis failure). This problem was solved in the framework of the concept 'leak before break' (LBB), developed in the USA, or the technology of preventing fracture, developed in Germany. Analysis of Western approaches shows that the application of these approaches to the nuclear power plant does not ensure the safety on the basis of the fracture criterion of MCP because it does not take into account the possibility of fracture of the pipeline without a leak[9].

Therefore, investigations were carried out to develop a Russian technology of ensuring the safety of the main circulation pipelines on the basis of the criterion of fracture resistance using the System concept of strength and the concept of the leak before break, based on the determination of the probability of failure of the pipeline. This technology has been subjected to international analysis in the framework of the project TACIS[9] and, subsequently, the safety systems LBB were applied for the units 3, 4 the Novovoronezh' NPP and the units 1, 2 of the Kola NPP. The technology is described in detail in Ref. 9, and some of the main results and conclusions, obtained using the method described in section 6.3, are given below.

Figure 6.12 shows the characteristics of service safe operation of the main circulation lines on the basis of the fracture criterion in the 'probability of break – operating time in the generalised heating–cooling cycles' coordinates. The curves 1, 2 characterise the safety level prior to the application of the 'leak before fracture' system technology; the curves 3, 4 indicate the actual level of safety in the units 3, 4 of the Novovoronezh' NPP and units 1, 2 of the Kola NPP.

Reliability and safety were improved as a result of specially organised non-destructive inspection and hydraulic tests (HT) (the effect of hydraulic tests on the probability of failure of the vessels and pressure pipelines was discussed in sections 8.1 and 8.2).

When using the 'leak before break' system technology, and the probabilistic evaluation of safety and reliability in the four units of the nuclear power plants in operation, the high level of safety and reliability was reached quite rapidly and the beginning of the maximum design-basis

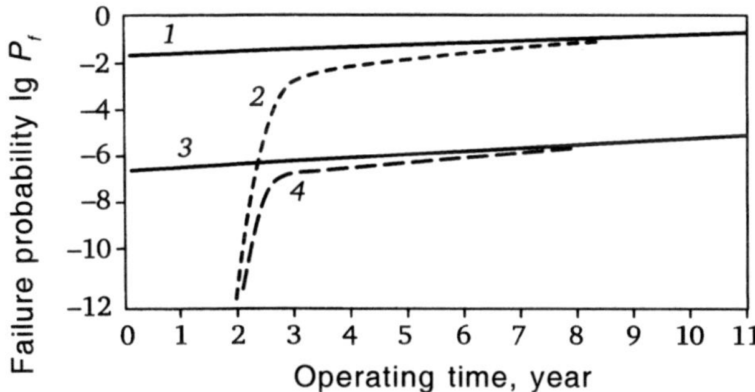

6.12 Variation of the probability of fracture of the main circulation pipeline of a VVER-440 reactor in relation to the measures taken: 1) initial reliability; 2) effect on the initial reliability of specially organised hydraulic test; 3) effect on initial reliability of special non-destructive inspection; 4) effect on initial reliability of non-destructive inspection and hydraulic test.

accident on these power units is completely prevented at this level.

The studies were used to reduce greatly the expenditure on the reconstruction of the power units 3, 4 of the Novovoronezh' NPP and units 1, 2 of the Kola NPP.

Optimisation of non-destructive testing

7.1 General

Optimisation is the selection of the best solution. Optimisation theory includes fundamental results and numerical methods which can be used to find the best option among many possible alternatives without exhaustive search and comparison.

In order to use the results and computational procedures of optimisation theory in practice, it is first necessary to formulate the considered problem in mathematical language, i.e. construct a mathematical model of the object to be optimised. The mathematical model is a more or less complete mathematical description of the process or phenomenon.

In most real situations, an exhaustive mathematical representation of the optimised system with all interactions of all its parts, interactions with the outside world and all the goals of functioning of the system is difficult or impossible. Therefore, in constructing a mathematical model it is necessary to separate and consider further only the most important aspects of the object to enable its mathematical description as well as the subsequent solution of the given mathematical problem. In this case, the factors ignored in the mathematical model should not significantly affect the final result of optimisation. Thus, the optimisation is a complex and demanding task, requiring from the researcher in-depth knowledge in the relevant region, expertise, intuition and critical analysis of the results.

Despite the fact that there is no general recipe for constructing mathematical models of optimisation, the process of mathematical modelling can be divided into the following main stages.

Definition of the boundaries of the optimisation object

The necessity of this stage is dictated by the fact that it is not possible to take into account and describe in detail all sides of most real systems. Highlighting the main variables, parameters and limitations, the system should be regarded approximately as some isolated part of the real world and to its internal structure should be simplified..

It may be that the original boundaries of the optimisation object were chosen unsuccessfully. This becomes clear in further analysis of the system and its mathematical model, interpretation of the search results of optimum solutions, comparing them with practice, etc. Then, in some cases, the system boundaries should be expanded while in others they should be made narrower. To facilitate the search for optimum solutions it is reasonable to explore each section as a separate selected system.

Generally, in engineering practice it is necessary, wherever possible, to try to to simplify the system to be optimised, break up complex systems into simpler subsystems, if it is believed that this will affect the final result within acceptable limits.

Selection of controlled variables

At this stage of modelling it is important to distinguish between those variables whose values can vary and be chosen with a view to achieving the best results (controlled variables), and the values that are fixed or determined by external factors. Determination of the values of controlled variables which correspond to the best (optimum) situation is the task of optimisation. Depending on the selected boundaries of the system to be optimised and the level of accuracy of system description, the same values can be either controlled variables or not.

Determination of restrictions on the controlled variables

In the real conditions, the choice of the controlled variables is usually restricted due to the limits of available resources, facilities and other features. When constructing a mathematical model, these restrictions are usually written in the form of equalities and inequalities or determine sets to which the values of the controlled variables must belong. The set of all restrictions on the controlled variables defines the so-called permissible set of optimisation tasks.

Selection of the numerical criterion of optimisation

The obligatory part of the mathematical model of the optimisation object is a numerical criterion, with the optimum variant of behaviour of the investigated object corresponding to the minimum or maximum value of this criterion (depending on the task). This criterion is completely determined by the selected values of controlled variables, i.e. it is a function of these variables and is called the target function.

Combining the results of previous stages of constructing the mathematical statistical model, the model is written in the form of a mathematical optimisation problem which includes building a target function and the defined constraints on the controlled variables. In a sufficiently general form the mathematical optimisation task can be formulated as follows: minimise (maximise) the target function with the constraints on the controlled variables taken into account.

Discussing the optimisation of in-service non-destructive testing (ISI), various authors have referred to various parameters and various optimisation criteria.

Thus, V.N. Volchenko[75] investigated the definition of economically optimum standard forms of defects q, based on the analysis repair costs. In this case, under the norms of the defects q refer to the allowable percentage of rejects in a batch of parts of the same type (or welded joints).

In Refs. 105 and 106, the optimumly organised operational of non-destructive testing is the system providing the required level of safety specified by the standard documents. The level of safety was estimated by the probability of failure of structural elements of the reactor, reactor core melting, and the release of radioactive elements outside the reactor[105]. The optimisation parameters, as well as the optimisation method are not defined accurately in Refs. 105 and 106. The advantage of Refs. 105 and 106 is that they attempted quantitative analysis of ISI as part of safety systems.

An optimisation method of ISI on the basis of the criterion of achieving maximum reliability of the controlled element of the structure of a nuclear reactor was proposed in Ref. 107. The optimised parameter is the time interval between the inspections. This formulation of the optimisation task is not entirely justified since it is obvious that the higher the frequency of inspection, the higher reliability can be ensured. Taking this apparent regularity into account, the time interval between inspections is often reduced in practice in order to increase the reliability of the given structure.

As discussed in Refs. 75, 105–107, ISI is optimised using probabilistic representation of strength. In reality, the optimisation can be carried out in both the deterministic and probabilistic formulations[89].

Recently, a number of papers have been published on the risk-oriented organisation ISI which can be called a semi-quantitative approach to optimisation of ISI. An example of this approach is given in the following sections 7.2 and 7.3.

A quantitative approach to the optimisation of ISI based on a generalised methodology for determining the probability of failure is described in section 7.4 and 7.5 as well as in section 9 in solving the problem of ensuring the integrity of the heat exchanger tubes of steam generators.

7.2   Overview of approaches to optimising ISI, based on information about risks: Semiquantitative approach

This section has been prepared following the documents ENIQ[108] and IAEA-TECDOC-1400[109,110] and a document prepared based on them[153].

The task of ISI based on information about risk is to form a rational strategy for managing nuclear plant safety based on risk analysis for each individual plant.

Basic principles of the ISI program based on risk information, are as follows:

• the possibility of determining the consequences, probability and risk associated with structural failures, so that the ISI programme focuses on an integrated strategy of defence in depth;

• development of the ISI programme which will reduce the risk of failure in high-risk areas, taking into account economic considerations (cost) and radiation doses received by the personnel at the nuclear power plant.

The advantage of this approach is the optimisation of operational inspection. In this context, the term 'optimisation' means the process of maintaining defence in depth which envisages:

• improvement or at least maintenance of the overall safety of the nuclear plants;

• reduction the radiation doses received by personnel involved in inspection;

• improvement of NPP reliability.

The main elements that make up the planning process of operational inspection based on risk information, are:

• participation of the operating organisation in the development and application of methodology based on risk information;

• formation of a group of experts to evaluate the ISI based on risk information;

• identification of equipment/structures or their elements for which the programme of operational inspection based on risk information will be used;

• gathering and analyzing information needed to perform risk assessment;

• determination of the level of assessment (i.e. for each equipment or its component or for the same type of group);

• assess the probability of failure for all the components;

• assess the consequences of failure for all the components;

• ranking of the risks associated with all components;

• selection of components to be controlled according to selected criteria;

• periodic certification of inspection;

• feedback of received information on the results of inspection by ISI methods and certification of inspection.

The first practical step in the development of the ISI programme, based on information about the risks is to determine its volume (areas of application) that will clearly define the scope of the programme, for example, which systems and structural elements (circular, longitudinal welds, fasteners, nozzles, etc.) should be included in the programme.

For successful development of this programme, based on risk information, it is necessary to solve several problems which can be described as follows:

• stability of the programme from a technical point of view – independent internal and external (e.g. by specialists of a regulatory body), checks should not be terminated as a result of changes to the inspection program (risk ranking, the number and type of inspections);

• programme requirements should be tailored to the specifications, based on the failure mechanism and the minimal conservatism of the model of probabilistic risk assessment;

• maintaining or improving plant safety – the use the integrated ISI programme, which includes the methodology based on risk information, must demonstrate that plant safety can be maintained or improved.

A comprehensive and consistent assessment of the consequences should be based not only on the understanding of probabilistic risk assessment and information specific to each nuclear power plant, but also on the traditional logic and structural probabilistic risk assessment.

As noted in the documents, it is essential to take into account the difficulties in assessing the results obtained after inspection based on risk assessment, as compared with the results obtained after inspection based on risk information. In this case, the documents note that the use of risk information in determining the probability of failure must be carefully coordinated, i.e. to determine how well this information can be entered into the process of identifying risks, despite the confirmation of the uncertainty associated with highly reliable components. It is noted that the conservatism used in the complex process of assessing the probability of failure should lead only to an increase in the number of inspections, whereas conservatism in other methodologies/processes may mask other important components and artificially reduce the number of inspections. It is also noted that when assessing the probability of failure it is important to take into account its instability with respect to defence in depth and unknown degradation mechanisms.

The risk matrix is used to reduce the influence of subjective opinion (as these documents do not contain methods for estimating the probability of failure).

The concept of ISI, based on risk information, can be presented as a scheme (Fig. 7.1). It is recommended to refer to the proposed scheme as the Main scheme of ISI risk information (the Main scheme herein). The Main scheme is considered in an IAEA document[108].

In order to give a short description of each scheme element, its values and also its impact on the overall process of ISI and improve its efficiency, implemented through the introduction of assessment methodology based on risk information, the individual elements of the scheme are given in a large scale with letter designations,

The basis of this process is the risk. Therefore, the element D (Fig. 7.2) of The Main scheme is the basic element for carrying out risk assessment. The basic factors are the consequences of failure and the likelihood of failure. These two factors are not independent of each other, they form the axis of the scheme of risk assessment. Both axes can be divided into five sections corresponding to five risk categories:

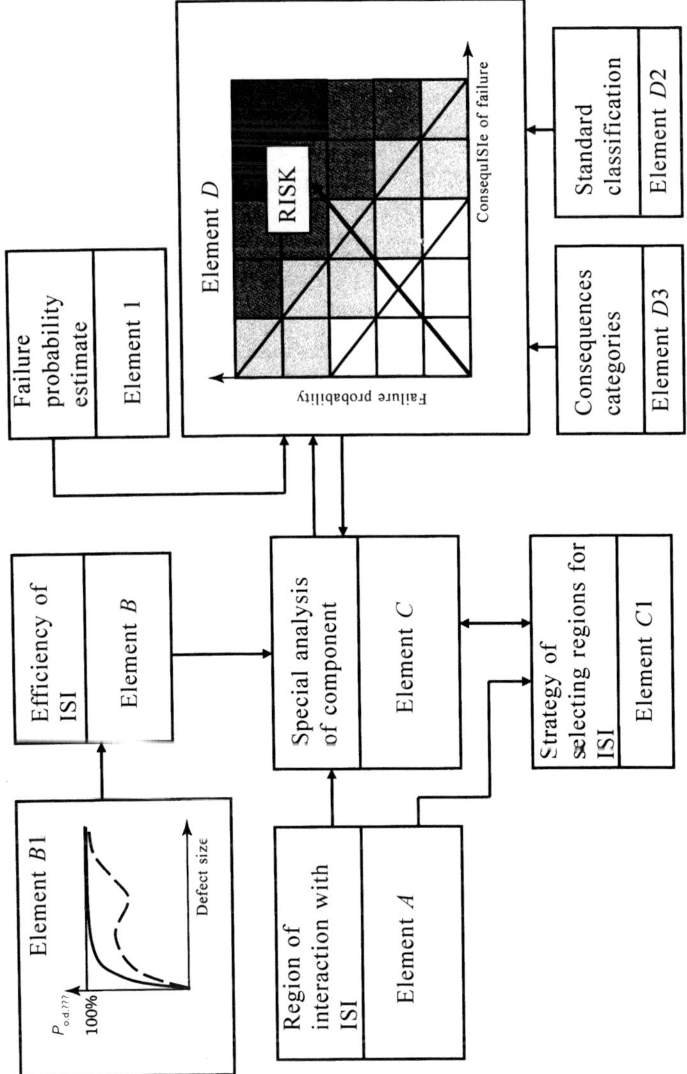

7.1 Main scheme of in-service inspection based on risk information.

- very high;
- high;
- average;
- low;
- very low.

Such a breakdown makes it possible to classify the various initial elements with the proper differentiation and evaluation. Three sites in the upper right corner of the risk matrix are generally excluded from

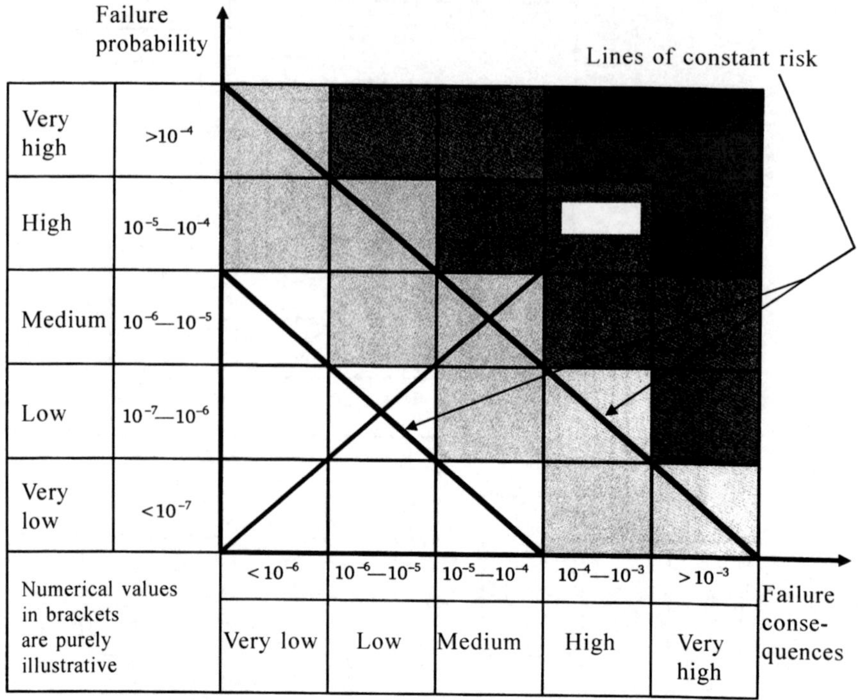

7.2 Risk matrix.

consideration. They symbolise the highest levels of risk and are not taken seriously. The remaining sections are divided into three risk categories and are marked with gray, light gray and white colours. 'Risk lines' represent two different levels of risk.

In element *D3* there are the key elements needed to build the abscissa of the risk matrix as a source of initial data, and the categories of potential consequences are listed:

- frequency of the maximum initial release of fission products;
- frequency of emission of fission products;
- core damage frequency;
- as well as the cost factor in the case of loss of operational readiness and the need for repairs.

The concept of defence in depth includes the probability of consecutive violations of safety barriers which, consequently, to a large extent affects the classification of the levels of consequences.

These categories of effects can be correlated with traditional regulatory classification of the components as shown in element *D2* of the Main scheme:

- absence of the failure probability of the reactor pressure vessel;
- class of equipment and pipelines 1;
- class of equipment and pipelines 2;

- class of equipment and pipelines 3;
- equipment and pipelines not included in classes 1–3.

However, this classification takes into account primarily the existing safety barriers so it does not always determine the levels of impact or correspond to them, although these levels should be clearly differentiated by using a model based on risk information and subsequent analysis of the consequences.

The ordinate of the risk matrix is the probability of an event or probability of failure (POF). The factors in element $D1$ important factors for this axis are:

- estimate of the probability of failure;
- databases;
- models;
- databases and models without inspection taken into account.

As component failures still occur, data on these failures at the two highest levels of probability (very high and high) may provide the basis for performance evaluation in this range. Since the probability goes beyond the accumulated experience, reliability models of structures depend greatly on the analysis of structural integrity. The lowest level of probability (very low) is considered as representing very distant events, relating to the category of consequences with the lack of probability.

Element A (Fig. 7.3) lists the subjects, criteria and methodologies, which work closely with ISI. It also shows promising possibilities of inspection, factors affecting the existing methodologies of ISI and its optimisation criteria derived from the above interaction. Particular attention is given to the introduction of an approach based on risk information. Considerable attention is paid to the process of selecting areas of the ISI and the selection of the initial data needed for this option, and also to the influence of this choice on the ISI.

On the left side of element A (Figure 7.3) there is a list of the basic data required for design which cover a large number of criteria and assumptions. During operation, these criteria, regardless of their values, may change, leading to the formation of defects, degradation, or damage that change the state of structural integrity, which in turn are very likely to lead to changes in the NDSI programme.

Issues related to manufacture are of particular importance for a re-evaluation completed after upgrades and repairs, as well as in the case of detection of defects by using a more efficient inspection methodology during operation.

Once again, it should be noted that during operation the values of these criteria potentially differ from their values at the time of manufacture. It is likely that these changes will affect the structural integrity of the component. Defects formed in the component in the manufacturing process can grow during operation, and new defects can form in the boundary conditions that differ from the design and manufacturing conditions.

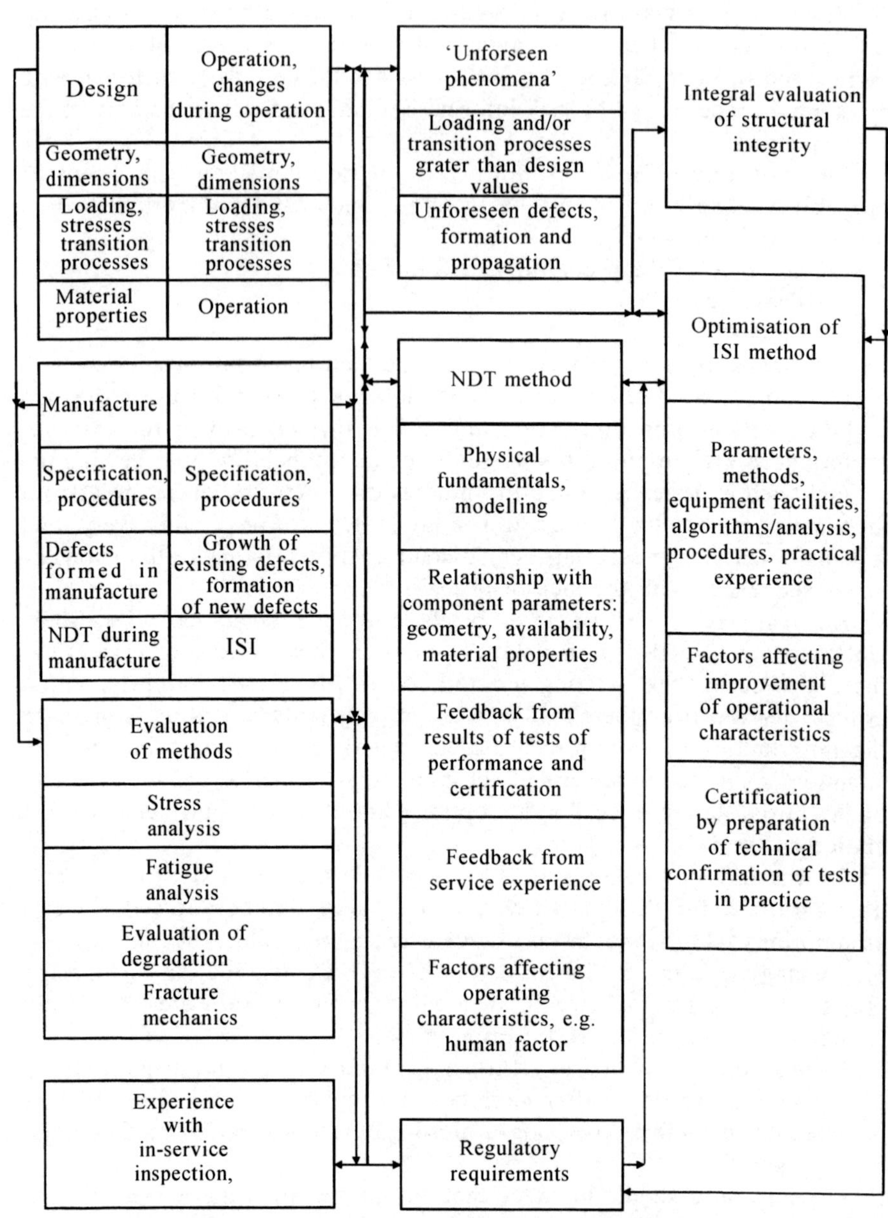

7.3 Areas interacting with ISI, for element A.

Actual values and criteria are evaluated using these methods to assess in terms of structural integrity. In this sense, the ISI with its specific criteria and performance is in addition to NDT, executed in the period of manufacture.

A specific criterion is the experience with ISI and received data, which will form a strong feedback with the majority of positions of the element B. In this context, it is important to considered unforeseen events affecting the structural integrity and also how can they be taken into account. For example, by considering the worst defect situations or the choice of specific additional areas of the ISI.

Element A is regarded as the actual method of non-destructive testing.

Element B (Fig. 7.4) shows performance of non-destructive testing for a specific area and indicates a number of reasons.

Non-destructive testing uses criteria that are not only simple but as informative as possible. When these criteria are considered as indications in NDT (in terms of ultrasound inspection), then this indication in most cases is the signal amplitude translated into time corrected again. Another example of indications is the dynamics of the sound path, etc. In this sense, the actual size determined by the appropriate method of measurement can also serve as an indicator. Indications used in eddy-current testing include

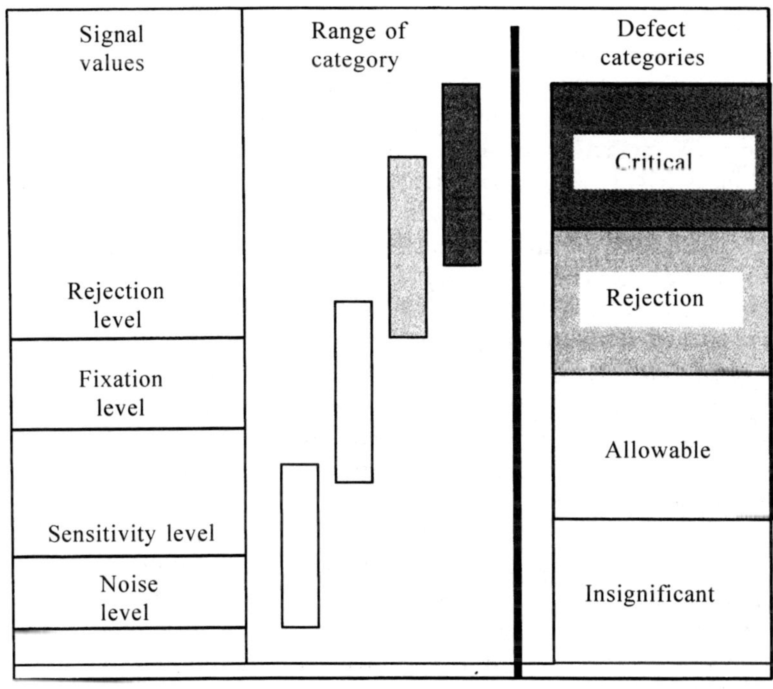

7.4 Efficiency of non-destructive testing for element B.

the signal amplitude and phase, and in radiography a catalog of reference defects.

This means that such indications are based on assumptions of models of defects which are used – through abstraction, generalisation and simplification – to describe the true state of the defects. Taking into account the determining intrinsic factors of defects, such as their orientation, structure, depth, and material (e.g. stainless steel) or geometry, it becomes clear that there may be discrepancies between the results obtained when using the indication and the actual defect state. In the case of low efficiency of non-destructive testing (e.g., based on identifying only reflectors of simple geometric shape for ultrasound inspection), these discrepancies can be significant. They may even mean the complete absence of any meaningful testing results.

Compared with element $B1$ (Fig. 7.1), representing the commonly used scheme for evaluating the effectiveness of NDT, the scheme presented in element B (Fig. 7.4) goes beyond simple ratios of the depth of the defect with detectability, recommending to consider the real importance of defects (depending on the types of defect) and distribute them to the major categories.

The scheme shown in Fig. 7.4, should read as follows:

The left side of the scheme to the part indicated by black vertical lines are the results of non-destructive testing, characterised by the degree of importance which may also be known as the indication. Different levels of indication are characterised by typical criteria relating to NDT, as the noise level of sensitivity, the level of fixation, etc. Vertical rectangles in the central part of the scheme (black, gray and white) correspond to different categories of defects, which are designated by the same tones and are placed in the right side of the scheme. The diagram shows that the change of indication values for the category 'allowable defects' is located within the 'level of fixation' and the 'rejection level' extending behind at both the lower and the upper boundary. The latter means over-rejection. If all allowable defects should be recorded, under-rejection should also be taken into account in the inspection method.

To determine whether a particular method and system of non-destructive testing are effective or not possible, it is necessary to take into account the degree of correlation between the indication obtained with ISI and the actual defect state. The ideal inspection system assumes the existence of direct correlation.

As stated above, the given correlation demonstrates acceptable but not a perfect inspection system, since as regards the allowable defects there would be some under-rejection as the range of indications for these defects falls below 'the level of fixation' indication. For the allowable defects the system also permits some over-rejection in the range of indications reaching the level of rejection.

The most objective information on the effectiveness of inspection can be obtained by using both approaches, as reflected in the schemes of the elements B and $B1$.

Element C lists the main criteria for selecting areas of inspection:

- special analysis of the component;
- identification of areas related to:
  - critical defects
  - acceptable defects
  - efficiency of NDT
  - frequency of NDT
  - certification criteria.

It should be noted that the areas of inspection are selected exercised on the basis of risk assessment presented in element D (Fig. 7.2). It is expected that in most cases this process will be iterative, especially during the initial phase of inspection based on risk information. This iterative process provides feedback to the ISI for optimisation and efficiency confirmed by the certificatin of inspection.

Element $C1$ (Fig. 7.5) shows the approximate strategy for selecting areas of ISI together with realistic values of some important criteria. As an example, the quantitative criteria are proposed for two main type of damage: fatigue defects and discontinuities. In the case of fatigue defects the zone

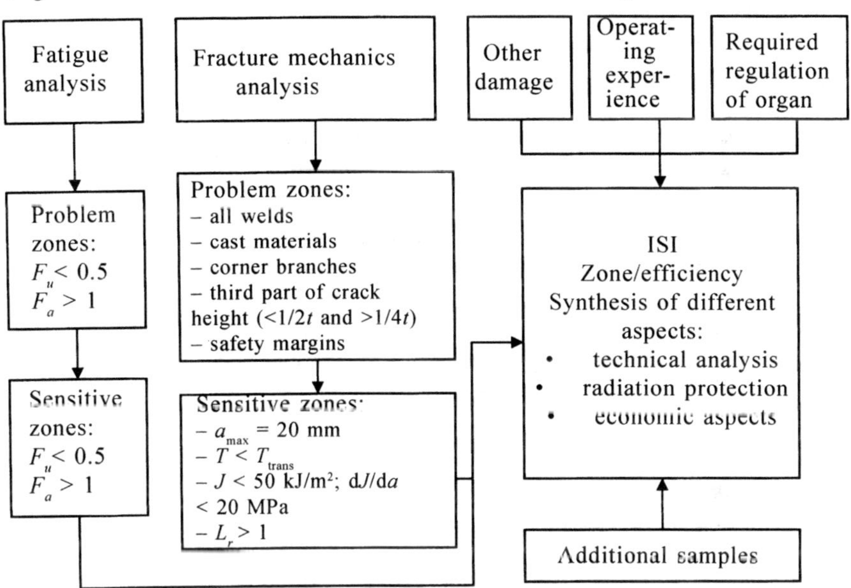

7.5   Example of the strategy for selecting areas for ISI (element $C1$): F_u – fatigue factor; F_d – fatigue factor for areas with crack-like defects; t – wall thickness of the element; a – the depth of the surface defect (crack); T – temperature; T_{trans} – ductile–brittle transition temperature; J – fracture toughness; dJ/da – modulus of ductile fracture resistance; L_r – reference stress – applied load/ limited load.

is considered sensitive, in accordance with the Code RCC-M, if the damage is more than 1. Additional assessment of the state of the element is carried out for those areas where the fatigue factor is in the range from 0.5 to 1.

Analyzing the permissibility of defects and the risk of rupture, it is assumed that the area is sensitive if the height of the crack does not correspond to the accepted safety factor (2 for level A, 1.6 for level C and 1.2 for level D).

Furthermore, the following zones are regarded as sensitive:
- zones which can operate in a brittle mode: operating temperature is lower than the ductile–brittle transition temperature of zone;
- zones where the fracture toughness characteristics are low: $J < 50$ kJ/m^2 and $dJ / da < 20$ MPa;
- zones of high mechanical loads: redistribution of load in the zones with cracks, taking into account the conditional yield strength, must be greater than 1 ($L_r > 1$).

The relationship between the volume of the ISI and its worth taking into account the risk factor is important. The qualitative nature of the approach does not require determination of failure probability in the absence of inspection and if inspection is carried out we consider only the reduction in the probability of failure as a result of effective inspection. Nevertheless, this qualitative approach can be transformed into a quantitative one, which will assess the real benefits of carrying out inspection whose volume has been defined properly.

If we consider the cost as a 'consequence' in a risk matrix, then risk assessment is the ranking of risks in the values of the potential losses from unplanned shutdowns. The first assessment should be done without any inspection. After that, a certain amount of inspection is introduced which should include some efficiency ISI, in this case we can evaluate how the risk will change with increased inspection volume. Subtracting the risk after the introduction of ISI from the risk prior to ISI introduction, we obtain a measure of risk reduction provided by ISI. If the accumulated risk is described graphically taking into account the risk reduction provided by each individual inspection with a given volume, we obtain a curve steadily decreasing with increasing amount of inspection (see Fig. 7.6). However, if the cost of each inspection is added, a graph of the total cost for nuclear power plants is plotted. This curve shows the minimum cost of inspection in the overall expenditure of the nuclear power plant. The value above it is the net cost, but the cost of inspection exceeds its benefits and this increases the total cost for the nuclear power plant in excess of this value for each additional inspection.

If the effectiveness of ISI is now represented as a parameter, the in the case of reducing the effectiveness of ISI the probability of failure increases. Therefore, the total cost of such inefficient inspection will increase compared to the cost of effective inspection of the same amount. Consequently, in the case of less effective ISI its lowest total cost

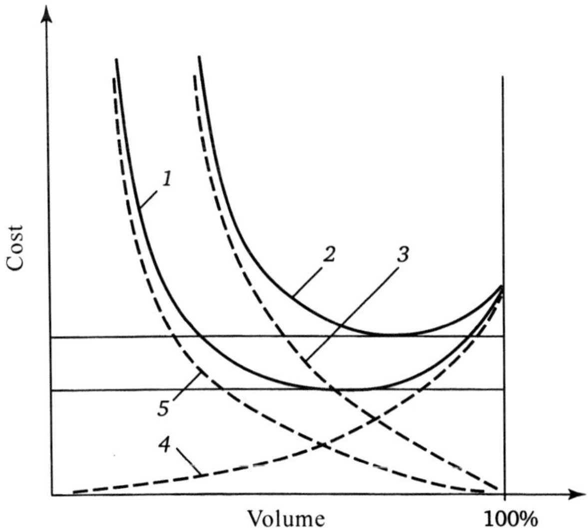

7.6   Relationship between ISI and its cost: 1) total cost in 'good' (correct) selection of inspection volume; 2) total cost at 'poor' (incorrect) selection of inspection volume; 3, 5) cost for nuclear power plant; 4) cost of ISI.

is displaced to the right and to a higher level (i.e. will increase) at which the minimum total cost of less effective inspection may well be greater than the total value of effective inspection. Thus, the higher the efficiency of the ISI, the larger the volume of inspection that can be substantiated and verified.

However, the radiation doses received by staff even with effective inspection are a limiting factor restricting the inspection volume. In-service inspection, based on risk information, offers a flexible approach to reduce radiation doses. In the case where several areas of inspection with the same level of risk are assessed for inclusion in the inspection volume, it is recommended to give preference to regions with the lowest possible radiation exposure. Consequently, the given level of risk can be achieved at the lowest dose rates.

Such a comprehensive concept that takes into account all factors affecting the cost and safety, considering the real value of inspection, should be used in developing standard programmes for ISI of equipment and piping, components and structures of nuclear power plants.

The relationship between the frequency of the ISI and its effectiveness should be supplemented in the light of the concept of inspection based on risk information.

As described in Ref. 110, the determination of the operating life of the component and the time periods between inspections can be shown schematically in the probabilistic aspect taking into account the required

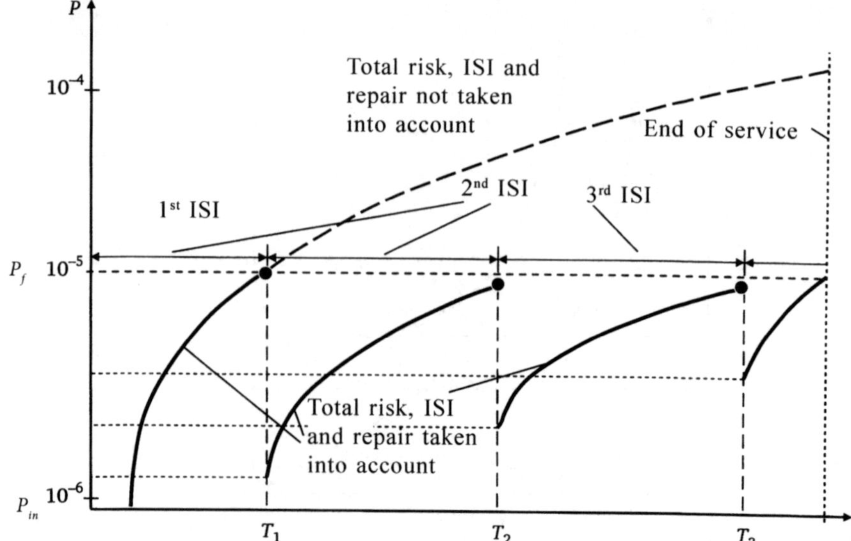

7.7 Frequency of ISI taking the risk factor into account (numerical values are purely illustrative).

values of risk during the operation of this component (Fig. 7.7).

If the initial level of component failure probability is P_{in}, then during operation due to the development of various processes of damage in the metal of the component the failure probability will increase. When the probability Pf reaches the permissible value [P] at which ISI and repair must be carried out, the component is either decommissioned or measures are taken to decrease the probability of its failure. Such measures may include ISI and subsequent repair of identified defects. After ISI and repair, the failure probability of the component decreases (in Fig. 7.7 this corresponds to times T_1, T_2 and T_3. However, as mentioned by the authors of the studies in question, the initial level of probability of failure is not reached because of the irreversible degradation of metal and possible residual defectiveness of the component. When the possibility of reducing the probability of failure due to ISI and repair is exhausted, the component reached the end of its operating life.

7.3   Optimisation of the risk-based oriented in-service inspection at the Ignalinsk nuclear power plant[111]

The scheme of the reactor RBMK-1500 of the unit 2 of the Ignalinsk nuclear power plant is shown in Fig. 7.8.

The Lithuanian nuclear regulatory authorities require:

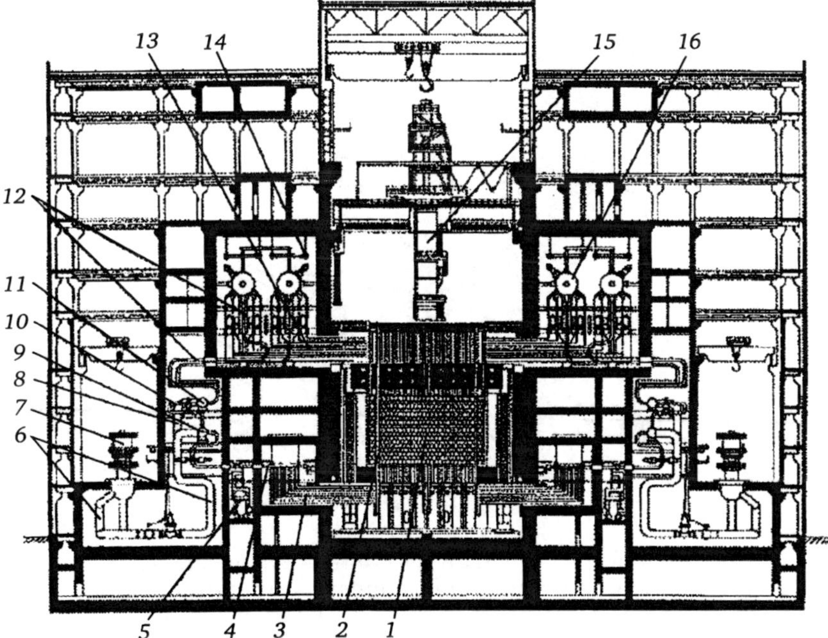

7.8   RBMK-1500 reactor of unit 2 of Ignalinsk nuclear power plant: 1) graphite stack; 2) fuel pipe of fee channel; 3) water pipeline; 4) distributor; 5) emergency core cooling pipelines; 6) standpipe; 7) MCP; 8) suction pipeline; 9) pressure header; 10) bypass pipe; 11) suction header; 12) risers; 13) steam pipelines; 14) steam pipeline; 15) refueling mechanism; 16) drum separator.

- To avoid the need for evacuation to a distance outside the zone established in the standards for nuclear installations, measures must be taken to ensure that the probability of worst case emergency release of radioactive material does not exceed 10^{-7} reactor·year.
- in addition, it is stated that the core damage frequency should be less than 10^{-5} reactor·year.

However, in Lithuania as well as in most other states members of the IAEA, there are still no regulatory requirements (even in the framework of PSA) to achieve the stated values of security.

As PSA shows, it is important to enhance the safety of the Ignalinsk nuclear power plant. One of the main risk factors was ageing of the pipeline DN300 by stress corrosion cracking. These pipelines are shown in Fig. 7.9.

In Fig. 7.10 shows an increase in the number of detected defects with operating time.

The main purpose of non-destructive testing (NDT) is to detect possible degradation in its early stages and prevent the damage that can lead to destruction. In this regard, the goal to develop a NDT programme based on risk information.

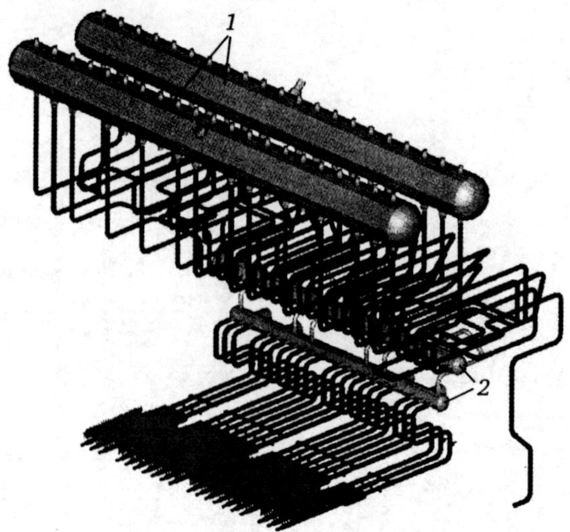

7.9 Pipelines of the circuit for multiple forced circulation o RBMK-1500 reactor with the total number of welded joints of 1241: 1) drum separators; 2) pressure and suction manifolds.

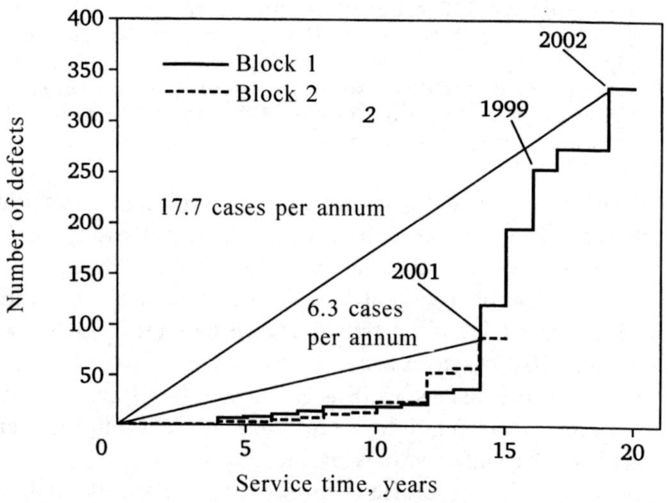

7.10 The number of detected defects caused by intergranular stress corrosion cracking.

The programme based on a study of the risk of the Ignalinsk nuclear power plant (IRBIS) was designed to:

- perform risk analysis of the chosen system of tubes at Ignalinsk plant, 2 units;
- use the results of risk analysis to identify inspection areas;

- assess the cumulative radiation exposure to inspection personnel of the plant;
- perform a sensitivity analysis of damage from the basic parameters, such as the intensity of the vibrations, the barriers of the system, etc.
- recommend improvements and further research that may increase operational safety of the Ignalinsk nuclear power plant, 2 blocks.

Risk assessment involves the following steps:

- analysis of the mechanisms of failure and degradation;
- assessing the consequences of failures C;
- evaluation of failure probability P;
- definition of the measure of risk $R = PC$;
- perform risk ranking for each section of the pipeline;
- define a new selection of inspection sites (ISI-programme), with the components with the highest risk;
- rate ΔR (new case of risk – old case of risk);
- assess the cost and radiation dose exposure to the new ISI-programme in comparison with on the existing ISI-programme;
- make appropriate recommendations based on research into risk in order to improve operation and maintenance.

The consequences of failure were assessed using the PSA. At the same time, the following were taken into account:

- extent of LOCA;
- location of LOCA.

In assessing the probability of failure, the availability to inspected areas was also taken into account (Fig. 7.11).

Effectiveness of inspection was assessed by the detection of defects (Fig. 7.12) using the equation of the probability of detection for different inspection efficiency:

$$P_{d.d}(s)=\Phi[c_1+c_2 \ln(s)], \ c_1 \in R, \ c_2 \geq 0, \qquad [7.1]$$

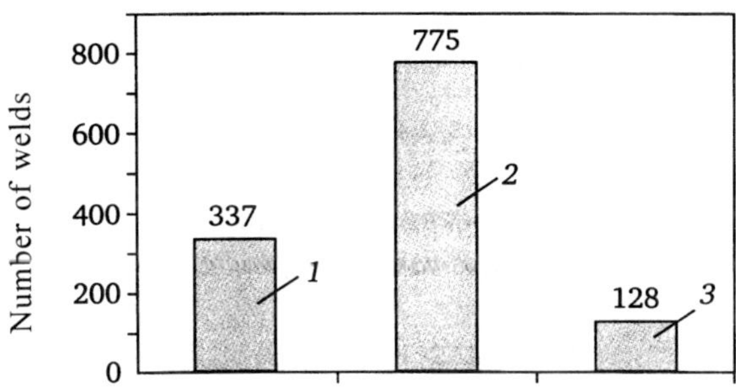

7.11 Number of welds with different access: 1) good availability (automatic and manual inspection); 2) normal availability (manual inspection only); 3 - bad availability (manual inspection with difficulties).

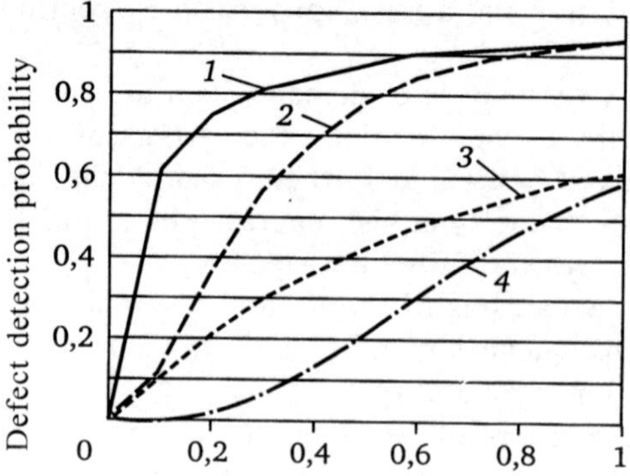

7.12 Detection of defects: 1) good; 2) normal; 3) poor; 4) very poor.

Table 7.1 Constants c_1 and c_2 for different defect detectability

The level of detection of defects	c_1	c_2
Good (good availability, automatic and manual examination by ultrasound inspection)	1.526	0.533
Normal (normal availability by using only manual examination by ultrasound inspection)	1.600	1.200
Poor (poor availability, using manual ultrasound inspection technique)	0.300	0.700
Very poor (radiography)	0.240	1.485

where the constants c_1 and c_2 take the following values (see Table 7.1).

The risk of failure is determined as follows:

The risk of i-th component P_f (probability of failure)·C_f (consequence of failure);

Probability of failure of welded joint $P = SL + LL + P_f$ where SL is the small leak; LL is the large leak, $P_f = PC$ is the probability of failure; P is the probability of failure as a result of annual degradation of the welded joints; C is the probability of reduction of the safety of the entire system as a result of degradation of the welded joints.

(The same procedure can be used to determine the risk for other degradation levels).

The risk for the entire pipeline system is equal to the sum of the risks of each welded joint.

Depending on the level of risk, the following intervals between inspections are selected (Table 7.2).

In addition to the previously described programme RBI-1, alternative RI-ISI programmes were developed (Fig. 7.13):

Risk category	Inspection intervals
Very high	1 year
High	2 years
Average	4 years
Low	4 years
Very low	4 years

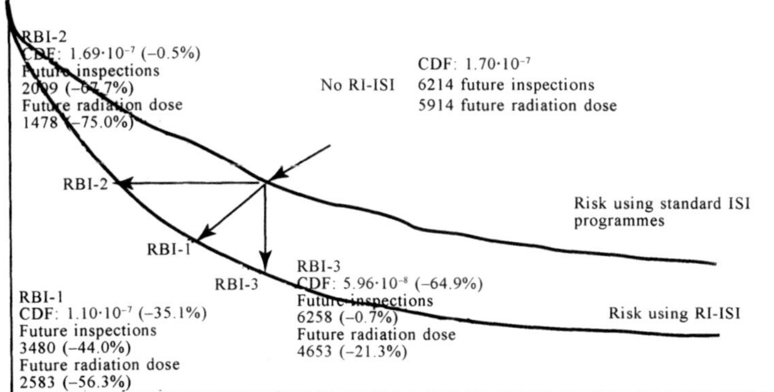

7.13 Dependence of risk on frequency and volume of inspection during operation obtained on the basis of the risk matrix (Fig. 7.2).

RBI-2: Minimise inspection at the given risk levels;

RBI-3: Minimise risk at the given inspection cost.

The results of these studies can be summarised as follows:

- using quantitative risk analysis for 2 units of the Ignalinsk nuclear power plant, the number of future inspections can be reduced by 44%, with the risk reduced by 35%;
- Radiation exposure of personnel can be reduced by 3300 mSv (by 56%) as compared to the tubes subjected to standard inspection.

After completion of the project RI-ISI, the Ignalinsk NPP has been using the results of preliminary research and has been preparing a new inspection programme, taking into account the welds with the highest level of risk.

The number of inspection has not been reduced but the risk was reduced significantly.

The Lithuanian regulatory authorities (VATESI) have agreed to use the RI-ISI programme for austenitic piping and are waiting for proposals of the Ignalinsk nuclear power plant.

In the case of reducing the volume of ISI it is also required to instal monitoring systems for leaks with high sensitivity.

At the same time, the requirements of the State Inspectorate for Nuclear Security specify 100% inspection.

7.4 Quantitative approach to optimisation of ISI

7.4.1 Optimisation of ISI based on the characteristics of probability of failure and a systemic approach

To achieve a more detailed presentation of the issue, the deterministic and probabilistic aspects of optimisation of ISI are described below.

The methodology and optimisation methods set out belo are based on economic criteria. The following optimised parameters are considered: norms of defectiveness, the time interval between inspections (frequency of inspection); inspection areas, inspection volumes over the entire length of operation of an inspected object, combination of inspection methods.

In general, the search for the optimum value of the investigated inspection parameter can be depicted graphically as the search for the extreme (minimum or maximum) value on the curve in the economic characteristic – optimised parameter coordinate (Fig. 7.14).

In this section, the considered optimised parameters are different inspection characteristics mentioned above. The optimisation criterion is to achieve the minimum cost at maximum possible positive technical result. The optimisation methods for ISI described below are based on taking into account the probabilistic relationships of detection of defects and failure.

With strict mathematical formulation of the problem of optimisation of

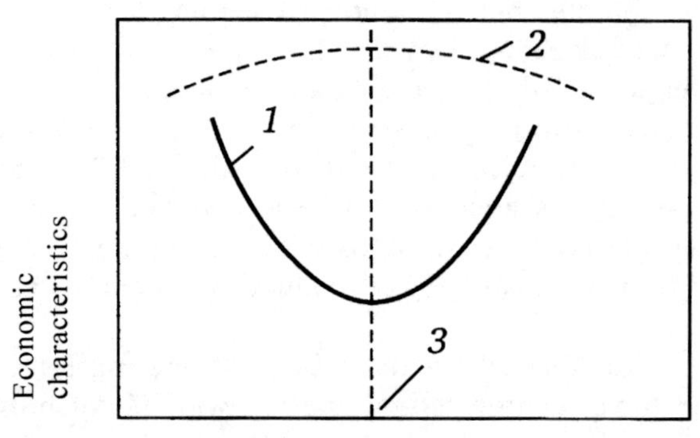

7.14 Graphical description of search for the optimum value of the optimised parameter; 1) damage; 2) earnings; 3) optimum value.

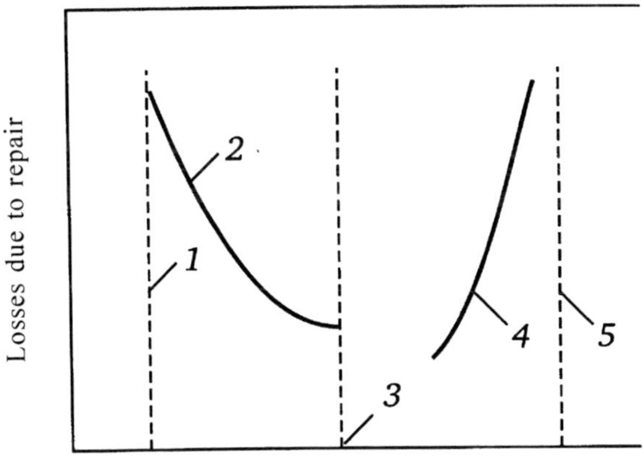

7.15 Justification of the economic optimum norm of defects in operation, identified using fracture mechanics methods: 1) defect norms for manufacture; 2) losses due to repair; 3) optimum normas of operational defects; 4) losses due to failures; 3) critical size of discontinuities.

ISI it is necessary to create a mathematical model of inspection in relation to reliability and economic characteristics of the object operation depending on inspection. An example of solution of this optimisation problem is given Sec. 7.4.4. However, often the best solution may be obvious and can be justified without constructing exact quantitative inspection models. Examples of this approach are given in Sec. 7.4.8.

7.4.2 Cost-optimum norms of defects in service (deterministic approach)

The optimum norms of operational defects are the norms based on fracture mechanics. Indeed, the left branch of the curve in Fig. 7.15 reflects the loss of the organisation as a result of repair of non-hazardous defects. The left branch has the flowing form since with increasing size the number of discontinuities decreases and, consequently, maintenance costs also decrease.

The right branch of the curve (see Fig. 7.15) is ascending in nature, since which norms of the defects approach the critical size of discontinuities (decreasing safety margins) the number of accidents caused by defects should increase and, consequently, losses from accidents should grow.

The minimum value of the curve of losses in Fig. 7.15 correspond to the norms of defects identified as permissible defects in operation (subject to properly safety margins).

As noted above, the Russian nuclear power plants use the norms of detects for manufacture. In connection with this the volume of repairs at nuclear power plants to ensure safe operation exceeds the required level by ten or more times. Introduction of economically optimum standards of defects to operating nuclear power plants reduces the labour content and exposure doses ten times while improving their safety (due to reduced maintenance, additional analysis of strength and operating life and resources and the use of freed-up resources to increase safety during operation).

7.4.3   Cost-optimum time interval between inspections (deterministic approach)

The time interval between inspections (or frequency of inspection) affects the economic factors.

Figure 7.16[106] shows the dependence of the income earned in the operation of structures (e.g. NPP) on the frequency of inspection for the entire period k.

Curve 1 corresponds to the case when inspection of the element of the structure (or the entire object) is economically not advantageous, i.e. losses from its unreliable operation are small and do not exceed the cost of inspection, and the optimum frequency of inspection k_1 is then equal to 0.

Curve 2 corresponds to the case when the income from operation in inspection at a low frequency increases but the inspection volumes increase dramatically and the total revenue from operation begins to fall. The maximum point on curve 2 corresponds to the economically optimum inspection frequency.

Curve 3 corresponds to the constructions whose failures have catastrophic consequences and are unacceptable for society. Any inspection costs in this case are much smaller than the losses due to the accident. The optimum inspection frequency k_3 at which the 100% reliability of the structure is reached (100% reliability can be unattainable and it can be replaced by another that is acceptable).

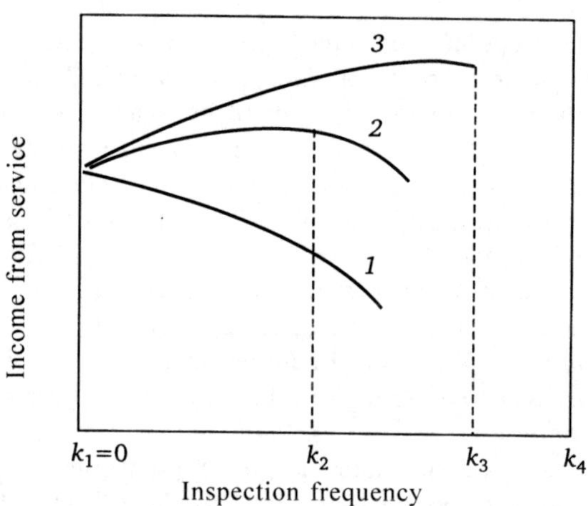

7.16   Determination of optimum inspection frequency.

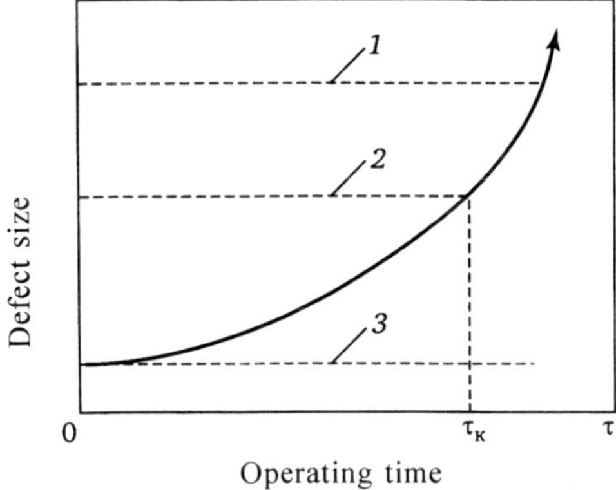

7.17 Determination of the economically optimum time period between inspections: 1) critical size of discontinuities; 2) allowable size of operational defects; 3) norms of defects in manufacture.

If the ISI methods or other methods of technical servicing in operation do not enable the desired level of reliability to be reached, it is economically reasonable to postpone operation and modernisation of the structure in order to enhance its reliability (safety).

It can be shown that for the cases described by curves 2 and 3 in Fig. 7.16, the economically optimum time interval between inspections is the time during which the size of a discontinuity increases from the size defined by defect norms in manufacturing to the permissible size of defects in operation (Fig. 7.17)[106,111].

The scheme shown in Fig. 7.17 for determining the economically optimum time between the inspections is only applicable if the inspection means and methods provide absolute inspection reliability. In fact, the detectability of defects both at the level of defect norms in manufacture and at the level of admissible sizes of defects in operation is not 100%. The actual detection is significantly below 100%, which means that the definition of the economically optimum time interval between inspections requires knowledge of the laws of probability of ISI or, more precisely, the probability characteristics of the detection of defects and inspection reliability.

Optimisation of inspection sites and inspection volumes throughout the entire operation and combination of inspection methods also requires knowledge of the laws of probability of ISI and failure and will be discussed below.

7.4.4 The optimum time between the inspections (probabilistic approach)

Since operational inspection is associated with material costs, the optimum organisation of inspection should include preserving a certain ratio between the total inspection costs and the benefits derived from inspection[106,112]

The concrete goal of optimisation is the maximum profit from operation of the nuclear power plant. The ratio between profit per unit time D, the number of inspections k, average duration of inspections τ_i, the average duration of repair x_r, the cost of repair per unit time Y_r, and the cost of inspection per unit time Y_i

$$D = D_0[1-\tau_r P(k)-\tau_i k]-Y_r\tau_r P(k)-Y_i\tau_i k, \tag{7.2}$$

where D_0 is the profit per unit time from the operation of equipment if there is no failure of equipment; $P(k)$ is the average number of breakdowns per unit time at inspection frequency equal to k.

The value of $P(k)$ is, in fact, the strength characteristic of the inspected element of the structure or, more accurately, the probability characteristic of failure according to the criterion of the appearance of a defect in the structure, or other criteria, for example, total destruction of the structure. It can vary in the range from some maximum value P_{max}, corresponding to the case when inspection is not performed, i.e. $k = 0$, to the minimum value equal to 0. In the latter case, k has its maximum value k_{max}.

Changing the function $D(k)$ in the range $0 < k < k_{max}$ can correspond to three cases (curves 1–3 in Fig. 7.2):

$$dD / dk > 0; \tag{7.3}$$

$$dD / dk = 0; \tag{7.4}$$

$$dD / dk < 0; \tag{7.5}$$

The meaning of the expressions [7.3]–[7.5] is as follows. Equipment which satisfies the condition [7.3], should operate with 100% reliability, i.e. $P(k) \to 0$. This is possible if the reliability of inspection is high.

For equipment with condition [7.4] a small number of breakdowns is allowed. For equipment in the third case inspection is disadvantageous.

Thus, the conditions [7.3]–[7.5] are criterial for the separation of equipment into three classes. Differentiation of expression [7.2] and application of [7.5] for the equipment of the first class:

$$-\frac{dP}{dk} > \frac{\tau_i\ (D_0+y_i)}{\tau_r\ (D_0+y_r)}. \tag{7.6}$$

If the inspection overlaps with refuelling, the expressions [7.2] and [7.3] have the following respective forms

$$D_i > D_0\left(1 - \frac{P(k)}{\tau_r}\right) - y_r P(k)\tau_r - y_r P(k)\tau_i; \qquad [7.7]$$

$$-\frac{dP}{dk} > \frac{\tau_i}{\tau_r} \frac{y_i}{(D_0 + y_r)}. \qquad [7.8]$$

For class 2 equipment:

$$-\frac{dP}{dk} = \frac{\tau_i}{\tau_r} \frac{(D_0 + y_i)}{(D_0 + y_r)} \qquad [7.9]$$

or if the inspection is combined with refuelling,

$$-\frac{dP}{dk} = \frac{\tau_i}{\tau_r} \frac{y_i}{(D_0 + y_r)}. \qquad [7.10]$$

For class 3 equipment, the condition of disadvantageous inspection takes the form:

$$-\frac{dP}{dk} < \frac{\tau_i}{\tau_r} \frac{(D_0 + y_i)}{(D_0 + y_r)} \qquad [7.11]$$

or if the inspection is combined with refuelling,

$$-\frac{dP}{dk} < \frac{\tau_i}{\tau_r} \frac{y_i}{(D_0 + y_r)}. \qquad [7.12]$$

The above equations include the probability of failure P. Since failure may occur by different mechanisms (fatigue, stress corrosion cracking, etc.) analysis using the formulae [7.6]–[7.12] should be performed for every possible failure mechanism. In this case, the conditions for optimising inspection on the basis of the the parameter of inspection frequency can be rewritten as:

$$-\frac{dP_i >}{dk_i <} = \frac{\tau_{ii}}{\tau_{ri}} \frac{D_0 + Y_{ii}}{D_0 + Y_{ri}}, \qquad [7.13]$$

where the index i stands for a specific failure mechanism (damage).

Equations [7.3]–[7.6], [7.8]–[7.13] are the conditions of optimumity of operational inspection with respect to the time interval between the inspections (or inspection frequency). They combine together such features of an integrated system of operational inspection as the quality of inspection means and methods, the properties and state of the metal, the level of reliability (strength) of equipment, the cost and time characteristics of inspection and maintenance of equipment, and these equations make it possible to optimumly solve a number of problems of operational inspection. The most important features are the definition of inspection sites and settings, inspection frequency, determination of requirements for the sensitivity of technical means of inspection, determination of norms of defects of metal, definition of the level of reliability of equipment after testing and repair, etc.

Here are two examples of determining the optimum inspection frequency: for VVER-1000 RPV and the casing of a pressuriser of the VVER-440 reactor.

7.4.5   ISI optimum frequency for VVER RPV

The purpose of design analysis is to determine the optimum inspection frequency of a cylindrical part of the reactor casing for a design life of 30 years. The probability of failure of the reactor pressure vessel was assessed by the criterion of the resistance to brittle fracture taking into account metal embrittlement due to radiation. The income from the operation of the VVER-1000 reactor in 30 years was taken to be $D_0 = 4 \cdot 10^8$ roubles (in prices of 1983 when the analysis was performed). The inspection costs for 30 years were defined as the wages of 30 people for 30 years with overhead costs taken also into account ($Y_i = 1.8 \cdot 10^7$ roubles). The cost of repairs was assumed to be $Y_r = 10Y_i$ (assuming that the LBB concept holds for reactor pressure vessel and complete destruction of the body will not take place). The ratio of inspection and repair times was $\tau_r/\tau_i = 10$. The analysis was performed for two cases: inspection combined with scheduled preventive maintenance and refueling; the reactor was shutdown for inspection.

The analysis results are reported in Fig. 7.18 in which the curves 1 and 2 correspond to the following values:

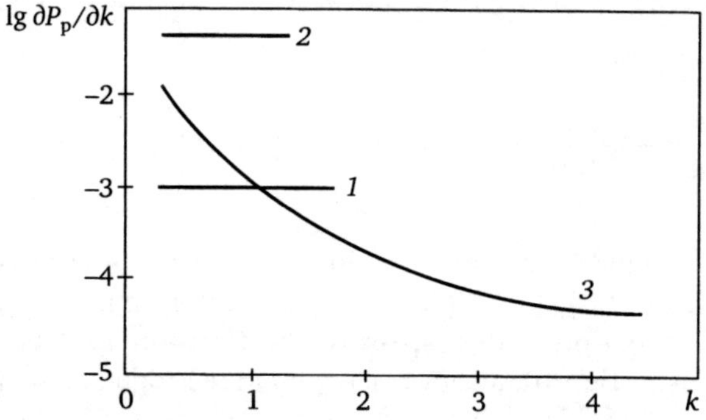

7.18 Results of determination of the optimum frequency of flaw inspection of a reactor pressure vessel for the case described by equation [7.4].

$$\frac{\tau_i}{\tau_r}\frac{Y_i}{D_0 + Y_r}; \quad \frac{\tau_i D_0 + Y_i}{\tau_R D_0 + Y_r}. \tag{7.14}$$

The point of intersection of curves 1 and 2 with curve 3 corresponds to optimum frequency of flaw inspection of the cylindrical part of the reactor pressure vessel for 30 years of operation. As seen from Fig. 7.18, in the case of a special reactor shutdown for the inspection of the reactor pressure vessel the inspection frequency is approximately 0, i.e. in this case, flaw inspection is economically impractical.

In the case of combining inspection with planned preventive maintenance (PPM) and refuelling, the optimum frequency of inspection is slightly higher than 1 which means that the optimum time between the inspection of the cylindrical part of the reactor is 10 years. The time interval between inspections adopted in Western countries (US, France and others) is also 10 years. The four-year interval between the inspections of the reactor pressure vessel, adopted at nuclear power plants in Russia and Germany (once in four years) is conservative and economically optimum.

7.4.6 ISI optimum frequency of the pressuriser

Analysis was performed for the most stressed sections of the pressuriser (PRZ) casing – jumpers between the holes in the bottom of the casing. In these places, refined analysis of stresses revealed elevated levels of total membrane stresses and a very low safety margin according the criterion of transition of the jumpers to the plastic state (the actual safety margin is equal to 1 at a regulatory value of 1.5). In this regard, the actual reliability of the casing and the conditions for the service inspection was analyzed in 1976–1977.

The PRZ casing was made from 22K steel. The wall thickness of the bottom was 145 mm. From the analysis of operating conditions it was found that the reliability of the PRZ should be evaluated by the criterion of resistance to: unaccetable plastic deformation; ductile, quasi-brittle and brittle fracture, as well as by the criterion of resistance to the nucleation of fatigue cracks.

As a result, computational and experimental studies have identified the main parameters determining the state of the PRZ and the requirements to control them during operation (Table 7.3). The characteristics of the probability of nucleation of fatigue cracks and optimum frequency of their inspection were obtained as an intermediate result. Figure 7.19a shows the variation of the probability of nucleation of fatigue cracks in the jumpers of the dished end of the PRZ, and Fig. 7.19b shows a graphical solution of the optimisation equation. The optimum frequency of inspection of the dished ends of the PRZ to detect fatigue cracks is 1 in 30 years. Since the inspection frequency of developing technological defects is higher, in practice high frequency of inspection was recommended.

Table 7.3 Requirements on in-service inspection of the pressuriser

Strength criterion	Inspection parameter	Permissible value of parameter		Inspection frequency	
Ductile fracture resistance	1. Pressure 2. Temperature 3. Tensile strength 4. Wall thickness	Nominal - According to actual state 142 mm	Nominal - According to actual state 160 mm	Continuous - > 30 years > 30 years	
Quasi-brittle fracture resistance	1. Pressure 2. Temperature 3. Tensile strength 4. Wall thickness 5. Crack length L 6. Number of cycles N	Nominal >100° in HT According to actual state 142 mm 8 mm 160	Nominal >170° in HT According to actual state 160 mm 14 mm 160	in HT conditions In HT conditions > 30 years > 30 years 1 in 4 years Continuous	in HT conditions In HT conditions > 30 years > 30 years 1 in 4 years Continuous
Brittle fracture resistance	1. Pressure 2. Temperature	Nominal	Nominal	During heating	During heating
Fatigue strength	1. Number of start-shutdown cycles 2. Number of hydraulic pressing cycles 3. Number of hydraulic tests	726 726 88	726 726 88	 Continuous	 Continuous

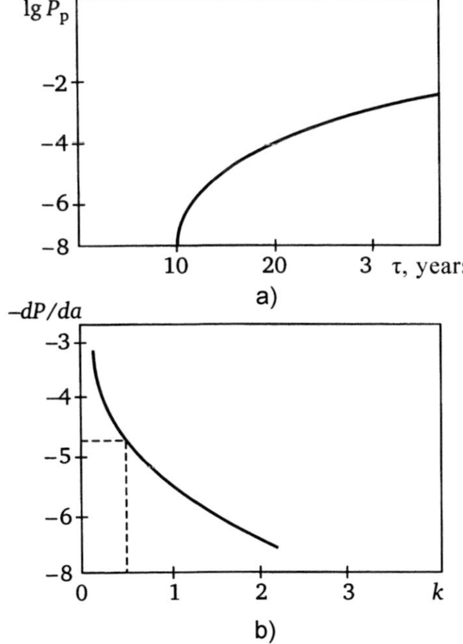

7.19 Variation of the probability of nucleation of a fatigue crack in the dished end of the PRZ (a) and graphical determination of the optimum frequency of inspection and reconditioning operations on the dished end of the PRZ (b).

The above example with the PRZ is interesting for two reasons. First, the analysis was carried out within the concept of complex inspection. The conditions for optimum inspection of not only defects but also other parameters that determine the condition and reliability of the PRZ were determined. Secondly, it is shown that flaw inspection can be optimised not only for residual defects but also in the risk of crack nucleation directly during operation.

7.4.7 Optimisation norms for defects during operation (probabilistic approach)

In section 7.4.2 it was shown that the norms of defects based on fracture mechanics methods are optimum. It was assumed that the strength margins used in the calculations were selected by the optimum procedure. In reality, the safety margins used in practice are evidently close to optimum (because they reflect a large amount of experience in strength calculations in various areas of technology), yet they may differ from the optimum values in one direction or another.

The laws of probability in flaw inspection and analysis of the strength can also be used to explore and define the optimum safety margins in a quantitative formulation. The use of the optimum values of the safety margins in the development of defect norms leads to strictly optimum normas of defects. In reality, the problem of optimising the norms of defects is reduced to finding the optimum safety margins. One of the possible schemes of simplified search for the optimum safety margins is described below.

In this case, the optimisation equation can be written by the following form:

$$\frac{d}{d[a]}(Y_r + Y_{fail}) = 0,$$ [7.15]

where Y_r are the losses from repair of non-hazardous defects; Y_{fail} are the losses from accidents due to loss of strength by the structure, $[a]$ is the characteristic allowable size of the defect (discontinuity).

Losses from repairs re directly related to defects in the construction and the reliability of inspection (since only detected defects are repaired):

$$Y_r = N_{det}C_r + D_0 \frac{N_{det}\tau_r}{\tau_{ser}},$$ [7.16]

where N_{det} is the number of detected defects; C_r is the cost of repairing one defect; D_0 is the revenue from operating the facility during τ_{ser} if no defects are found; τ_r is the repair time of a defect.

Losses from accidents due to loss of strength can be written as written as

$$Y_{fail} = P_f C_{fail} + D_0 \frac{P_f \tau_{fail}}{\tau_{ser}},$$ [7.17]

where P_f is the probability of structural failure, C_{fail} is the cost of mitigation of the consequences of an accident; τ_{fail} is the time of the mitigation of the consequences of one accident.

The values of N_{det} and P depend on the allowable (normalised) size of a discontinuity a. In particular, for N_{det}

$$N_{det} = \int_a^s N_{res}(a) \cdot W(a) da,$$ [7.18]

where s is the wall thickness or the maximum possible size of the defects.

Taking equations [7.15]–[7.18] into account gives

$$\frac{d}{d[a]}\left(C_p \int_{[a]}^s N_{res}(a)W(a)da + D_0 \frac{\tau_p}{\tau_{ser}} \int_{[a]}^s N_{res}(a)W(a)da + \right.$$

$$\left. + P_f([a])C_{fail} + D_0 P_f([a])\frac{\tau_{fail}}{\tau_{ser}} \right) = 0.$$ [7.19]

Equation [7.19] includes the characteristic of the detectability

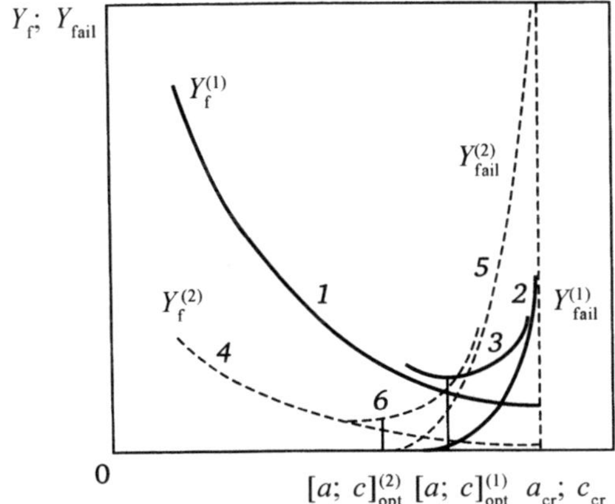

7.20 Graphical determination of the optimum norms of defects during operation: 1) the losses from repair of defects at inspection detectability W_1; 2) the losses from failures at inspection detectability W_1; 3) the sum of losses described by the curves 1 and 2; 4) the same as curve 1 but at inspection detectability W_2 ($W_2 < W_1$); 5) the same as curve 2 but at inspection detectability W_2 ($W_2 < W_1$).

of inspection $W(a)$. This means that the optimum allowable sizes of discontinuities (and therefore the safety margins) depend on the detectability of inspection.

Obviously, the lower the detectability of the inspection, the fewer defects will be discovered and the lower will be the cost of repair. This means that with decreasing detectability the optimum allowable size of the discontinuities will decrease. Conversely, with increased detectability of inspection the optimum allowable size of the discontinuities will increase (see the schematic solution for two values of inspection detectability in Fig. 7.20).

Increased detectability of quality inspection also increases the reliability of the structure and, consequently, reduces the losses from accidents, which in turn will lead to even greater shifts of [a] values to larger values. Thus, analysis shows that the optimum norms of defects in operation depend strongly on the detectability of inspection. The increase of the detectability of inspection is a cost-effective direction of work not only because it reduces the risk of structural failure but allows less conservative norms of defects to be used in operation. This also gives a positive economic effect, since it reduces the volume of repair and the corresponding costs associated with repairs.

7.4.8   Optimisation of the volume of inspection in various stages of the life cycle of nuclear power plant life cycle (factory inspection, installation and commissioning, operation). Superinspection

The overwhelming number of defects identified and repaired during the operation phase formed at time of manufacture or assembly and are of the technological nature. The reasons for this situation are both objective and, apparently, subjective factors. The subjective factor can include the desire of the manufacturers or assembly organisations to deliver the product to the customer as soon as possible and the absence of material and legal responsibility for factory rejects (technological nature of the defects) identified during subsequent operation. The nuclear power plant pays for almost all repairs of these defects. Input inspection in nuclear power plants is often carried out by inexperienced NDT inspectors, since the object is new and the team at the plant has just been formed. The objective factors include almost complete absence at the present time of NDT methods and tools allowing with 100% certainty to detect all defects of the technological nature directly at the manufacturing plant or after installation.

Detection of defects can be significantly increased (2–3 times or more) even without changing the methods and means of verification. To achieve this it is necessary to:

- carefully select NDT inspectors on the basis of the psychophysical, moral and professional characteristics;
- inspect repeatedly the same location inspected by different NDT inspectors.

Because of these two measures the number of operational defects missed can be reduced by 5–10 times or more.

Given the above the following tentative scheme can be proposed for the inspection of responsible elements of vessels and pipelines.

1. Before inspection, NDT inspectors should be selected on the basis of the results of 'blind inspection' conducted on full-scale test samples of hidden defects;

2. Following the stage of factory output inspection, representatives of the nuclear power plant (customer) in the form of two or three independent teams of NDT inspectors should carry out 100% inspection of all welded joints. Repairs should then be carried out at the plant using the inspection results;

3. 100% inspection of all welds should be carried out in the framework of input inspection. Equipment should then be repaired according to the inspection results;

4. During the commissioning tests (after hot and cold running of equipment, including hydraulic tests) all welds should be subjected to 100% inspection.

5. During the first 20 thousand hours of operation 100% inspection of all welds should be conducted;

6. The next inspection should be carried out after a period defined on the basis of quantitative analysis of the inspection results, equipment reliability, rate of damage of the metal in service, as well as taking into account other factors which define the state of the structure. When using this inspection schedule the number of inspections of the important elements of the reactors can be reduced to 1–2 for the entire period of operation (and in some cases to 0) without prejudice to the reliability of the inspected structural element.

A similar inspection scheme can be applied to inspect the field welds and can be adjusted depending on the type of structure and operating conditions. The advantages of this inspection control scheme are obvious.

Firstly, the total number of inspections does not exceed the number of inspections for the entire lifetime according to the current normative documents (5–7 inspections). In this case, the cost of inspections will be lower as inspection is carried out on a clean (not contaminated with radioactive elements and corrosion products) construction. For this reason, the efficiency of inspection must also be higher. In addition, the quality of repairs conducted at the manufacturing plant and(or) on clean equipment, will be higher and the cost lower.

Secondly, the detected defects can be removed at the expense of the manufacturer or installation company as the technological nature of such defects is obvious.

Thirdly, the volume of repairs during operation will be sharply reduced (5 times or more).

Fourthly, the losses associated with the outages of the nuclear power units for will be lower.

Fifthly, the reliability of pressure vessels and pressure piping of nuclear reactors will improved because of more efficient inspection and repair conducted at the factory or on clean elements of structure for nuclear power plants.

NPP operating experience provides an example proving the validity of the above proposals. This case was analyzed in one VNIIAES reports (Scientific Research Institute of Atomic Power Stations) drawnup under the guidance of one of the authors of this book (A.F. Getman).

In the mid 80-ies, the newly commissioned units of nuclear power plants with RBMK reactors showed an increased number of defects in welds of the DN800 pipeline of the repeated forced circulation circuit. In this regard, experts at TsNIITMASh (Central Scientific Research Institute of Heavy Engineering) proposed to conduct the so-called superinspection. The results of factory inspection are shown in Fig. 7.21. More extensive inspection was also conducted before operation of the nuclear power plant. The subsequent results of in-service inspection of welds on DN800 pipelines of unit 3 of the NPP which had been subjected to superinspection prior to operation,

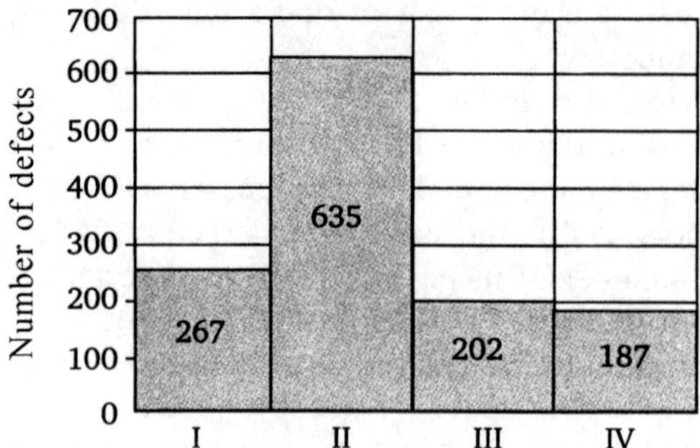

7.21 Number of defects in welded joints in DN800 pipelines and repeated forced circulation collectors of RBMK reactor detected by NI methods and repaired at the manufacturing plant (Dzhuro Dzhakovich): I) inspection after welding (radiography); II) inspection after welding (ultrasound); III) inspection after heat treatment (ultrasound); IV) superinspection prior to delivery of pipeline (ultrasound).

did not reveal any defects (see below):

1990 – 78 welds inspected (9 field, 18 shop, 51 longitudinal). Defects were not detected;

1991 – 160 welds inspected (26 field, 34 shop, 100 longitudinal). Defects were not detected.

1992 – 371 welds inspected (89 field, 71 shop, 211 longitudinal). Defects were not detected.

1993 – 182 welds inspected (89 field, 32 shop, 61 longitudinal). Defects were detected.

1994 – 191 intersections inspected. Defects were not detected.

Table 7.4 Results of ultrasound inspection of DN800 pipelines of repeated forced circulation circuit of unit 1 of the nuclear power plant shown as the ratio of the number of identified discontinuities to the number of tested welds

Year of control	Field welds			Shop welds					
	circumferential			circumferential			longitudinal		
	1	2	3	1	2	3	1	2	3
1989	71	4	0.056	46	1	0.08	–	–	–
1990	48	13	0.27	25	3	0.12	100	2	0.02
1991	86	18	0.21	48	11	0.23	97	2	0.021
1992	156	25	0.16	46	13	0.28	251		0.032
1994	112	25	0.22	34	0	0	109	9	0.08
Total	524	85	0.16	199	28	0.14	557	21	0.037

Note. 1. Inspected number of welds. 2. Number of defective welds. 3. Ratio of the number of defective welds to the number of inspected welds.

Table 7.5 Results of ultrasound inspection of DN800 pipeline of teh repeated forced circulation circuit of the nuclear power plant given as the ratio of the number of identified discontinuities to the number of tested welds

Year of inspection	Field welds			Shop welds					
	circumferential			circumferential			longitudinal		
	1	2	3	1	2	3	1	2	3
1989	19	0	0	35	2	0.057	114	6	0.053
1990	24	3	0.125	20	5	0.25	96	10	0.104
1991	17	0	0	23	7	0.212	99	4	0.04
1992	43	2	0.047	38	3	0.079	112	3	0.027
1993	35	0	0	–	–	–	–	–	–
1994	16	0	0	–	–	–	–	–	–
1995	25	2	0.08	36	1	0.028	100	–	0.01
Total	179	7	0.08	162	18	0.11	251	24	0.046

Note. 1. Inspected number of welds. 2. Number of defective welds. 3. Ratio of the number of defective welds to the number of inspected welds.

1995 – 179 welds inspected (31 field, 38 shop, no longitudinal). Defects were not detected.

A total 902 welds were inspected (214 field, 193 shop, 533 longitudinal), in addition, 191 intersection areas of the longitudinal joints with a circumferential weld.

For comparison, Tables 7.4 and 7.5 show the results of inspection of the DN800 pipelines for blocks 1 and 2 of the same nuclear power plant, for which more extensive inspection was not performed before the start of operation.

7.4.9   Optimum combination of inspection methods

The results of the probabilistic strength analysis of structural elements, carried out to ensure correct organisation of NDT during operation can be used to develop recommendations for optimum combinations of various physical methods of NDT, NDT methods with other inspection methods, including inspection of the parameters of operational loading of the structure, NDT and hydraulic tests, NDT and inspection of leaks within the LBB concept. Unfortunately, this line of work has not been sufficiently developed, but some aspects of the problem and expected results can be discussed.

ISI and inspection of the parameters of operational impact

In the case of superinspection at the stage of input inspection and the initial stage of operation and the reduction of the probability of existence in the structure of technological defects to an acceptable required level, the subsequent inspection of the structure can be carried out by controlling the parameters of operational impacts on the structure (mechanical, thermal, thermomechanical, corrosion, radiation). Non-destructive inspection should be carried out only on the basis of the indications of the control of the parameters of operational impact. This organisation scheme of inspection has already been discussed in Ref. 97 within the framework of the Integrated System of Operational Control (ISOC) where it was named the 'resource control'.

The advantages of such an organisation of in-service inspection are evident, since the control of the parameters of operational impacts can often be organised in a simple and less expensive way in comparison with NDT. In addition, staff are exposed to smaller radiation doses.

Organisation of ISI in aviation can serve as a prototype of the above-described organisation of inspection where the volume and frequency of NDT is determined depending on the number of hours spent by aircraft in the air, the number of takeoffs and landings, etc.

Combination of different physical NDT methods

The results of strength analysis carried out with the residual defectiveness of the structure taken into account can be used to define conditions for optimum combinations of various NDT methods. For example, new cracks can nucleate in pressure vessels and piping of nuclear power plants, oil and gas industry and other similar structures only by the mechanism of fatigue, thermal fatigue, corrosion fatigue, corrosion cracking, intergranular or crevice corrosion. Obviously, such a crack can nucleate in the vast majority of cases only at the surface of the metal structure. Therefore, dye penetrant, magnetic particle, eddy current inspection and other methods for detecting defects on the surface of the metal of the structure should be the principal methods used during operation.

Methods for inspecting the state of subsurface layers of metal of structures should be used in conjunction with the methods of surface inspection at the stages of input and preoperational inspection to reduce the probability of the existence of technological defects to an acceptable level. During operation, these methods should be used only on the basis of the results of strength analysis conducted with the data on actual operational effects and the residual defectiveness of the structure when there is a risk of rapid crack propagation from the technological defects remaining in the structure.

NDT and hydraulic strength testing

The issue of the usefulness of hydraulic tests (HT) of strength by the pressure exceeding the working pressure is the subject of frequent discussions. The negative impact of HT is evident, as each cycle of mechanical stress leads to a decrease in the operating life by the criterion of fatigue resistance. At the same time, HT can be helpful in some cases. Analysis of the strength of the structure using methods of fracture mechanics and taking into account the residual defectiveness of the structures allows accurate assessment of the usefulness of HT. Figure 6.12 shows the calculated probability of rupture of DN500 pipelines after ISI as well as after ISI combined with HT. In the case of successful HT the pipeline in operation has a high level of reliability (probability of fracture of the pipeline is zero).

In general, it can be argued that the hydraulic tests fail to detect defects smaller than the critical dimensions for the HT regime. In some cases, successful hydraulic tests provide a high level of the reliability of structures and can serve as a substitute for ISI.

The effect of HT is discussed in more detail in sections 8.1 and 9.

8

Optimisation of hydraulic tests, technical certification and planned–preventative maintenance

8.1 Method for determination of the optimum pressure of hydraulic stress tests to ensure reliability and operational safety

Hydraulic (pneumatic) tests (HT) are widely used in practical operation to check the strength and leaktightness of vessels and pipelines, their parts and assembly units loaded with pressure.

Hydraulic (pneumatic) tests are performed:
– After manufacture of equipment or elements of the pipeline supplied to installation;
– After installation of equipment and pipelines;
– In service of equipment and pipelines, loaded pressure water, steam and water mixture or other liquids and gases.

Hydraulic (pneumatic) test of pressure vessels (pipelines)[6] are conducted at intervals established empirically in accordance with applicable regulations. Vessels and pipelines working under pressure, are loaded with the pressure of water or air p_{HT} which is higher than the working pressure P_w by the value $K \geq 1$. The vessels and pipes are tested under pressure for no less than 10 min, after which the pressure is reduced and the pressure vessels (pipelines) are inspected during the time required for inspection. The pressure load is selected on the basis of existing regulations to reflect current practices. The minimum allowable temperature during tests and exposure is determined by the strength calculation standards. Vessels and pipelines pass the hydraulic (pneumatic) test if testing and inspection reveals no leaks and ruptures of metal and during the time lag the pressure drop remained within the limits specified in the regulations, and no new residual strain is detected after the tests.

The disadvantage of this method is that the pressure p_{HT} and the frequency of HT are determined without regard to the actual level of reliability of the pressure vessel or pipeline achieved in HT and its changes over time. Therefore, despite carrying ot hydraulic (pneumatic) tests pressure vessels or pipelines can fail prior to the next HT and its operation can be aborted because of leaking or damage that usually leads to great economic losses and, possibly, loss of human life.

Technical results achieved using the technology described below are based on the correct choice of the frequency of HT conditions, pressure and test temperature which in turn will lead to guaranteed reliability and safe operation of pressure vessels and piping.

These technical results are achieved by subjecting the pressure vessels (pipelines) to periodic loading with the pressure of liquid or gas which exceeds the working pressure, followed by holding the vessels (pipelines) under pressure. The pressure then drops and the vessels (pipelines) are inspected and they are considered to have passed the hydraulic (pneumatic) test if testing and examination reveal no leaks and ruptures of tmetal in holding and the pressure drop does not exceed the established limits in the regulations and no new residual strains are detected after testing. The test conditions are selected to ensure that the size of a family of critical size cracks, corresponding to the appropriate test regime, is smaller than that of the family of the critical size cracks, corresponding to the operating mode $(a,c)^{cr}_{HT} < (a,c)^{cr}_{w}$, where $(a,c)^{cr}_{HT}$ is the size of the family of the critical size cracks corresponding to the test conditions, $(a,c)^{cr}_{w}$ is the size of the family of critical size cracks corresponding to the working conditons. The frequency of such tests is selected so that the time between hydraulic (pneumatic) tests does not exceed the duration of increase (growth) of the size of the family of critical size cracks, corresponding to the conditions of the hydraulic (pneumatic) tests, to the size of the family of the critical size cracks corresponding to the working conditions (temperature, pressure and other influences): $T\left[(a,c)^{cr}_{HT} \rightarrow (a,c)^{s}_{w} \right]$.

The method for selecting the optimum pressure and periodicity of HT is shown in Figs. 8.1–8.15.

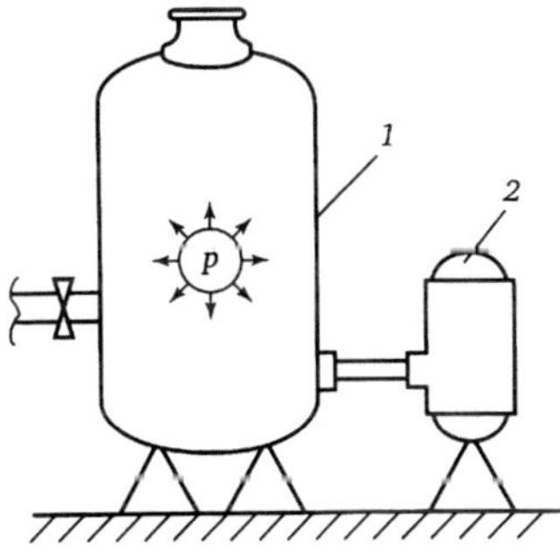

8.1 Diagrams of loading of a pressure vessel: 1) vessel; 2). pump.

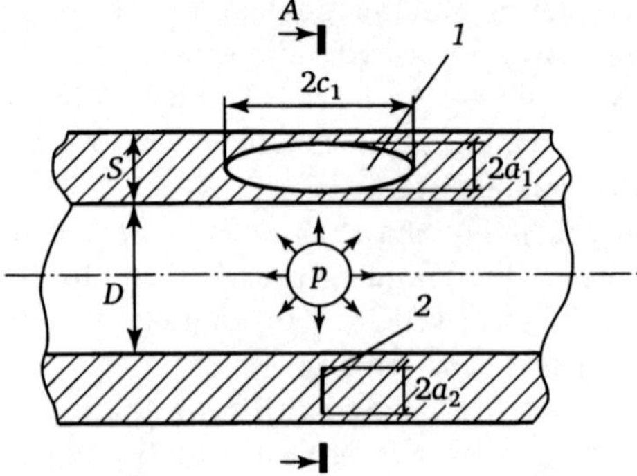

8.2 Location of subsurface cracks in a pressure vessel: 1) a longitudinal crack, width $2a$ and length $2c$; 2) a transverse crack.

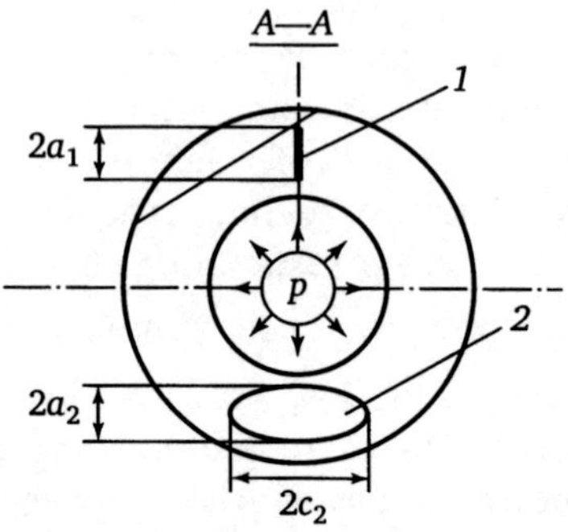

8.3 Location of subsurface cracks in a pressure vessel: 1) a longitudinal crack; 2) an annular crack, width $2a$ and length $2c$.

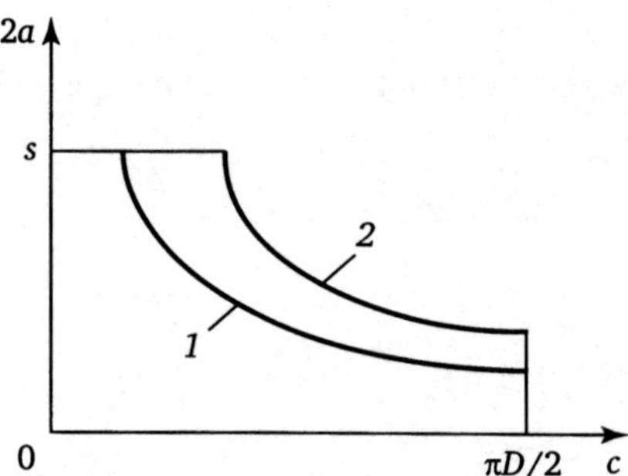

8.4   Families of all possible transverse subsurface critical size cracks in the HT mode (1) operating mode (2).

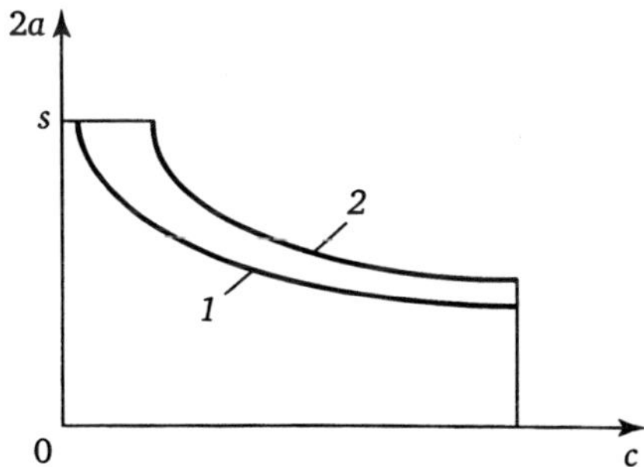

8.5 Families of all possible longitudinal subsurface critical size cracks in the HT mode of hydraulic and operating mode (2).

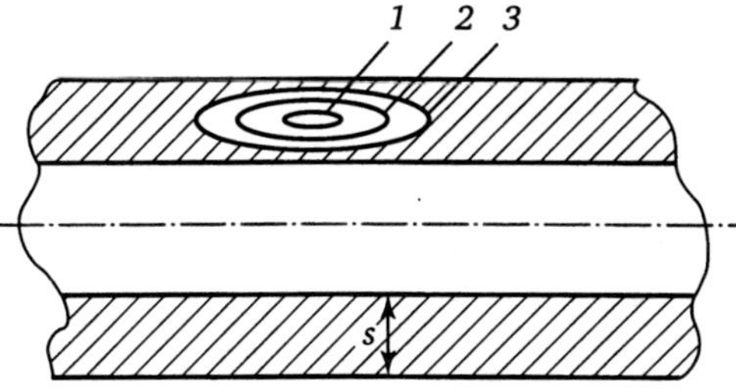

8.6 Development of cracks in time: 1) initial crack; 2) crack which propagated during operation; 3) crack at the end of operation.

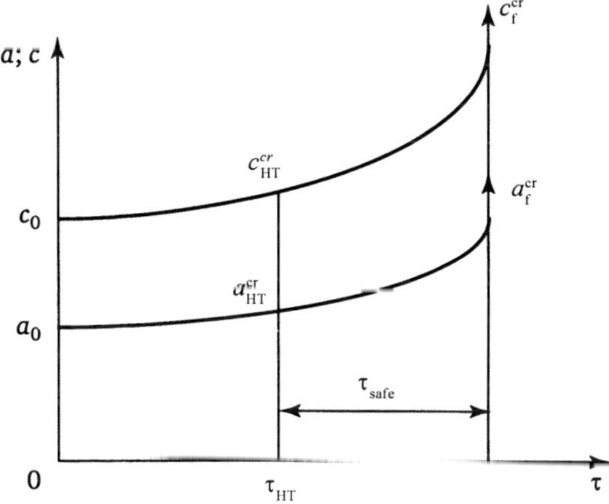

8.7 Graph of crack propagation in time.

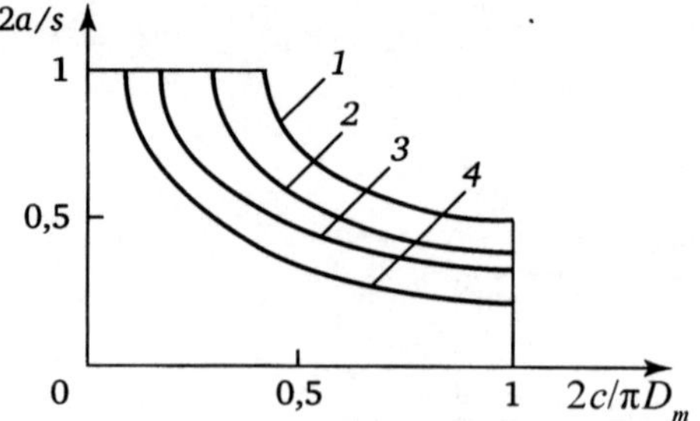

8.8   Change of the crack size, critical for the HT mode, as a result of growth during operation.

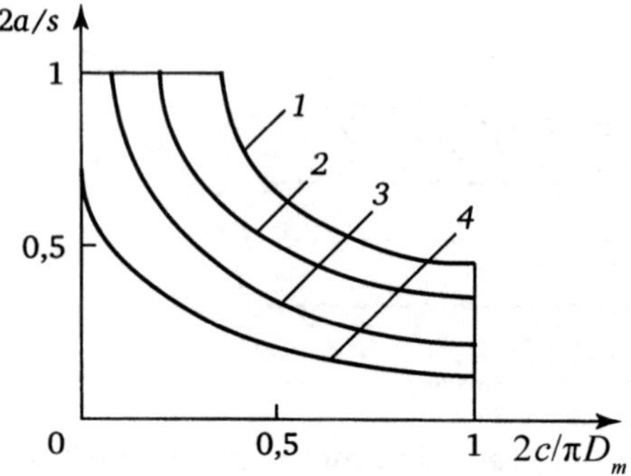

8.9   Curves of the critical size of subsurface transverse (tangential) cracks in a pressure vessel under the influence of different pressures $p_1 < p_2 < p_3 < p_4$.

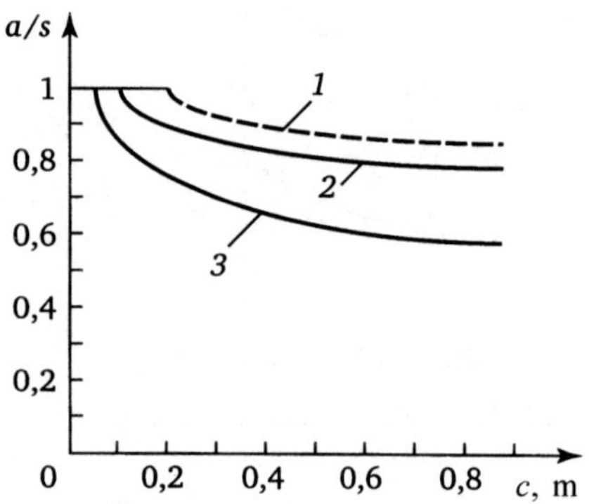

8.10   Curves of the critical size of subsurface longitudinal cracks in a pressure vessel under the influence of different pressures $p_1 < p_2 < p_3 < p_4$.

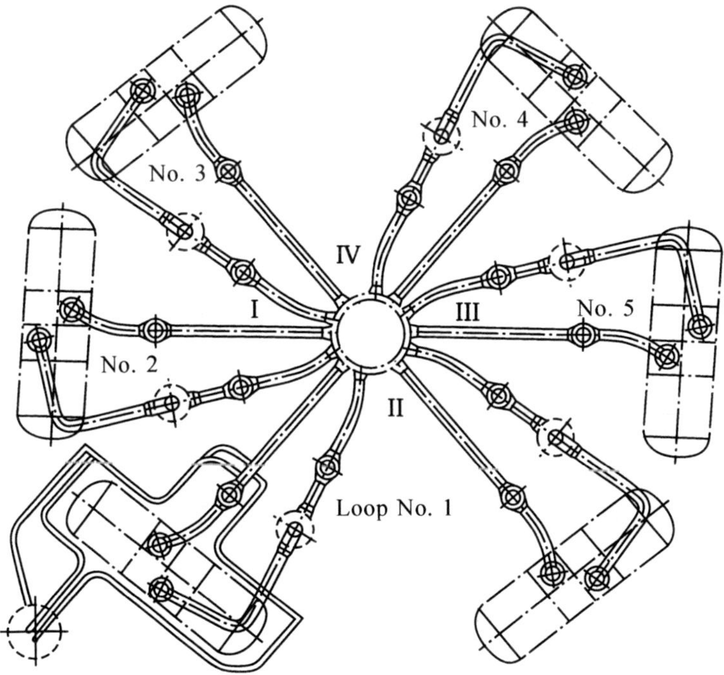

8.11 Diagram of energy equipment with a tested pipeline with an internal diameter of 500 mm (DN500).

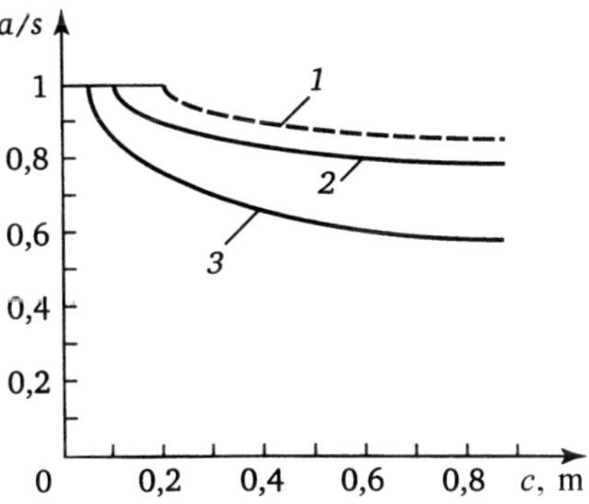

8.12 Results of calculations of critical crack size for test pressures of 140 and 175 kgf/cm² and operating conditions (longitudinal cracks): 1) normal operating mode; 2) operating mode with a failure situation; 3) hydraulic test.

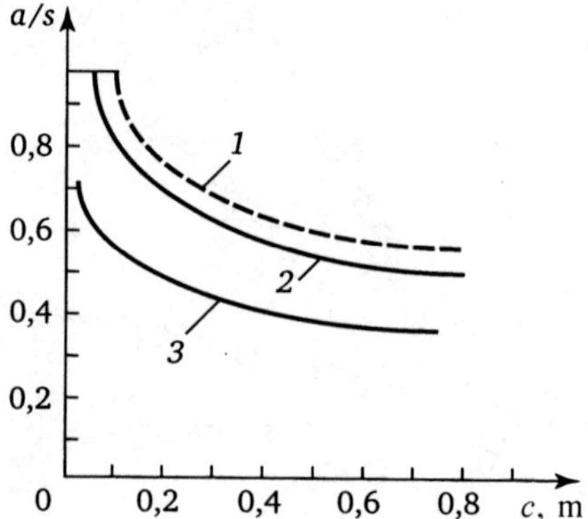

8.13   Results of calculations of critical crack size for test pressures of 140 and 175 kgf/cm² and operating mode (transverse cracks): 1) normal operating mode; 2) operating mode with a failure situation; 3) hydraulic test.

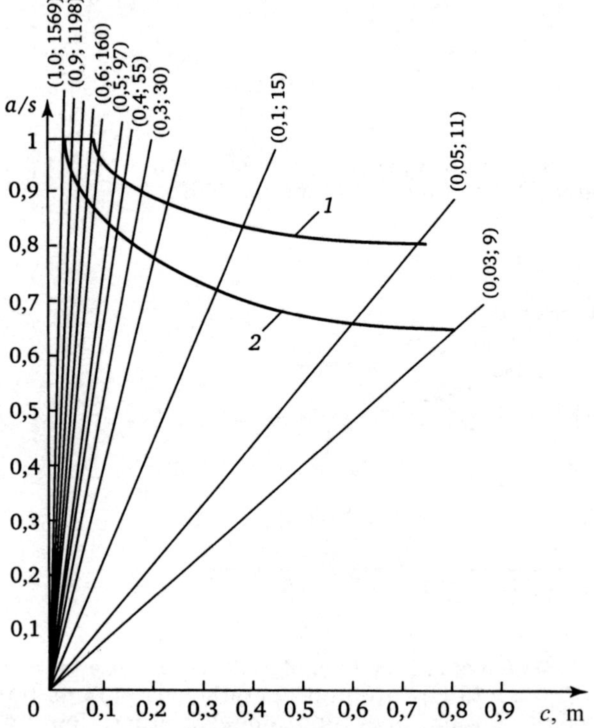

8.14   Results of calculation of the safe number of loading cycles for a longitudinal crack (the ratios of the half axes of the crack and the number of operating cycles are given in the brackets): 1) operating mode; 2) hydraulic test.

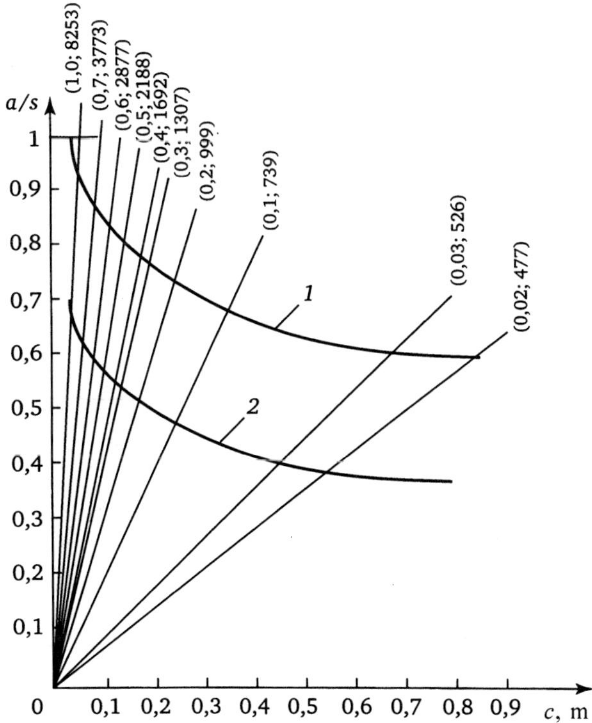

8.15 Results of calculation of the safe number of loading cycles for a transverse crack transverse orientation (the ratios of the half axes of the crack and the number of operating cycles are given in the brackets): 1) operating mode; 2) hydraulic test.

According to fracture mechanics, unstable (rapid avalanche-like destruction that cannot be stopped) destruction occurs when the defect size reaches the critical value. Any defect can be modelled conservatively by a crack, and any crack can be described by an ellipse with minor a and major c semiaxes The higher the pressure and lower the temperature, the smaller the critical crack size.

A family of cracks is a set of all possible cracks in the structural element including the orientation and location of this element, size, shape. For this particular loading the family of cracks can be divided into two parts – the cracks that do not lead to destruction for the given loading and cracks that lead to destruction. The boundary between these two sets of cracks is formed by the critical size cracks. The family of critical size is the set of all possible critical size cracks classified on the basis of type: the orientation and location in the given element and shape. Critical size cracks whose plane passes across the cylindrical part of the pressure vessel or piping, are referred to transverse (or tangential), if the plane of the crack passes through the centerline of the pipeline or pressure vessel, then such a crack is called axial. Figure 8.2 and 8.3 show the arrangement of subsurface cracks in a pressure vessel. The figures show a longitudinal crack $2a_1$ wide ad $2c_1$ long [s is the wall thickness of the vessel; D is the inner diameter of the pressure vessel (pipeline)],

Since the pressure vessel or pipe wall are in the plane stress state with principal stresses oriented along the axis or in the tangential direction, then every possible set of critical size cracks can be schematised by axial and transverse cracks and represented graphically in the coordinates (a; c), Figs 8.4 and 8.5. The ratio between the critical size a and c is such that as a increases c decreases, and vice versa (these relationships are known from fracture mechanics[113,etc]). In addition, the higher the load, the smaller the critical crack size.

In order to ensure safety after conducting hydraulic tests, the pressure and temperature during HT shall provide the conditions in which the critical sizes of cracks in the HT mode are lower than during normal operating conditions. Schematically, these conditions are shown in Fig. 8.6. In this figure: 1 – a source-defect; 2 – crack of critical size for the pressure in HT mode, 3 – a crack of critical size for loading pressure vessels and piping in the operating mode.

8.2   Optimisation of the frequency of hydraulic tests

During operation, the cracks can grow and increase in size. Figure 8.7 shows the propagation of cracks over time. The initial sizes of the defect are denoted by c_0 and a_0. During operation, if the crack grows, the dimensions of a and c change, when the critical size is reached, the crack becomes unstable and total or partial destruction of the structure takes place. In Fig. 8.7 the period of time from the moment of successful HT τ_{HT} to the moment when the crack size becomes equal to the critical size of the crack in the operating mode of operation corresponds to the time of safe operation τ_{safe}.

The growth of the entire set of cracks with a particular orientation can be represented graphically (Fig. 8.8). If at the beginning of operation the cracks are characterised by curve 1 (a family of critical size cracks to HT pressure), then at certain points in time due to crack growth the curves occupy the position of curves 2, 3 and 4. Curve 5 characterises the critical crack size for the operating mode of operation. Curves 2 and 3 reflect the safe operating conditions (the probability of bursting of a pressure vessel is zero), and at the time of contact of curve 4 with curve 5 the probability of rupture is different from zero.

The size of all the critical cracks depends on the load, such as pressure. Figures 8.9 and 8.10 show the curves of the family of the critical size subsurface transverse (tangential) and axial cracks at different pressures $p_1, p_2, p_3, p_4,$ and $p_1 < p_2 < p_3 < p_4$. Curve 1 corresponds to pressure p_1, curve 2 to p_2, curve 3 to pressure p_3, and curve 4 to pressure p_4. The graphs show that the critical crack size decreases with increasing pressure.

If the loading conditions are such that the dimensions of critical cracks in the HT mode are smaller than in the operating mode, HT ensures the safety of subsequent operation. This situation for longitudinal and transverse cracks is shown in Fig. 8.4 and 8.5. However, if the critical size of the cracks in the HT mode – all cracks or some cracks from the entire possible set of cracks (families of cracks), is equal to or greater than the critical size of cracks in the operating mode of operation, HT does not provide in this case complete safety of subsequent operation (safety according to the fracture criterion), i.e. partial or complete failure of the structural member can take place in this case.

Thus, for HT to ensure safety in subsequent operation, the pressure should be chosen so that all possible critical size cracks during HT are lower than all the possible critical size cracks in the operating mode (this situation is indicated on Fig.8.4 and 8.5). In this case, the greater the difference between the critical crack sizes in the operating mode and in HT, the longer is the duration of safe operation.

At the same time, it is important to ensure that the vessel or pipeline does not fracture in the HT mode. These successful HTs show that the tested structure does not contain cracks whose size is greater than or equal to the critical size of cracks in the HT mode. But at the same time, the structure may contain smaller cracks, including cracks with the size close to the critical size in the HT mode. However, these cracks are not dangerous for the operating mode.

These cracks can grow during operation. However, whilst the cracks are smaller than the critical size in the operating mode of operation, no failure can take place. But once the cracks reach the critical size in the operating mode of operation the structural member can fail.

Thus, the time during which the family of critical cracks in the HT mode can grow to the critical size in the operating mode is the time of safe operation. On this basis, the frequency of HT which provides full security is chosen as the time of crack size growth of the family of critical cracks, corresponding to the pressure in the HT mode, to the size of the family of critical size cracks, corresponding to the operating mode (Fig. 8.7 and 8.8). The crack propagation time is determined by the well-known methods of fracture mechanics, for example, Paris' formula for cyclic loading.

When selecting HT pressure, it is necessary to define the family of critical size cracks, corresponding to the pressure in the operating mode (using the fracture mechanics formulas), and this is followed by determining the HT pressure which must exceed the pressure in the operating mode. Subsequently, a family of critical size cracks is formed. If all the cracks for the pressure in the HT mode are smaller than the critical crack size for the pressure in the operating mode, this pressure can be achieved in practice. If all the dimensions of critical size cracks, or parts thereof, for the pressure in the HT mode are equal to or greater than the critical crack size for the pressure in the operating mode, then the HT pressure is increased and the critical crack size is calculated for this higher pressure. And so on, until all the critical crack sizes for the pressure in the HT become smaller than all of the critical crack sizes for the operating mode. This pressure is then chosen as the HT pressure.

The time of the next HT test is then defined for the HT pressure determined by this process by the time of the growth of critical cracks for the pressure in the HT mode to the critical crack size for the pressure in the operating mode of operation.

In addition to the stresses caused by pressure, pipes and pressure vessels are also loaded by general or local bending stresses whose magnitude in operation and in HT differs. The magnitude of bending stresses is determined by weight and thermal loads, as well as by the possible time-dependent thermal effects in operation.

A pipeline of a power plant is shown in Fig. 8.11 will be taken as an example. The inner diameter of the pipe is 500 mm (DN500), wall thickness $s = 32$ mm. The pipeline has straight sections and bends, is produced from 0Cr18Ni10Ti austenitic steel, operating temperature $= 300°C$, HT temperature $=20°C$, working pressure $= 125$ kg/cm^2. The stress during operation from internal pressure in the axial direction is $\sigma_p = 451$ kgf/cm^2, in the tangential direction $\sigma = 902$ kg/cm^2, while in the axial direction there are general bending stresses of $\sigma_b = 60$ kg/cm^2. The yield strength of the material at room temperature $\sigma_y = 220$ kgf/cm^2, at the service temperature $\sigma_y^p = 190$ kg/cm^2.

To determine the HT pressure $p = 140$ kgf/cm^2 is selected, and the stress in the axial direction $\sigma_p = 506$ kgf/cm^2, in the tangential direction $\sigma = 1012$ kg/cm^2, and bending stresses are equal to 5 kgf/cm^2.

It should be noted that the material is in the viscous state in both the test regime and the operating mode.

We define the critical crack sizes for the operating mode and for the HT mode:

– for transverse cracks (the safety margin is equal to 1; if there is no safety margin, the size is critical, and if there is a margin, the cracks do not cause failure):

$$\sigma_B = (2/\pi) R_F^T \{2\sin\gamma - n_a [a/s]_1 \sin(n_\varphi [\varphi]_1)\};$$

$$\gamma = 1/2 (\pi - n_a [a/s]_1 n_\varphi [\varphi]_1 - \pi\sigma_m / R_F^T);$$

– for axial cracks (safety margin 1):

$$\sigma_m + 0.67\sigma_B = R_F^T (1 - a/w),$$

where σ_B in the general bending stress, R_F^T is the half sum of the yield limits and ultimate strengths of the material, φ is the length cracks in radians, S is the pipe wall thickness, σ_m is membrane stress, n_a, n_φ are the safety margins for the crack size (in this case $n_a = n_\varphi = 1$);

$$a/w = \frac{n_a \left[a/s\right]_1}{1 + \frac{2}{\pi} \frac{\left[a/c\right]_1 n_a / n_0}{\left[a/s\right]_1 n_a}}.$$

The results are presented in Figs. 8.12 and 8.13, where 1 – critical size cracks the operating mode, 2 – for p_{HT} = 140 kgf/cm², 3 – for p_{HT} = 175 kgf/cm². Figure 8.12 shows that the size of the family of critical size cracks for HT mode is smaller than the size of the family of critical size cracks for the operating mode.

Safety in the test with such pressure is guaranteed because all the critical size cracks in the HT mode are smaller than the critical size cracks in the operating mode.

The time of safe operation in the case of successful tests at the selected pressure will be determined. For this steel and the given operating conditions the crack grows only under the effect of cyclic loads. Cyclic loading is produced by loading the pipeline operating pressure and temperature.Crack growth is defined by

$$da/dN = C_1 \left(\Delta K / \sqrt{1-R}\right)^m,$$

where C_1, m are material characteristics and depend on the loading conditions.

In this formula, the coefficients C_1 and m are chosen from Table 1 in Ref. 113, and the stress ratio is zero (R = 0), i.e., the cycle is pulsating. The calculations were performed for various ratios of a to c. The calculation results are presented in Figs. 8.14 and 8.15.

The rays diverging from the origin characterise the family of cracks with the same ratio of a and c. The first number indicates the ratio of a to c and and the second number is the number of cycles during which the critical crack in the HT mode grows to the crack size, critical in the operating mode.

The calculations show that the minimum number of cycles during which crack grows to 'dangerous' dimensions in the operating mode is 9. The duration of safe operation is the time during which the loaded component enters and leaves the operating mode nine times. For this pipeline two such loading cycles occur per year, so the time for safe operation is four and a half years. The following loading with the HT pressure regime is carried out after 4 years with the pressure being the same as in the initial loading after the 9th cycle of reaching and leaving the working regime andprior to transfer to the 10th working load cycle.

8.3   Optimisation of the technical inspection and scheduled preventive maintenance

8.3.1   Optimisation of scheduled–preventive maintenance

Optimisation of scheduled–preventive maintenance (SPM) using probabilistic methods of structural strength can be performed on the basis of both the composition of and periodicity of operations.

The simplest approach is to monitor the actual level of reliability of components and pipes which are important for safety. SPM of these elements should be carried out on the basis of the criterion of reaching an unacceptable level of reliability.

Figure 8.16 shows schematically the change in the probability of the existence of discontinuities with the unacceptable size in equipment in operation (curve 2), the probability of leaking (curve 4) and the probability of failure of the structural element (curve 6). SPM should be carried out when unacceptable values of probabilities are recorded.

The case of expulsion from the SPM of eddy-current testing of tubes of steam generators reactors VVER-440, units 3 and 4 of the Novovoronezh NPP, is described in section 9. The volume of these operations was reduced on the basis of analysis of the reliability of the tubes which, after

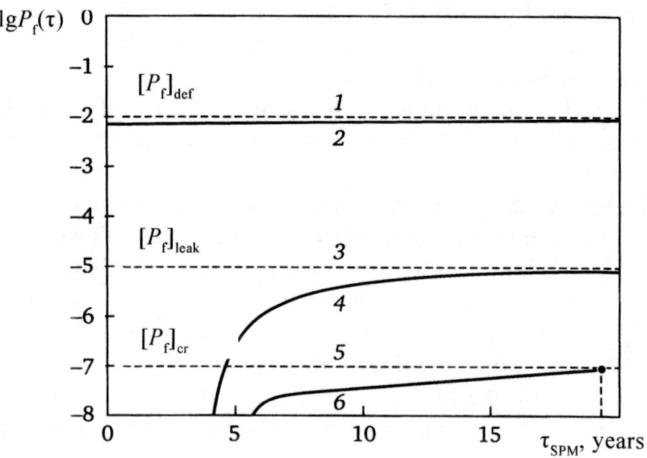

8.16   Time of next scheduled–preventive maintenance determined using the reliability characteristics on the basis of the criteriaof maximum detectability of unacceptable discontinuities in operation lg $P_f(\tau)$ (curve 1 – normative and 2 – actual), the maximum probability of leakage (curve 3 – regulatory and 4 – actual) and the maximum probability of failure (curves 5 – regulatory and 6 – actual).

completing the necessary measures, reached acceptable levels.

The second direction of optimisation of SPM is to ensure the optimum stock of spare parts and accurate timing of repairs. Repair is the most common and easiest way to restore the operational efficiency of the structure to an acceptable level. In most cases, repair is carried to restore the operational efficiency of structural members and equipment that have defects of technological nature (cracks, cavities, etc.). In this case, the repair can be accomplished either by welding or without it when the defect can be removed by mechanical grinding.

More effective repair procedures can be justified by the optimisation method in Ref. 138.

As a rule, maintenance is efficient if technological defects are removed. If the reason for appearance of the defect is poor design, reconstruction is inevitable.

The reliability characteristics are used to create a maintenance reserve. This technique is well developed and tested and thus it is not discussed here.

The third area of optimising SPM is associated with the optimisation of technical inspection of equipment and pipelines

8.3.2 Optimisation of the frequency and composition of works on technical inspection

Technical inspection is an integral step in assessing the state of equipment and pipelines (EP).

The purpose of technical inspection is to determine the actual status of the EP, confirmation of their compliance with the project, existing rules and standards, other technical documentation, confirmation of their serviceable condition, the possibilities and conditions for their further operation on nuclear power plants.

Technical inspection is carried out according to the results of:
– verification of documentation;
– visual survey of EP in accessible locations inside and outside;
– hydraulic test;
– previous non-destructive testing;
– prior testing of the mechanical properties, microstructure, chemical composition, geometrical dimensions and operational impacts.

Using the probabilistic methods of structural strength, described above, to determine the actual values of residual defects, strength, resource, reliability, safety and durability it is possible to optimise the composition of operations and frequency of technical inspection. In this case, the results of technical inspection are used to:
– define the original actual state of EP before operation (if technical inspection is carried out during commissioning);
– determine the state of EP at the time of technical inspection (if it is

carried out during operation);

– determine the time of next technical inspection; deadline for the next technical inspection is determined depending on the actual characteristics of residual defects, strength, useful life, safety, reliability and survivability of EP;

– Determine the conditions of operation, maintenance and operational control up to the next technical inspection, which would provide the level of residual defects, reliability and safety in accordance with the established requirements.

Below are the results of studies of the effect of periodicity of technical inspection (TI) on the reliability of the main circulation line (MCL) and the pressurizer (PRZ) of the VVER-1000 reactor. The inspection method was described in Ref. 115. Initial data for calculation (defectiveness of metal, mechanical properties, operating conditions, etc. correspond to power units of the Balakovo NPP on the condition of compliance with the standard terms of design, manufacture and operation, including inspection, maintenance and repairs).

Figures 8.17 and 8.18 show the results of analysis aimed at determining the effect of periodicity of MCL and PRZ technical inspection in the range of up to 10 years. Reliability was determined by the criteria of resistance to formation of leaks and damage during operation. The solid lines on the graphs show the change in the probability of a leak or fracture during the time between two adjacent TIs. It is evident that in both the MCL and PRZ, the probability of leakage or failure varies only slightly and at a 10-year

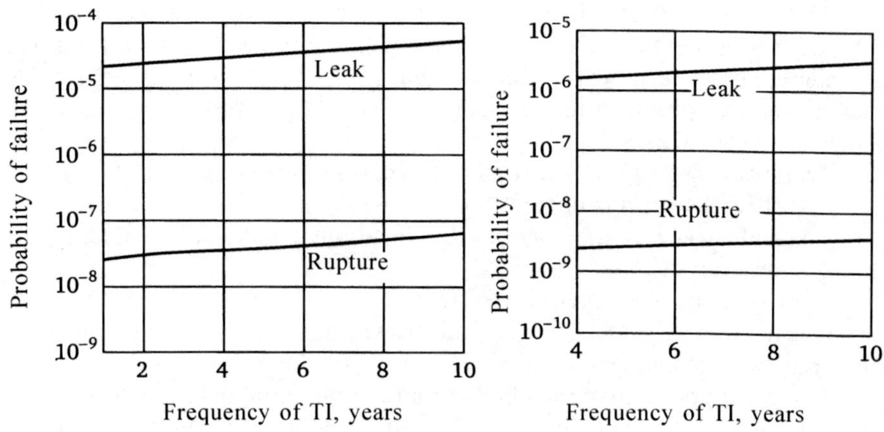

8.17   (left) Dependence of probability of leakage and rupture on frequency of TI under the conditions of NI and repair of identified defects for MCP Du 850.

8.18   Dependence of probability of leakage and rupture on periodicity of TI under the conditions of the NC and repair of identified defects for the casing of the pressure compensator.

periodicity is less than 10^{-7} (failure) and 10^{-4} (leakage) for MCL, and 10^{-8} (failure) and 10^{-5} (leakage) for the PRZ.

Analysis shows that the frequency of TI of the VVER-1000 reactor may increase to 10 years while maintaining a high level of reliability and safety in which the destruction of MCL or PRZ is almost impossible and leakage is unlikely. Practical realisation of this transition requires the unit-by-unit analysis of reliability and safety of all items of equipment and pipelines, including the reactor pressure vessel, the collector and piping of steam generators, and the main circulation pump housing.

Using probabilistic methods for solving the problem of ensuring leak tightness of heat exchanger tubes of nuclear power plant steam generators

The problem of the integrity of heat exchanger tubes (HET) of steam generators (SG) for the above-described probabilistic methods is of considerable interest because:

– there are a number of similar elements; for example, the number of HETs in a VVER-1000 reactor is 44 000;

– there is a large amount of damage, for example, the number of damaged HETs reaches hundreds and thousands of tubes;

– the dynamics of damage is very high and the effectiveness of measures taken to ensure integrity becomes noticeable very quickly (within a few days to several years of operation);

However, the large amount of statistical material and its rapid change makes it possible to determine the effectiveness of an approach.

This chapter provides three examples of practical application of probabilistic methods:

– Monte Carlo method (section 9.2);

– method based on the binomial distribution law (section 9.3);

– a generalised method that takes into account the residual defects (section 9.4).

9.1 Ensuring the leak tightness of tubes of vertical and horizontal steam generators

Nuclear power plants used steam generators (SG) of fundamentally different designs: horizontal (HSG) – in Russian projects, vertical (VSG) – in Western countries, the USA, Japan, Korea and in other countries. These designs have evolved considerably over many years and in most cases the evolution was stimulated by problems arising during operation. One major problem was the problem of HET plugging as a result of the formation of very large numbers of defects.

Comparison of special features of the formation of defects on differently oriented surfaces of heat exchangers with different flow pattern may affect the mechanisms and reasons for the formation and development of defects. Different materials of HET, their diameters and wall thickness are also important for various aspects of the HET integrity, but not a fundamental obstacle for comparative analysis.

Mill annealed alloy 600 (600 MA alloy) was used in the U.S. and most other countries for the manufacture of CPS and I&C nozzles for reactor covers and some other primary circuit nozzles. All these nodes have problems. Covers of PWR reactors are replaced at nearly the same rate as SG. Moreover, if the U.S. carried out the replacement after the discovery of defects (to 2004, 10 covers were replaced, and 23 planned to be replaced), in France covers are being replaced continuously (44 of 54 were replaced by 2005).

Replacement of VSG because of problems with alloy 600MA is now a routine procedure. There are around 350 SG of this type in the world. According to the data from US NRC and METI (Japan) websites by mid-2006 there were around 70 SGs of this type.

The initial approach to choosing material for SG HET for nuclear power plants in the USSR and the United States was the same. However, the use of stainless steel type SS304 in the first commercial nuclear power plant in the USA (Shippingport) gave negative results (leakage through two HETs of the steam generator 150 hours after start) in early 1957[116].

Negative experience with stainless steel for HET of steam generators was confirmed in other nuclear power plants such as Yankee Row and Indian Point-1, as well as non-commercial reactors Savannah River, Hanford and Nautilus[116]. Many of the problems with the HET in vertical steam generators were associated with the wrong choice of a new material for HET (high-nickel alloy 600MA) in the U.S. and all other countries. In these countries, nuclear power plant equipment (or SGs for them) were manufactured by US firms or under US license. The only country which quickly and independently stopped using Alloy 600MA was Germany. After the experience gained at the Obrigheim NPP, where HETs of SGs were also made of alloy 600MA, the new structures used 800NG alloy. Because of the special features of design and efficient choice of material there were only a small number of problems with the vertical HETs with pipes made of 800NG alloy (related to corrosion). Although, if we consider not only the corrosion causes of breakage of HET, the comparison of the extent of HET plugging in all German SGs and Russian PGV-1000 shows the practical equality in this parameter (about 0.6%)!

USA and all other countries were forced to start from 1980 (Surry-2 NPP) the replacement of SG with HET made of the alloy type 600mA by SG with HET made of alloy 600TT (in the U.S. 17 nuclear power plants, with 281 000 HET). Prior to 2002, 1400 HET of the SGs made of this steel were plugged in USA (up to 2005 more than 1700, by mid-2006 1884) – most of them due to wear in the zone of bending grids. Stress corrosion cracking (SCC) in the HET made of this alloy was detected in the USA only in 2002[117].

In France and South Korea, corrosion defects in HET made of this alloy were observed previously[118]. In 2002, a HET fractured in one of the SGs of Ulchin-4 NPP in Korea[119]. Since 1988 (Cook NPP) most of the SGs taken out of operation were replaced by new ones made of 690TT alloy. Corrosion problems were noted, most of the plugging up took place as a result of vibrational wear on a very small scale. At the end of 2005, 395 HETs (data for 78 SGs and 26 units) were plugged in U.S. SGs with HET made of the alloy. At the end of 2005, 395 heat exchanger tubes (data for 78 steam generators and 26 units) worked in the steam generators of the USA with the heat exchanger tubes produced from the same alloy. However, in the middle of 2005, the first inspection of the steam generators at the Osonee-1 nuclear power plant (14 months after replacement) showed further problems. Vibrational wear took place in almost all spacing zones in the 3200 heat exchanger tubes[120]. 48 heat exchanger tubes were plugged. The reasons have not as yet been established. However, as shown in the presentation material of the nuclear power plant (at the meeting in NRC), 200 tubes are being prepared for pluggage by the unit No. 2 shutdown in October 2005. According to the most recent data, only five heat exchanger tubes were plugged up in the unit No. 2. Three leakages were also detected in the heat exchanger tube of the steam generator made from this alloy. The selection of the material for spacing plates was also not very efficient. The initial selection of carbon steel and of the structure with drilled holes for these elements in the first models was regarded as the reason of denting which is a phenomenon typical only of the vertical steam generators. At the moment, the elements of the vertical steam generators are produced from stainless steel.

It is well-known that in the horizontal steam generators the stainless steel was used for spacing elements from the very first steam generator. The important design solution was the selection of the diameter and wall thickness of the heat exchanger tube. The vertical steam generators used several standard dimensions of the heat exchanger tube. There are two main standard dimensions (the vertical steam generators manufactured by Westinghouse): 19.05×1.05 mm ($k = 0.055$) and 21.43×1.23 mm $k = 0.0$ 57). In the most recent models (F, D75, etc.), the dimensions of the heat exchanger tubes are 16.86×1.02 mm ($k = 0.060$) and 19.05×1.09 ($k = 0.0$ 57). The vertical steam generators produced by the Babcock and Wilcox company use the heat exchanger tubes with the size of 15.80×0.86 mm ($k = 0.054$). The vertical steam generators manufactured by CE company use the heat exchanger tubes with the size of 19.05×1.22 mm ($k = 0.0$

64) and 19.05 × 1.03 (k = 0.054) mm. In Japan, there is only one standard size 22.23 × 1.27 mm (k = 0.057), and in Germany 22.0 × 1.23 (k = 0.056).

There are only two standard dimensions for the horizontal steam generators: 16.0 × 1.5 mm (k = 0.0 94) and 16.0 × 1.4 (k = 0.088). At this selection greatly increase the resistance of the heat exchanger tubes of the horizontal steam generators to corrosion and fracture.

One reason for the heavy damage to the HET of the SG was not the best choice of a design solution the spacing zone. According to Ref. 117 in 2002 of the 237 nuclear power plants (worldwide) 42% had SGs with spacer plate with drilled holes. In about 60% of nuclear power plants SGs were replaced, and new SGs use different spacing structures, 27% grid/eggcrate, 73% broached – with three or four contact zones). All the replaced SGs use the drilled hole-type spacers. It should be noted that simultaneously with the change in grid design the material of the elements of the spacer was also changed. Carbon steel was replaced with stainless steel. The change of the configuration of this zone of the HET should have significantly changed the situation with the effect of corrosion. And so it happened. After nearly 20 years of operatione no denting was detected in the VSGs with HET made of 600TT alloy with new design of the spacing zone and no stress corrosion cracking with intercrystalline corrosion as well as SCC (stress corrosion cracking) with intergranular corrosion. The first defects in the spacing zones appeared in the SG in the US plants only in 2002.

Here, apparently, it is worth mentioning that such a development of corrosion processes in the HET made of the new alloy was predicted in many studies by one of the leading U.S. specialists on corrosion - R.W. Staehle. Critically evaluating the methodology for selecting new materials (assuming that a mistake was made by choice of the alloy 690TT), he stated that it takes time (for a variety of factors – 10–20 years) to create suitable conditions for intensive corrosion processes. This position is described in detail in, for example, Ref. 121. According to this author, the lack of corrosion in the HET of the new alloys is largely the result of structural changes in the spacing zones and fixing areas, and not merely the result of selecting a new material.

In the HSGs the spacing between HET using spacing strips (option to which the U.S. came through trial and error – grid/eggcrate - is very close to the traditional method for HSG). This is one of the most dangerous places of localisation of all types of corrosion in VSG. Again, it should be noted that one of the main reasons was the wrong design solution: in the first models rolling (mechanical) was carried out only on a small part of the sealing area, so that an annular gap almost half-meter high remained. In the SGs of different models with HET made of the 600MA alloy this problem has not been solved. In models 51 manufactured by Westinghouse rolling of the HET in the tube plates was carried out at full length, first by an explosive method (to eliminate narrow gaps[122]), then mechanically, and, finally, with the model D5 by the hydraulic method. A number of other measures were taken to avoid re-rolling, reducing the stress state in the transition zone, etc.

Table 9.1 **Replacement of vertical steam generators because of problems with heat exchanger tubes made of 600 MA alloy**

Parameter	USA	France	Japan	Others
Total number of steam generators (SG)	209	198	68	175
Of these:				
with HET made of 600MA	176	63	29	59
with HET made of 600TT	33	111	20	—
with HET made of 690TT	0	24	19	—
Total number of replacements	121	45	29	53
Of these:				
with SG with HET made of 600TT	25	0	0	—
with SG with HET made of 690TT	93	45	29	26
with SG with HET made of 600MA	3	0	0	—
with SG with HET made of 800NG	0	0	0	10
Mean life, years	18.1	17.9	20.4	—
Maximum life, years	31	25	26	—
Minimum life, years	8	10	11	—

Basic data on changes of VSGs are shown in Table 9.1.

In Russian nuclear power plants only three SGs were replaced at nuclear power plants with VVER-1000 reactors (Unit 2 of the Balakovo nuclear power plant, 1999).

Of 176 (57 units) VSGs with HET of the 600MA alloy 127 SGs on 43 units were replaced in 2005 (the state at 01.01.06). By 2009, according to published plans[122], there would be only 5 SGs in units. Since plans often change, the data given here refer to 55 VSGs in 20 units (4 of them were marked for replacement in 2005 but they passed technical inspection of the HET).

As can be seen from Table 9.2, the era of replacements of SGs with HET made of the 600MA alloy is close to completion. Apparently, by 2010 there will be no such SGs anywhere in the world (only a few in the U.S.). There were nearly 350 of them. Replacement of one SG costs about $100 million.

The analysis shows that there are in fact two defect formation mechanisms in HSG and VSG: stress corrosion cracking (SCC) and crevice corrosion (CC). Their action is quite specific because of different HET materials as well as their location due to vital differences in the structure.

All the mechanisms for 600MA alloy and 08Cr18Ni10Ti steel are considered separately below.

Figure 9.1 show a localisation scheme for the mechanisms of defect formation in the HET of the VSG in the U.SA[124]. The black box indicates the area of occurrence of SCC in the VSG in the primary circuit. In the USA this type of corrosion was detected for the first time at the beginning

Table 9.2 Number of plugged HET on the existing HSG in the USA

Start year	Nuclear power plant	Output, MW	Company, type of SG	Number of SGs	Number of HET in SGs	Year replaced	Number of plugged HET
1974	Arkansas One-1	836	B&W, FA177	2	15531	2005	1704
1976	Beaver Valley-1	810	WEST, M51M	3	3388	2006	1952
1987	Beaver Valley-2	820	WEST, M51M	3	3388	—	440
1990	Comanche Peak-1	1150	WEST, D4	4	4578	2007	776
1977	Crystal River-3	818	B&W, FA177	2	15531	2009	1213
1978	Davis Besse-1	873	B&W, FA177	2	15531	—	823
1985	Diablo Canyon-1	1073	WEST, M51	4	3388	2009	825
1986	Diablo Canyon-2	1087	WEST, M51	4	3388	2008	890
1974	Fort Calhoun	476	CE, S67	2	5005	2006	753
1990	Palisades	762	CE, M2350	2	3219		823
1980	Palo Verde-1	1270	CE, S80	2	11012	2005	2076
1988	Palo Verde-3	1270	CE, S80	2	11012	2007	1500
1974	Prainie Island-2	512	WEST, M51	2	3388		500
1981	Salem-2	1106	WEST, M51	4	3388	2006	1100
1983	San Onofre-2	1070	CE-3410	2	9350	2009	2383
1984	San Onofre-3	1080	CE-3410	2	9350	2010	1355
1982	Sequoyah-2	1106	WEST, M51	4	3388		507
1983	St. Lucie-2	839	CE, S3410	2	8519	2007	4268
1973	Three Mile Island-1	786	B&W, FA177	3	15531	2009	2532
1985	Waterford-3	1075	CE, CE70	2	9350	2005	1524
1996	Watts Bar-1	1170	WEST, D3	4	4674	2006	1158

of the 70s holds a special place in their analysis of US speciliasts. It is this type of corrosion that besides SG HET (from inside) affects primary circuit equipment. This type of corrosion taking place in the US SGs was first recorded in the early 70's. Its contribution to the number of defects, which required repair, has changed over time, reaching 10%. This type of corrosion has led to the rupture of HET in 1976 at Surry-2 NPP in the USA, in 2000 at Indian Point-2 NPP (and also in 1979, at Doel-2 NPP in Belgium). In VSG HETs in the United States this mechanism is localised in a few places (notably on bends and in the fitting zone). In 1994, experiments started at the U.S. Farley-2 NPP with addition of zinc into the primary circuit coolant to reduce the intensity of SCC. A few more nuclear power plants in the U.S. and Germany joined these experiments later. The largest amount of experience was obtained at the Diablo Canyon nuclear power plant, where zinc was added during three cycles. It is concluded in Ref. 125 that the intensity of SCC slightly decreased, but it was stressed that the results of laboratory studies are contradictory. This type of SCC may appear in the form of circular cracks. Defects of the primary circuit in HSG are extremely rare. The PGV-1000 system did not show a single case of plugging due to these defects. In PGV-440, these defects led to plugging of several HET (less than 10). They were found in several units, and not in all SGs. No apparent localisation was observed and no defects were not found in the bends.

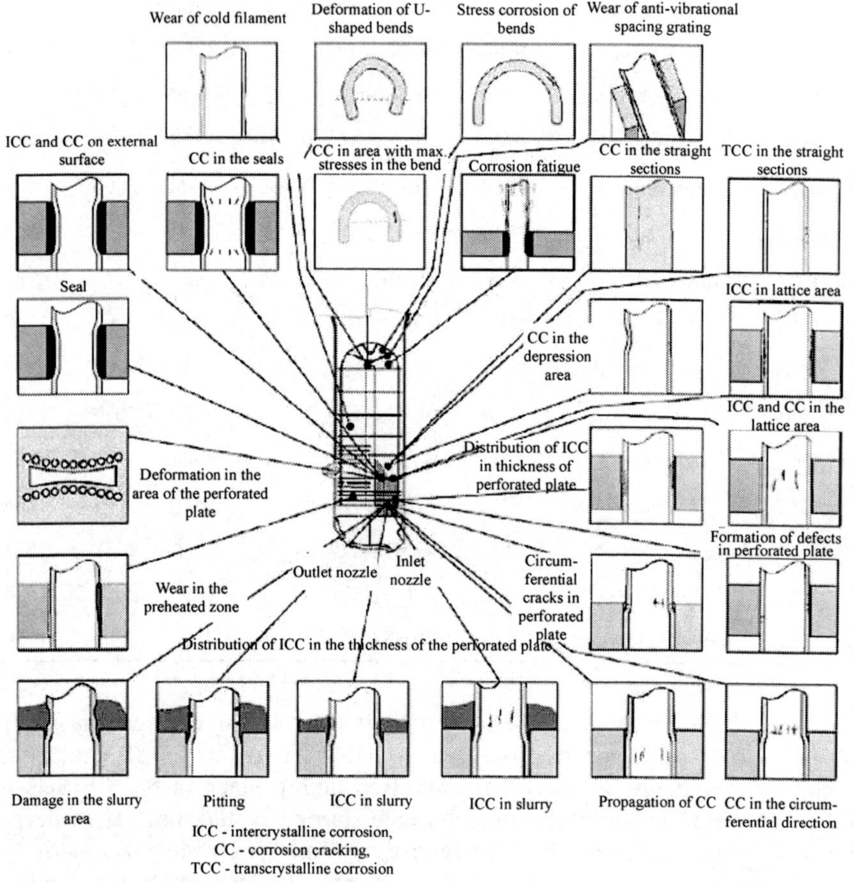

9.1   Scheme of localisation mechanisms for 600MA and 08Cr18Ni10Ti steels.

Heat exchanger tubes of SGs in the power unit No. 2 of the Balakovo nuclear power plant are discussed below(1996).

The case study described below shows the quantitative relationship of the lifetime, reliability and safety. From the methodological point of view, the case is interesting because it shows the use of system technology for assessment and lifetime provision, reliability and operational safety of nuclear power plant elements.

The previous section provides a general description of the SG.

Schematic representation of SG with HET is shown in Figs. 9.2–9.4. These schemes show the numeric designation of the HET spacer grids in the direction of the 'hot' end face of SG. The tubes are made of steel 08Cr18Ni10Ti, tube diameter 15 mm, wall thickness 1.5 mm. The properties of the steel conform to specifications TU 14-3-197–89.

Steam generators Nos. 1, 3 and 4 were put into operation in the unit No. 2 of the Balakovo nuclear power plant in March 1987. Low-temperature treatment and removing of residual deformation of the collectors were carried out in 1987. The non-failure operating time on 1/1/1996 was 50 200 hours.

The project provides for large strength margins and HET life. The normative and limiting numbers of loading cycles of the HET are shown in Fig. 9.5.

In April–May 1996, inspection of SG1, SG3 and SG4 on the unit 2 of the Balakovo NPP showed leaks from the primary to secondary circuit of up to 5 l/h which in accordance with the instructions for service of the SG required shutdown of the unit, search and elimination of leaks.

Flaw inspection, first by the aquarium and then eddy current (EC) methods, showed damage to the heat exchanger tubes.

The operating organisation Concern Rosenergoatom and Balakovo NPP organised large-scale activity to define the size of damage, causes of damage, eliminate the causes of injury and the possibility of further operation, repair and commissioning of steam generators. A research program was drawn up as part of these studies and its aim was to:

1) analyse the causes of damage to the HET in SG II of unit 2;

2) investigate the possibility of further operatio of SG HETs after plugging, cleaning surfaces and other prescribed activities;

3) justify HEt plugging criteria;

4) more accurately determine residual strength, residual life and reliability of the HETs of SG1, SG3 and SG4;

5) determine the extent of technical measures in the SPR-96-97 needed to ensure safe and reliable operation of SG with the existing damage in the HET until the next scheduled–preventive maintenance;

6) development and substantiation of the required scope of work in the next SPM (after SPM-96).

As part of this research program, the above methods and techniques (Chapters 1–3) were applied to develop and implement technical solutions

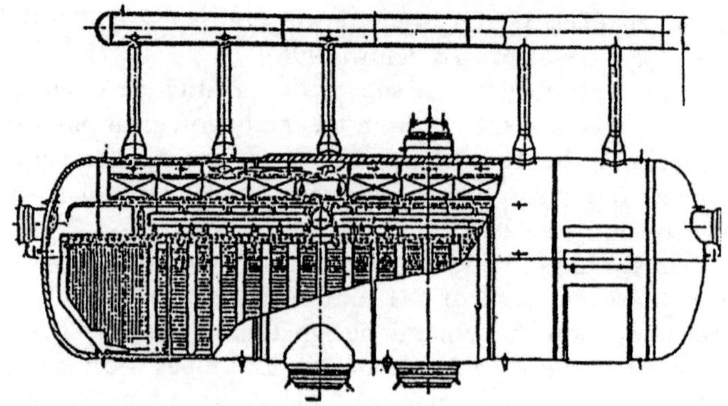

9.2   Steam generator PGV-1000, longitudinal section.

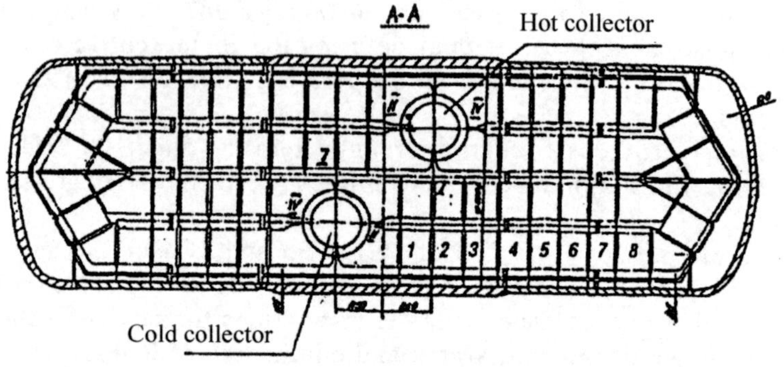

9.3   Steam generator PGV-1000, horizontal section.

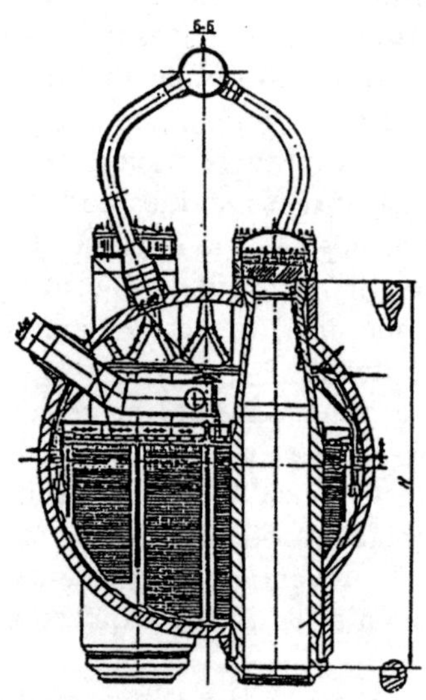

9.4   Steam generator PGV-1000, cross-section.

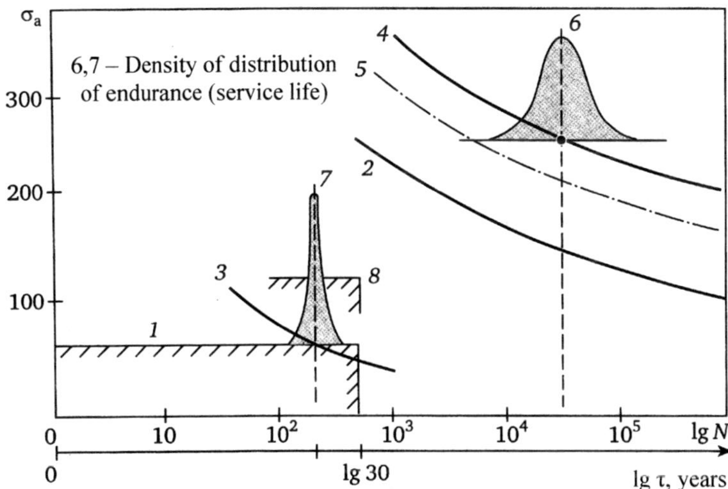

9.5   Fatigue curves for 08Cr18Ni10Ti steel: 1 – the projected area of HET operation, 2 – normative curve for temperatures up to 450°C, 3 – life curve as defined by the fact of damage to the tubes; falling part of the curve is determined by the quality of boiler water the day before SPM-96; 4 – results of tests in the air; 5 – tests in water with Na and Cl of 30 g/l;]8 – stress range, taking into account stress concentration from a corrosion pit.

providing an acceptable safety level of steam generator tube bundles operation, taking into account the provisions of the LBB concept in the system setting.

The operating conditions prior to SPM-96 were generally in compliance with regulatory requirements. Thermomechanical operating conditions in respect of the number of loading cycles were in the normal range. The data on water chemistry regime suggest occasional deviations from the requirements of OST 34-37-769-85 and temporary Water Chemistry Standard the Second Circuit, introduced on 01/01/1991.

The total time of deviation of pH, Na, Cl and electrical conductivity from the normalised values was less than 1% of operating time. According to the nuclear power plant data, the deviations are associated with increased suction of cooling water in turbine condensers and turbine pumps.

On April 24, 1996 unit 2 was shut down due to leaks in steam generator No. 4. After removal of leaks by plugging tubes, leakage was recorder in SG1 and SG3 during heating reactor installation. The leaks were removed but subsequent pneumatic–hydraulic trials revealed new leakingtubes in SG1, SG3 and SG4, a total of 67 tubes.

Under the assumption that the leaks in pipes will continue to emerge, the leadership of Rosenergoatom appointed a commission that stressed the necessity to carry out eddy current testing (ECT) of SG tubes and suggested possible causes of damage.

Programmes have been developed for ECT, cutting out and inspecting pipes and similar mesures were proposed to verify the corrosion state of SG and the water chemistry regime (WCR), as well as cleaning of SG piping to remove sediments by chemical cleaning.

A number of leaking tubes (i.e., pipes with through-wall defects) were identified: SG1 – 45, SG3 – 39, SG4 – 42.

Leaks were identified mainly in the tubes of the lower rows: the SG1 from the 106th to 31st, the SG3 from the 110-th to 99th, the SG4 from 110th to 89th.

Eddy current testing was performed using automatic equipment manufactured by Interkontrol with a rotating probe. The technique allowed to define only one size of damage – in the direction of wall thickness. In addition, the coordinate of damage along the length of the tube (from the collector) was also determined.

The scope of inspection was determined according to 'The working program of inspection of HET metal and SG collectors lintels of unit 2 by Interkontrol equipment'.

The rejection damage level was set at the level of 70% of the tube wall thickness.

The results of monitoring showed:

1) the total number of defective tubes (with the depth of the defect greater than 70%) for SG1 was 941 out of 10 947 inspected tubes, SG3 – 431 out of 10 380 inspected tubes, PG4 – 351 out of 10 942;

2) the percentage of damaged pipes (in relation to the inspected tubes): PG1 – 8.6%, PG3 – 4.15% SG4 – 32%;

3) the density of defective tubes in the lower part of SG is significantly higher than in the upper part, as evidenced by the results of the first stage of inspection: SG1 – 23.2%; SG2 – 18.0%, SG3 – 22.9 %.

4) through-wall defects detected by the pneumatic–hydraulic method, also located in the HETs of the lower rows;

5) defects are closer to the hot collector and most of them are located in the second semicircle;

6) defects between spacer grids are located primarily in the lower rows;

7) defects in the spacer grids are located in both the lower and the upper rows of HETs.

The distribution of defective HETs over the rows of tubes is shown in Fig. 9.6. After SG drying visual inspection revealed that approximately 10 of the lower rows are immersed in the sludge.

On August 7, 1996 a tube 60-96 of the second semicircle in the region of the hot collector was cut out from SG3 by the technology developed at the Balakovo nuclear power plant (PR-026)

The surface at a distance of 70, 90 and 130 mm from the collector was examined on the cut-out tube. At a distance of 70 mm there were deposits,103.6 g/m^2, black-brown, loose, easily removed by mechanical means. Under the deposits there was a grey oxide slick, strongly bonded

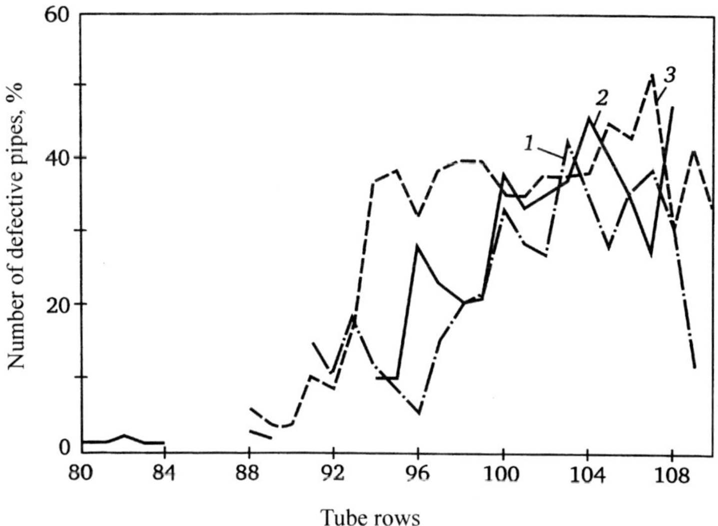

9.6   Distribution of defective tubes in rows: 1) SG1, 2) SG3, 3) SG4.

to the surface. The chemical composition of the deposits: FeO_3 – 91.35%; CuO – 8.6%. Pitting and corrosion traces were not found.

At a distance of 90 mm: 226.9 g/m² deposits, porous sediments, beneath a small amount of Cu deposits, distributed unevenly over the surface of the individual spots with a diameter up to 1.5 mm 8–10 deposits per 100 cm. Under the Cu sediment there were pits with a diameter up to 0.5 mm and 0.3 mm deep. The chemical composition of the sediments: FeO_3 – 82%; CuO – 17.1%.

At a distance of 130 mm: deposits of 632.3 g/m² were detected. Under the layer of unconsolidated sediments the whole surface was covered with Cu deposits, strongly bonded to the tube metal. Under the Cu deposits there were a lot of small pits of up to 1 mm in diameter and up to 0.5 mm deep. There was a penetrating pit with a diameter of 3 mm from the secondary circuit and with a diameter of 1 mm from the primary circuit. The composition of sediments: FeO_3 – 63.44%; CuO – 30.32%.

Studies were conducted on other tubes cut from the SG3 (60-96, 60-102). The following conclusions were drawn on the basis of the results:

1) the chemical composition of steel of the damaged HET 60-96 (SG3) corresponds to the steel 08Cr18Ni10Ti according to TU 14-3-197-89;

2) the mechanical properties satisfy the requirements of TU 14-3-197-89;

3) damage typical of ICC was not detected (inspected by the colour flow method);

4) non-metallic inclusions – within the range of TU 14-3-193-89;

5) the inner surface of the tube showed corrosion or other damage (except traces of the passage of the probe);

6) damage to the tube 60-96, SG3:

– no defects were found between the spacing grids 1 and 2;

– traces of pitting corrosion were found between the spacing grids 2 and 3 in the form of clusters of pits with a diameter of 0.3 to 2 mm. One of them is a through-wall pit; the diameter of the outer surface 3 mm, on the inned surface an ellipse 0.1 × 8 mm. Dead-end cracks propagated from both the surface and from pits, one through-wall crack from a pit;

– pits 0.2–0.3 mm deep were found below the second spacer grid;

7) damage to tube 60-102, SG3:

– below the spacer grid 3 – pits and cracks up to continuous, emerging from both the pits and on the surface;

– below the spacer grid 4 – corrosion pits and stress corrosion cracking on the outer surface. A crack developed parallel to the surface.

8) damage to tube 61-1, SG1:

– under the spacer grid 4 there were small pits with a diameter of 0.1 mm with a density of 2 pits at 1 mm, and also cracks developing from both the surface and from the pits. The maximum depth of cracks was 51% of the thickness of the tube wall;

9) all the tubes contained deposits, including those under the spacer grids. Sulphur (0.1–1%), copper, traces of chromium, chlorine, sodium and other elements were found in the cracks;

10) the crack trajectory is wave-shaped which indicates that it is determined mainly by the influence of corrosive environment and microheterogeneity of the steel;

11) in general, cracks are oriented along the axis of the tube which indicates that the tangential stresses exceed the axial stresses

Photographs of different types of defects are shown in Fig. 9.7–9.11.

Causes of damage in the HET of SG1, 3, 4 unit 2 of Balakovo NOO analyzed by retrospective network modeling.

At the first stage the main characteristics of HET damage of SG-1, 3, 4, unit II Balakovo NPP were determined, and significant information from other units was collected. One of the main criteria for the correct analysis of the causes of damage is that these (this) reason(s) should explain all the features of damage and the process of damage.

At the second stage assumptions about the causes (mechanisms) of damage were identified. All assumptions suggested at meetings of the working group and other meetings dedicated to discussion of SGs on unit 2 of Balakovo NPP (of 8 hypotheses) were taken into consideration.

At the third stage facts, data received as a result of reviews, calculations and experimental analysis as well as other circumstances that could prove or disprove a hypothesis were subjected to analysis.

As a result of tests carried out at stages 3 and 4, it was found that the root cause of damage such as pits and pits with cracks is the transfer of corrosion products from the condensate feed circuit and their accumulation in the form of deposits and sludge.

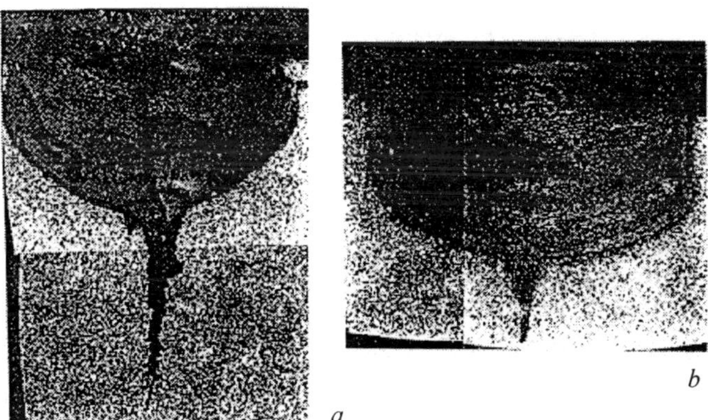

9.7   The pipe section (60-96) between spacing grids 2 and 3 (a) and the microstructure of a pit with a crack in the transverse microsection, x100 (b).

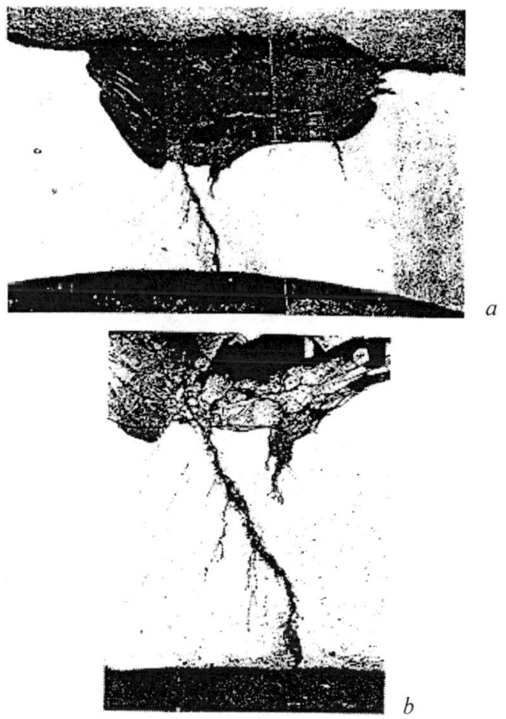

9.8   The microstructure of sediments and through defects on the outer surface of the tube, x100: a) section of a tube (60-102) below spacing grid 3, a general view, b) the zone with the crack.

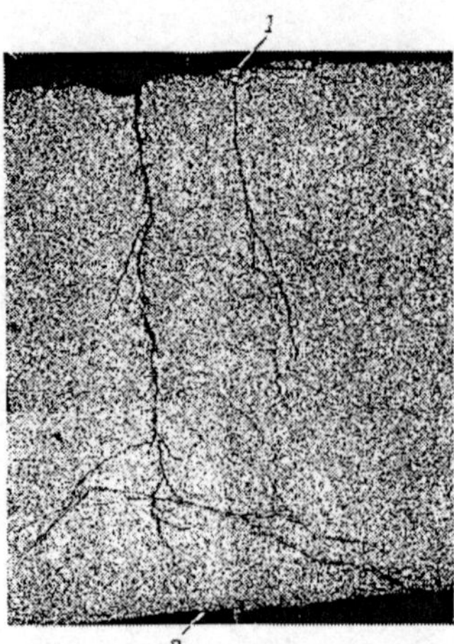

9.9   Cracking of a heat exchange tube (61-1, second semicircle) in the zone of contact with spacing grid 2, ×100: 1) outer surface of the heat exchange tube, 2) the inner surface.

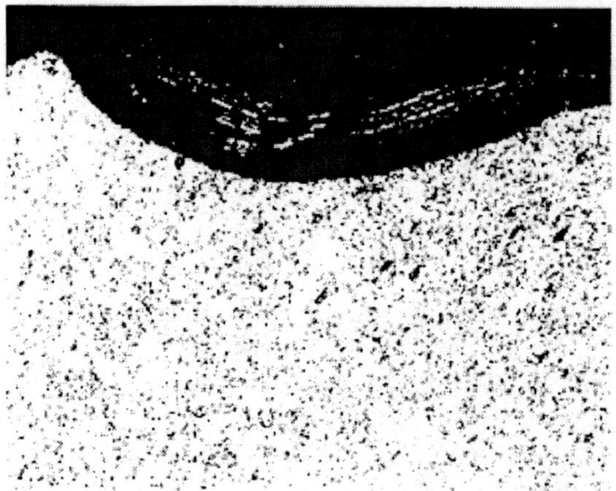

9.10   Corrosion defects on the outer surface of the tube section under spacing grid, ×100.

Accumulation of corrosion products from the condensate feed circuit in combination with products of suction, as well as short-term violations of water chemistry, were the cause of damage such as cracks.

9.2   Application of the Monte Carlo method to the problem of ensuring the integrity of HETs of VSG

This analysis was carried out in Ref. 29. Growth of stress corrosion cracks in steam generator tubes is estimated using statistical approaches and Monte Carlo simulation method. Statistical parameters that define the

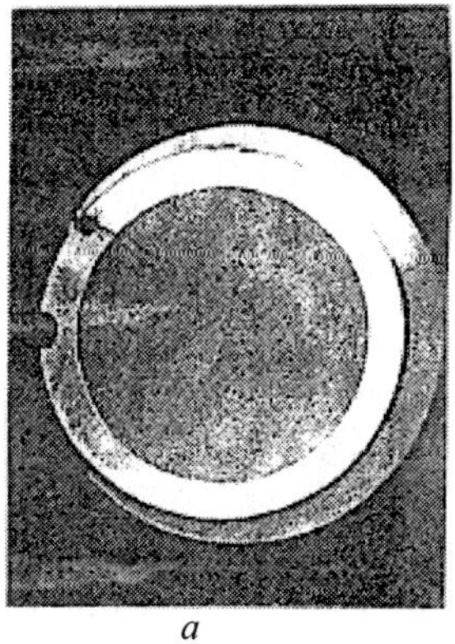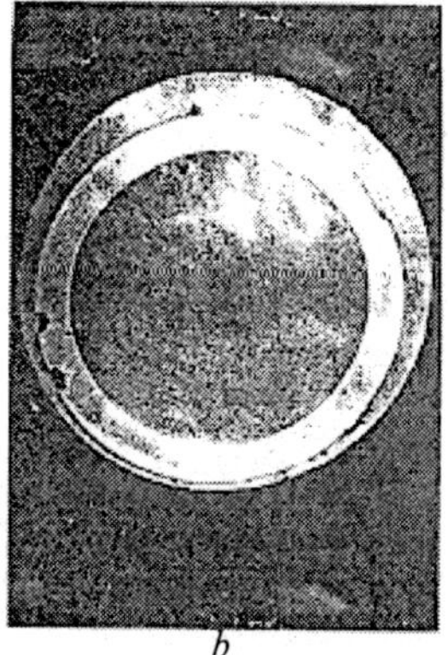

9.11   A pipe section (60-96 between spacing grids 2 and 3 (a)
and the form of corrosion defects on the microsection, ×5 (b).

characteristics of initiation and crack growth were obtained from the results
of regular periodic in-service inspection (ISI). A model of crack growth
and size distribution of cracks in the final cycle of operation (FCO), based
on statistical approaches, was proposed. Because the ISI data differ from
physical data (i.e. actual defects), a simple method of predicting the actual
number of cracks on the basis of the ISI results was proposed. The actual
number of cracks and crack growth can be easily determined on the basis of
simulations using the Monte Carlo method. At the same time, the probability
distribution of the number of cracks and their maximum size in FCO are
determined. Satisfactory results were obtained in comparing the forecast
with the FCO data.

Model for estimating crack growth

The two approaches used for statistical evaluations of structural
strength were combined in Ref. 29. The first approach is based on the
deterministic laws of fracture mechanics in which elements of statistics are
introduced[126]. This approach is useful for analysis of some local issues. But
its use is associated with difficulties when considering large structures due
to the large number of variables (load, mechanical properties, crack size,
etc.) and the complexity of their theoretical description.

Another approach is based on statistical distribution functions for the
most important variables[127]. Parameters of distribution functions can be
obtained from the analysis of real data for some of the most important
variables (crack size, time of initiation of cracks, crack growth rate,
etc.). This method cannot predict the impact of changes in non-destructive
testing systems because it does not address the physical mechanisms of ISI
on the test results.

Combination these two approaches allows us to develop a modified
methodology that uses statistical distribution functions of the crack growth

rate, the time to crack nucleation and the maximum size of the crack with a simple equation for the crack growth rate.

To calculate the probability of failure or estimate the size of a leak one must know the number of actual defects and the distribution of their sizes. However, these data differ from regular ISI data for the defects in steam generator tubes which results in insufficient reliability of inspection, detection of defects and their sizes. A simple method, which includes the use of the Monte Carlo method, was developed for estimating the actual number of cracks on the basis of the ISI results.

Figure 9.12 illustrates the procedure used in the proposed statistical model to estimate crack growth. The model is used to estimate the number of cracks and their size distribution at the end of the i-th cycle of operation using data from the i-th cycle of ISI. Statistical features of crack nucleation and crack growth rate are represented by statistical distribution functions whose parameters are derived from statistical analysis of ISI.

The number of actual cracks in steam generator tubes is estimated using the results of ISI and data on the reliability of quality control, more precisely, the probability of detecting a defect (PDD). Since the water chemistry regime (WCR) may vary during operation, WCR assessment and determination of the actual number of cracks were performed for each cycle of operation. The actual number of cracks was determined and their numbers at the end of this cycle predicted at the beginning of the i-th cycle of operation. The forecast was then compared with the results of ISI carried out at the end of the i-th cycle.

Rate of crack growth

The crack growth rate, used in the simulation, was calculated statistically according to the ISI data. Figure 9.13 shows data on the rate of crack growth in the pipes of the steam generator, model F (FPG) in Korea. These steam generators were in operation for 12.38 effective operating years (EOY 12.38 during which 12 full-scale inspections were carried out. The crack growth rate for the i-th inspection was obtained by dividing the increment of crack length during the time between the i-th and $(i +1)$-th inspections. The units of increment of crack length were in millimetres, time in EOY, respectively. Many point in Fig. 9.13 show a negative crack growth rate which is physically impossible. The negative growth of cracks is due to unreliable ISI.

Figure 9.14 shows the dimensions of cracks measured in each ISI. It may be noted that the crack length does not increase in some case. To obtain non-negative and more precise estimates of the crack growth rate it was necessary to carry out regression analysis of the crack length for each cycle of ISI using polynomial approximation. Some regression analysis results are given in Fig. 9.15. The linear regression equation is used when the number of points is less than 4. However, when the number of points is

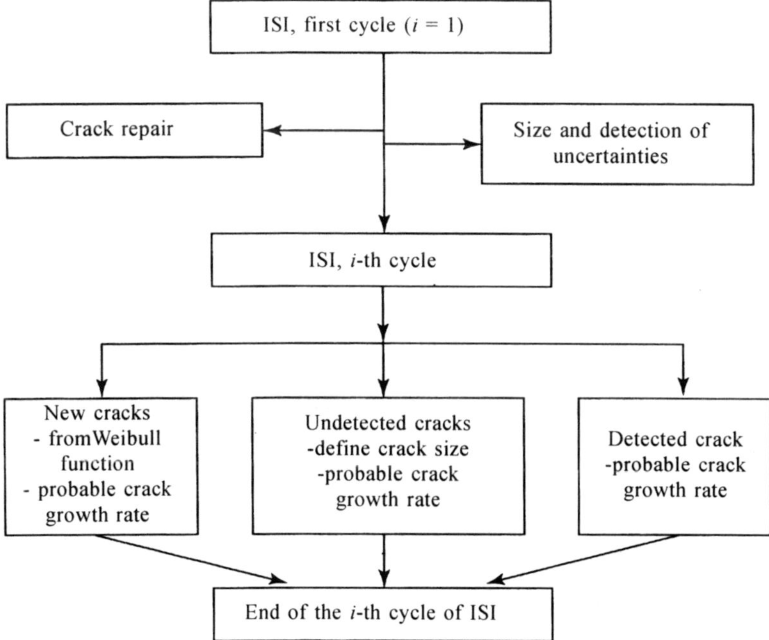

9.12 Statistical analysis of the model of crack growth.

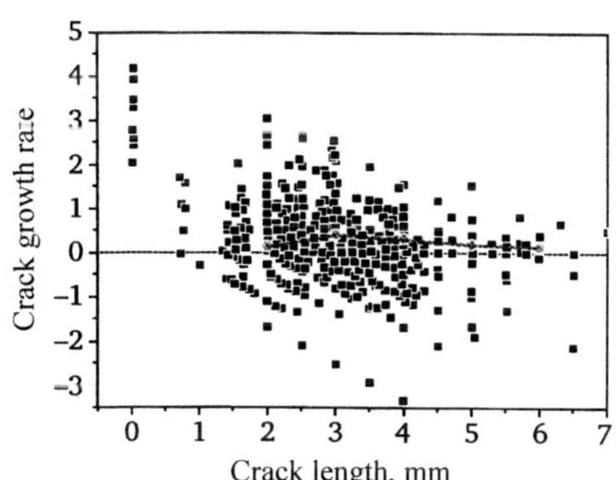

9.13 ISI data for crack growth rate (for steam generators model F).

equal to or greater than 4, the second-degree polynomial must be used. The crack growth rate is determined by the slope of the regression curve to the time axis.

Figure 9.16 shows the crack growth rate obtained from the results of regression analysis. All data show positive crack growth. If there is a relationship between crack length and its growth rate, the crack growth rate can be represented as a function of crack size. To find the relationship

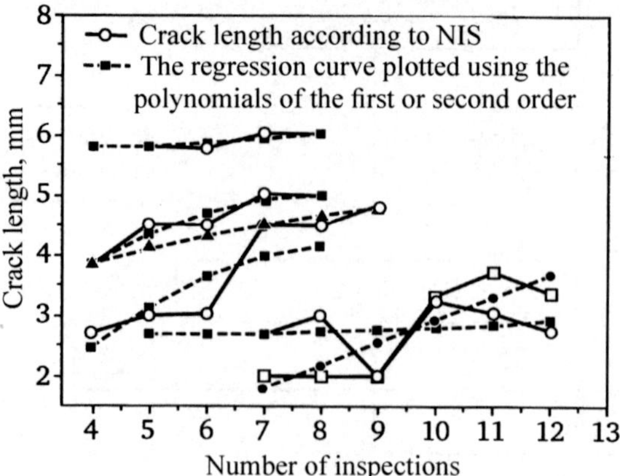

9.14   Description of data on crack length by polynomials.

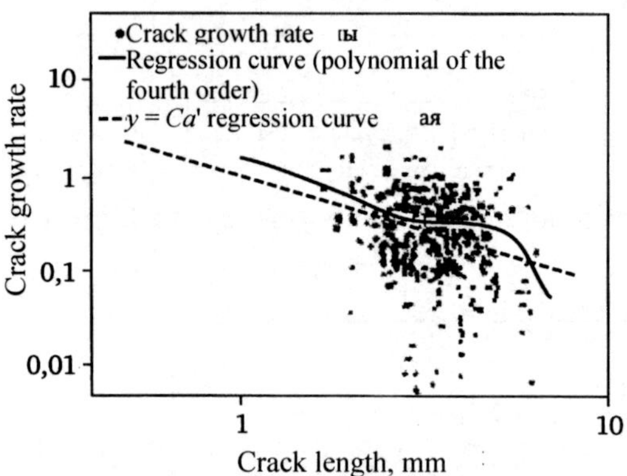

9.15   Regression analysis of the crack growth rate.

between the length of the crack and its growth rate, using a fourth degree polynomial. The results show that the growth rate decreases with increasing crack size. This pattern is consistent with earlier data[128].

As expected, the crack growth rate satisfies the following equation [127]:

$$\frac{da}{dt} = f(a)z, \qquad [9.1]$$

where a is the crack length, t is time, $f(a)$ if the function of crack length a, and z is a statistical random variable, taking into account the random error.

One of the simplest forms of equations [9.2] is as follows:

$$\frac{da}{dt} = Ca^{\lambda}z, \qquad [9.2]$$

where C and λ are constants. Logarithmic transformation of both sides of the equation [9.2] leads to the following:

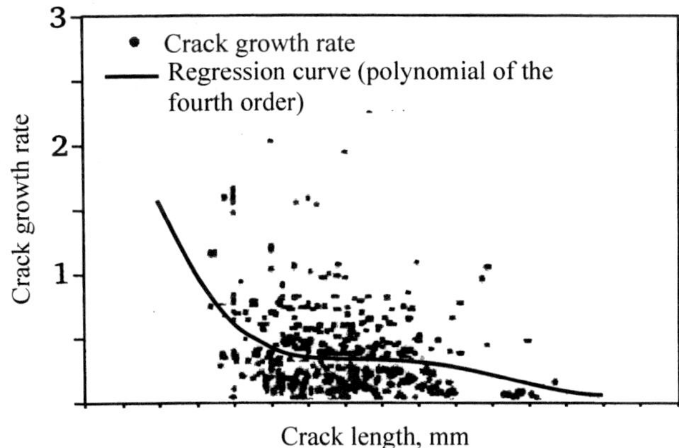

9.16   Crack growth rate data and the polynomial curve.

$$\log\left(\frac{da}{dt}\right) = \log C + \lambda \log a + \log z.$$ [9.3]

The constants C and λ can be obtained from the linear relationship between the variables $\log a$ and $\log (da/dt)$. Figure 9.15 shows the regression analysis results. The logarithmic scale is used for both horizontal and vertical axes. The graph shows a linear dependence and, for comparison, the fourth-degree polynomial. It may be noted that the difference between the straight line and the polynomial curve is small.

The value of $\log z$ was evaluated by statistical analysis of deviations from the mean *RSME*. Calculations were carried using the following formula[130]:

$$RMSE = \sqrt{\frac{\sum (Y_i - \hat{Y})^2}{n-2}},$$ [9.4]

where Y_i is the crack growth rate determined in the observations; $\hat{Y}$ is the rate calculated by the regression equation; n is the number of data (experimental points).

Distribution of $\log z$ can be described by a normal distribution which can be used to obtain the crack growth rate in the statistical representation [9.3].

The probability of detecting cracks and the number of actual fractures

The means and methods of nondestructive testing have been developed for detecting small defects. In reality, both small and large defects can remain undetected during inspection. This is due to insufficient reliability of the existing inspection methods.

Unreliability of ISI can be divided into unreliable detection of the defect size and insufficiently accurate detection of defects. Unreliability of estimates of the defect size can be accounted by statistical analysis of the measurement error. Insufficient detectability of the defect is characterised by the probability of defect detection (PDD) which is a function of defect size.

Currently, the PDD can be determined, according to the authors, only by experiments on special samples with artificial defects whose dimensions are known beforehand. Statistical analysis of nondestructive testing of this model produces a PDD estimate.

In Ref. 131 it is porposed to describe the PDD function by the functions shown graphically in Fig. 9.17

$$P_{d.d}(a) = \frac{\exp(\beta_0 + \beta_1 \ln a)}{1 - \exp(\beta_0 + \beta_1 \ln a)}; \qquad\qquad [9.5]$$

$$P_{d.d}(a) = \left\{ 1 + \exp - \left[\frac{\pi}{\sqrt{3}} \left(\frac{\ln a - \mu}{\sigma} \right) \right] \right\}^{-1}. \qquad\qquad [9.6]$$

In [9.5] and [9.6], values β_0, β_1, μ and σ are constant and can be determined by statistical analysis. Interdependence of the constants has the following form:

$$\mu = -\beta_0 / \beta_1; \qquad\qquad [9.7]$$

$$\sigma = \pi / \left(\beta_1 / \sqrt{3} \right). \qquad\qquad [9.8]$$

It is easy to estimate the physical dimensions of the cracks using the ISI data if the detectability of inspection is known. In single inspection it is also easy to estimate the number of actual cracks using the curve of defect detection probability. Nevertheless, some problems may be encountered in this case. For example, cracks can be detected only after they reach a certain minimum size. However, some cracks are detected immediately after reaching this size, while others are detected only after a few inspections. To count the number of actual cracks, it is necessary to know the history of inspection of each crack. This requires a special technique. It should be

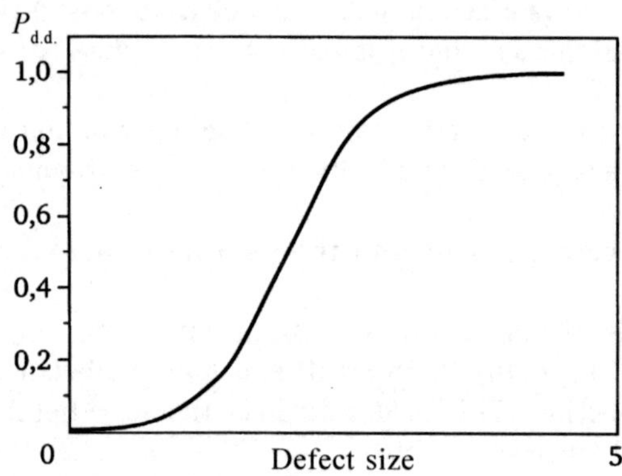

9.17   Generalised form of the PDD curve.

noted that in Ref. 132 the number of actual cracks was determined without consideration of repeated inspections using only function $P_{d.d}$.

The actual number of cracks is defined as the aggregate number of cracks found in ISI and the number of cracks missed due to imperfect ISI. The method of calculating the amount of actual cracks is based on determining the probability of detecting defects:

$$P_{d.d}(a) = \frac{\text{Number of detected defects } N_{det}}{\text{Actual number of defects } N_{act}};$$

[9.9]

$$N_{act} = \frac{N_{det}}{P_{d.d}(a)}.$$

[9.10]

The number of actual cracks is calculated using equation [9.10][132]. According to the authors of Ref. 132, this equation cannot be used for periodic ISI inspection because it does not account for multiple inspections.

To count the number of actual cracks using the data from periodic ISI, it is important to consider both the probability of detection of defects and of non-detection of defects. The probability of detecting a crack in n inspections is calculated by the probability of non-detection of cracks..

The probability of detecting a defect x times in n inspections is calculated using the binomial function:

$$P_{d.d}(x) = \binom{n}{x} p^x q^{n-x} = \frac{n!}{x!(n-x)} p^x q^{n-x},$$

[9.11]

Here, p is the probability of detection in a single inspection, and q $(= 1 - p)$ is the probability of missing the defect.

The probability of detecting a crack at least once in n inspections is:

$$P_{d.d}(x-1) = 1 - (1-p)^n.$$

[9.12]

To apply equation [9.12], we must know the exact history of crack growth and the number of inspections for each crack. However, these parameters are unknown.

Since each crack may be subject to inspection several times, the actual probability of detection is different from $P_{d.d}$. Let the actual probability of detection be the effective probability. The number of actual cracks is then expressed as:

$$N_{act} = \frac{N_{det}}{P_{d.d.ef}(a)}.$$

[9.13]

To know the history of crack growth and the number of inspections for each crack, we use the statistical modelling approach.

Determination of the effective probability of detection

Figure 9.18 shows the statistical modelling procedure. Let a crack 2.7 mm long be detected for the first time at the i-th ISI. The history of growth of this crack can then be obtained by inverse modelling of crack growth. Since PDD is a function of crack size, the value of PDD can be calculated for each ISI. From these values we can calculate the actual PDD, i.e. effective PDD. Repeating the simulation many times, the average value of the effective PDD can be determined.

For a particular inspection method the effective PDD is regarded as a function of crack size for each ISI. The number of actual cracks of a given length or length range can be obtained from equation [9.13].

The curve effective PDD for each ISI is plotted using the following procedure in which crack growth is simulated by the Monte Carlo method. The curve of effective PDD in the k-th ISI is obtained as follows:

1) a crack with length a_i in the k-th ISI is considered;

2) reversed crack growth back to the size of the minimum detectable crack length is simulated. A random variable Z is renegerated during simulation and, using equation [9.1], the crack growth rate is calculated;

3) the size of the crack and the corresponding values of PDD for each of the previous cycles of ISI are determined;

4) the resultant PDD values are used to calculate the effective value;

5) many estimates of the effective PDD are produced, repeating steps (1) to (4), and the average effective PDD is calculated.

6) steps (1) to (5) are repeated for increasing length $a_i + \Delta a$ to the k-th ISI and the average value of effective PDD are determined for different crack sizes.

PDD values used in the simulation are presented in Table 9.1. Modelling was carried out using discrete PDD values, as it is more convenient to calculate the number of actual cracks.

Discrete and continuous PDD curves, used in the simulation in each ISI, are shown in Fig. 9.19. Since the non-destructive testing system was changed after the 9th ISI, the PDD curve also changed after the 9th ISI.

The effective PDD was used to estimate the actual number of defects with the inspection results taken into account. Figure 9.20 shows the

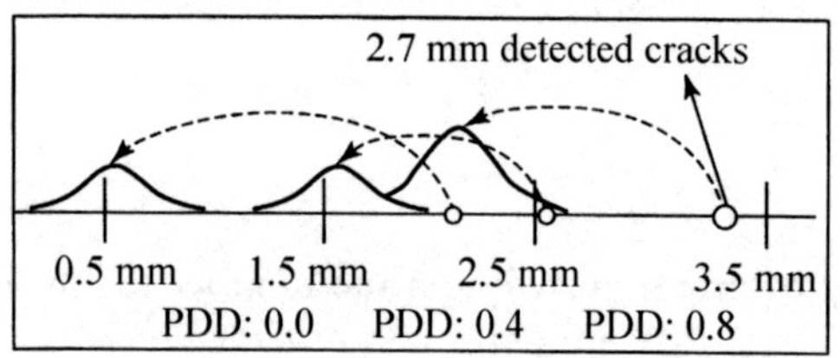

9.18   Scheme of statistical modeling to assess the history of the crack growth.

Table 9.1 PDD values for different crack length ranges and different ISI

Number of ISI	Probability of detection			
	0.5–1.5	1.5–2.5	2.5–3.5	more than 3.5
4–9	0	0.4	0.8	1
1012	0.4	0.8	1	1

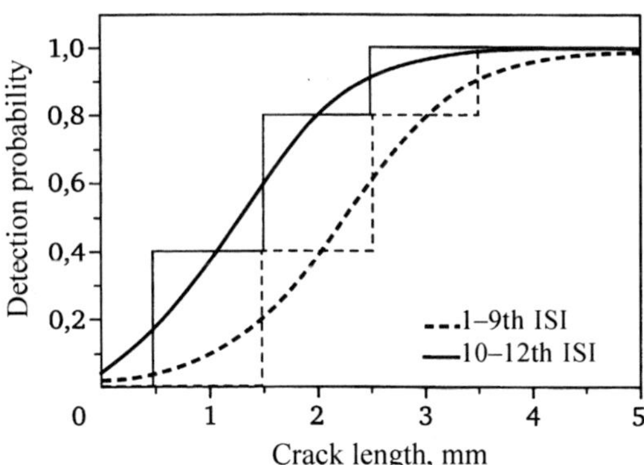

9.19   Continuous and discrete curves of probability of detection.

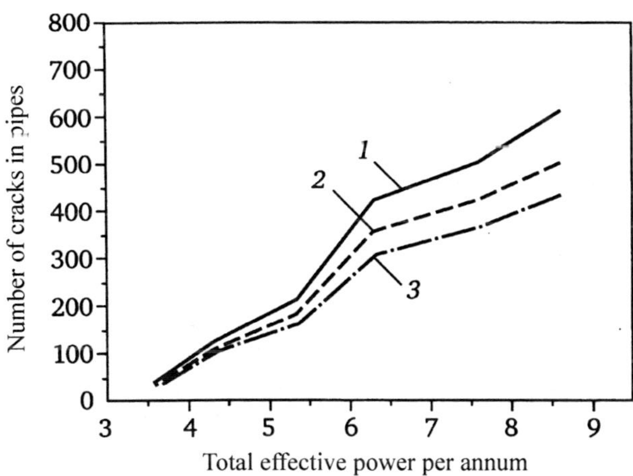

9.20   Number of cracks detected by ISI and effective ISI: 1) detection probability; 2) effective detection probability; 3) ISI data.

comparison of the actual number of defects obtained by using equations [9.10] or [9.13].

The number of actual defects, determined using PDD, is greater than those obtained using the effective PDD. For a conservative estimate we can the PDD instead of the effective PDD.

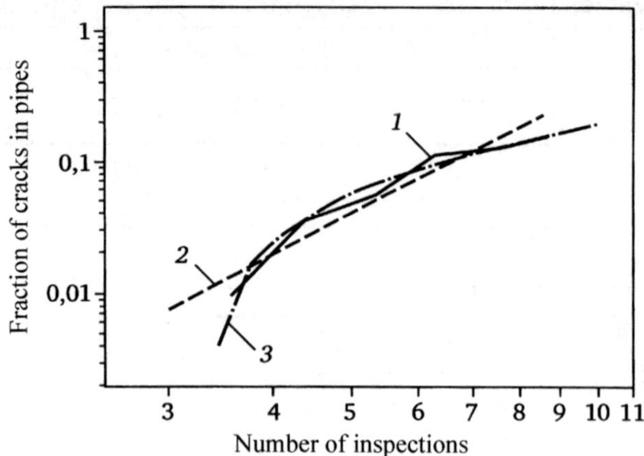

9.21   Formation of new cracks on Weibull curves with two and three parameters: 1) fraction of cracks in tubes (physical); 2) Weibull curve with two parameters; 3) Weibull curve with three parameters.

The exact number of actual defects can be determined only from by destructive experiments and it not possible to say which method is more accurate. Given that the simulation using the effective PDD value is more logical, it should be assumed the number of actual defects identified by this method is closer to the real values.

Crack formation

Statistical analysis of ISI data allows to estimate the number of cracks formed during the 1st cycle of operation. The Weibull distribution function is used in this case. Figure 9.21 shows the cumulative proportion of cracks occurred in each ISI and their Weibull functions with two or three parameters. The combined proportion of the cracks is calculated in the actual field, i.e., the number of actual cracks is calculated. Note that, as shown in Fig. 9.21, the Weibull curve with three parameters agrees better with the experimental points than the curve with two parameters. Thus, to estimate the number of cracks it is necessary to use the Weibull function with three parameters.

The results of simulation

Figure 9.22 shows the results of the forecast of the number and size of the cracks obtained by statistical modelling using the Monte Carlo method. The ISI data for model F steam generators from the 8th ISI is used to predict the number of cracks formed and the maximum size of cracks at the 9th ISI. Simulation is repeated 1000 times and the results are statistically analyzed. The results were obtained as probability distributions.

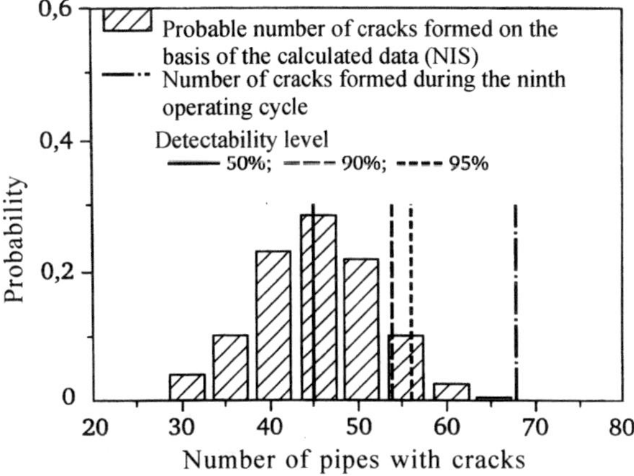

9.22 Modelling of the probability distribution of the number of new cracks detected in the 9th ISI.

Table 9.2 Estimated number of newly formed actual cracks in the 9th operating cycle for different levels of confidence intervals

Detectability level, %	Number of new physical cracks
50%	95
90%	109
95%	112
Data for 9th ISI	1988

The projected number of newly formed cracks in the 9th operating cycle are shown in Table 9.2 with a confidence level of 50%, 90% and 95%. Forecasting is performed using the three-parameter Weibull function, as described above. The value corresponding to the 50% confidence level, shows 88 cracks, which is similar to the number of actual cracks, calculated from the results of the 9th ISI.

It is also necessary to predict the number of newly detected cracks as well as the number of actual cracks. Figure 9.22 shows the probability distribution of the number of newly detected cracks in the 9th ISI. The estimates, corresponding to 50%, 90% and 95% confidence level, are also shown in Fig. 9.22 and in Table 9.3. Comparison of 68 cracks, obtained during the 9th ISI, with the simulation results shows that the simulation gives a slightly smaller number of cracks. The reason for this may be that the values of PDD are higher than the true values.

The discrepancy between the predicted number of cracks and the ISI data is associated with shorter cracks. For long cracks the predicted number of detected cracks is almost always the same as the number of cracks obtained

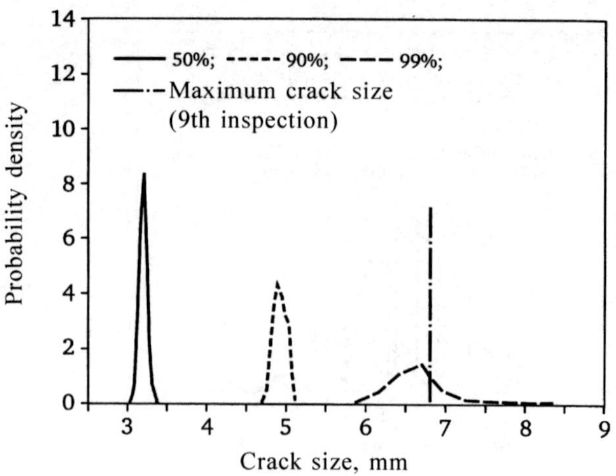

9.23   50, 90 and 99% crack depth.

Table 9.3 Estimated number of newly detected cracks in the 9th ISI for di0fferent confidence levels

Confidence level	Number of detected cracks
50%	45
90%	54
95%	56
9th ISI	1968

Table 9.4 Confidence level for cracks with the size of 50, 90 and 95% of wall thickness

Confidence level, %	Crack size, mm			
	50%	90%	95%	99%
50%	3.2	4.9	5.4	6.6
90%	3.3	5	5.5	7
99%	3.3	5.1	5.6	7.6
9th inspection cycle	6.8	6.8	6.8	6.8

by the ISI because for these cracks PDD and effective PDD have values close to 1.

Figure 9.23 illustrates the distribution of the probability density of crack sizes corresponding to 50%, 90% and 95% for the 9th ISI. Distributions were obtained on 1000 data. The distribution of the maximum size of the crack can also be obtained but in this case the maximum crack size can be unusually large. This is due to the probabilistic nature of the crack growth rate and large errors z in equation [9.3]. Thus, a more accurate value for the maximum crack size can be obtained if the size distribution of cracks shown in Table 9.4 is used. The maximum size of the crack measured in the 9th ISI is 6.8 mm.

Thus, the growth of cracks due to corrosion in steam generator tubes was estimated using Monte Carlo methods and statistical approaches. The input data for evaluating the statistical characteristics of crack growth and initiation of cracks were the ISI data. The proposed analysis method was used to predict crack propagation at the end of the life cycle. The efficiency of the method was tested by comparing the predicted data with the known data.

9.3 Application of a probabilistic method based on two-parameter distribution

Ths section considered the construction of a probabilistic model of failure in the analysis of efficiency of heat exchanger tubes[56].

Since 1990 the Russian nuclear power plants have been using the eddy current method for testing the quality of metal. Application of this type of inspection is usually accompanied by the detection of a sufficiently large number of defects. This led to the need to justify the performance of tubes of steam generators and prepare recommendations for preventive plugging of these tubes in the presence of defects. The eddy current method allows to detect a large number of defects of different length and depth and also the scatter of the values of the mechanical properties of materials of heat exchanger tubes in the certificate data. This indicates the need to use it analysing the performance.

Summary results of the probabilistic assessment of the permissible and critical defect sizes according to various failure criteria

The probability of formation of through-wall defects and large-scale failure is determined using the approach described in section 4.2, based on the consideration of the three failure mechanisms: the onset of stress corrosion cracking, establishment of the limiting plastic state (ductile fracture); corrosion fatigue growth of defects.

If the failure probability of tubes depends on their locations, then the failure probability for each group of tubes during time interval t for i heat exchanger tubes inthe given set n is determined by the formula:

$$P_n^i(t) = C_n^i (P(t))^i \left[1 - P(t)\right]^{n-i},$$ [9.14]

where $P(t)$ is the probability of failure of one of the tubes in time interval t which is regarded as the same for a group of tubes; C_n^i is the binomial coefficient, defined as

$$C_n^i = \frac{n!}{i!(n-i)!}.$$ [9.15]

The cumulative probability of leakage or large-scale failure in n heat exchanger tubes is

$$P_n = \sum_{i=1}^{n} P_n^i. \qquad [9.16]$$

The criterion of reliability of heat exhanger tubes is selected in accordance with the principles[132] described in section 4.2.

On the basis of the first principle, the PSA data for the given energy unit and the aggregated IAEA data[134] are used to determine the allowable probability of 'fracture of the heat exchanger tube'. As part of the PSA, IAEA publishes periodically summarised data on the probabilities of initiating events compiled on the basis of the analysis of the reliability and operating experience with various types of nuclear power plant in different countries, including Britain, Germany, Russia, France, USA and Japan. Summarised data[134] from the IAEA include the probability of a steam generator tube rupture. For nuclear power plants this probability of the investigated initial event ensures the reactor core safety, i.e., in accordance with domestic and foreign safety requirements the probability of core damage is less than 10^{-5} for reactor per year, i.e., the probability of 'steam generator tube rupture' which characterises the reliability of the steam generator corresponds to the (acceptable) level of reliability at which operation of the reactor core and nuclear power plant as a whole is safe.

The second principle of design and specification documentation is used to define the acceptable level of probability of formation of through-wall cracks and leaks not less than 5 kg/h based on the need to fulfil thermal engineering function for steam generators (the principle of safe failure).

The third principle is based on the availability factor of the steam generator. In general, the principle of minimising the number of outages of a power unit during operation is used. In this case, the availability factor of equipment in accordance with GOST 24722-81 should be at least 0.972[135].

If the resulting level of reliability parameters for heat exchanger pipes does not meet the reliability criterion, the criterion can be satisfied, for example, by establishing requirements on flaw inspection in the part of the allowable defect sizes. The development of recommendations on the admissibility of detected defects is reduced to determining the maximum admissible sizes for these defects.

The developed approach to performance analysis was used to prepare recommendations for preventive plugging of the heat exchanger tubes of steam generators of power units No. 3 and 4 of the Novovoronezh NPP power units No. 1 and No. 2 of Kola NPP.

For example, the results of eddy current testing, presented by the Kola nuclear power plant, were used for preliminary analysis and determination of the size distribution of the detected defects (Table 9.5). The number of inspected pipes, the number of defects found in the depth of more than 20% of wall thickness and defect density distribution of in the steam generator are presented in Table 9.6.

Table 9.5 Parameters of defect distribution

Number of steam generators	Number of defects			Number of inspected tubes	Density of defects per tube
	MM	CE	Total		
1	50	–	50	1608	0.0311
2	71	13	84	2838	0.0295
3	43	87	130	5518	0.0235
4	30	13	43	819	0.0525
5	30	12	42	5511	0.0076
6	5	2	7	883	0.0079

Comment. MM – 'lack of metal' defect'; CE – 'corrosion below the grid'-type defect

Statistical analysis of the defect size was performed for the variants of set restriction of the maximum depth of the defect – 100%, 80%, 60%, 40% in relation to wall thickness of the heat exchanger tubes.

Table 9.7 presents the results of calculations of the mathematical expectations and variances (standard deviations) with the uncertainty of the initial data on the statistical characteristics of defects, such as 'lack of metal' (MM < 100%) for the elements of tubes of the steam generator of power unit No. 1 of the Kola nuclear power plant, taking into account the uncertainty of initial data.

Statistical analysis of defect sizes showed that the Weibull distribution is the most suitable theoretical law for describing the depth distribution of defects.

The law of distribution of the defect length was used to analyse the propagation of defects in the metal of heat exchanger tubes based on the experiments carried out at the Novovoronezh NPP. In this case, the length of defects was described by a normal distribution with the mathematical expectation of 13.5 mm and a standard deviation of 4.55 mm.

Performance analysis of heat exchanger tubes assumes that the rate of nucleation of new defects in future will be the same as in 2002. This premise requires mandatory testing, either through application of the model of nucleation of defects described above, or by obtaining and processing new data resulting from eddy current testing of tubes.

The following indices and parameters of the Paris equation were used for the analysis of corrosion-fatigue defect growth: the exponent for steel 08Cr18Ni10Ti m_0 = 3.53; experimental factor for a given material in fatigue testing in air C_0 was assumed to be a random variable obeying the log-normal distribution law with parameters q = –26.86 and w = 0.6215; coefficient A_0 takes into account the influence of corrosive environment on the fatigue crack growth rate. The value of this coefficient for the contact zones of the heat exchanger tubes with spacing elements is accepted as equal to 20 (higher concentration), and for the remaining sections of the tube as 10.

Table 9.6 Distribution of defects in the heat exchanger tubes of steam generator unit ⊠ 1 at the Kola nuclear power plant

Event	Criterion	Calculated estimate	Permissible and critical depth and length of defects*, mm					
			4 years			30 years		
			Sv-10GN1MA		UONI-13/55	Sv-10GN1MA		13/55
			According to criterion	Acerage		According to criterion	Average	
Corrosion cracking		–	3.2 (16)	3.2 (16)	5.5 (27.5)	3.2 (16)	3.2 (16)	5.5 (27.5)
Formation of through-wall cracks	Corrosion-fatigue crack growth**	First	15 (75)	–	–	7 (35)	–	–
		Second	12 (60)	10 (50)	10 (50)	less than 3 (less than 15)	2.5 (12.5)	2.5 (12.5)
		Third	2.5 (12.5)	—	—	less than 3 (less than 15)	—	—
Large-scale failure	Critical crack opening***	—	19 (95)	19 (95)	21.5 (107.5)	14 (70)	14 (70)	17 (85)

*The results for each criterion include: the permissible or critical depth of the defects, and the values in the brackets are the permissible or critical crack length with defect density taken into account.
**The permissible defect sizes were determined on the basis of corrosion-fatigue growth of defects with corrosion cracking taken into account.
***Critical defect sizes were determined on the basis of the criterion of critical crack opening with corrosion cracking and corrosion-fatigue growth of defects taken into account.

Table 9.7 Results of the statistical analysis of 'lack of material' for heat exchanger tubes of steam generator unit No. 1 Kola NPP

Parameter	Straight section of pipeline					Sealing area
	SF1	SG2	SG3	SG4	SG5, 6	
$\tilde{q}$	0.949	0.875	0.968	0.968	0.940	0.987
$\tilde{w}$	0.211	0.226	0.247	0.232	0.227	0.227
$\tilde{D} = w^2$	50	80	119	43	43	17
$D(x)$	0.054	0.06	0.069	0.067	0.064	0.073
$w = \sqrt{D(x)}$	0.233	0.245	0.264	0.258	0.253	0.270

The values of critical stress intensity factor K_{1SCC} in a corrosive medium for the analysis of stress corrosion cracking were estimated using the data on crack resistance of austenitic steels under the effect of corrosion[136]. Statistical analysis was performed using the normal law of distribution of K_{1SCC} with the mathematical expectation of 12,00 MPa·m$^{0.5}$ and a standard deviation of 3.03 MPa·m$^{0.5}$ in conditions of a corrosive boilin medium in the secondary circuit.

The criterion which provides the necessary level of reliability of heat exchanger tubes was selected as follows. On the basis of the first principle of reliability according to the PSA data for the Kola NOO the frequency of the initiating event 'rupture of heat exchanger tubes' was adopted on average equal to $2 \cdot 10^{-3}$ for reactor per year. This does not pose any significant danger to the reactor core. Given that unit No. 1 of the Kola NPP consists of six steam generators, the permissible level of probability of large scale destruction of at least one heat exchanger tube is assumed to be $3.3 \cdot 10^{-4}$ per steam generator per year. This value agrees well with the IAEA data[134]. Then the condition that ensures the implementation of the first principle of reliability of heat exchanger tubes for safe operation of the reactor core, can be written as

$$\sum_{i=1}^{5136} P_{crn}^{i} < 3.3 \cdot 10^{-4}.$$

[9.17]

On the basis of the second principle of reliability it is assumed that in case of leaks up to 10% of the total number of HET can be plugged so that the reactor can reach 100% of its capacity. At the same time, for each of the six steam generators tested it is necessary to determine the individual reliability criterion due to different levels of damage and, consequently, the difference between the numbers of heat exchanger tubes plugged in the 30 years of operation. According to these data, the average number of failed tubes to be plugged per year should not exceed the value determined from the condition of available reserves of the heat transfer surface and of extend the life of the reactor for a further 15 years. The conservative variant in which the number of plugged tubes is 5% was also considered.

Enforcement of the second principle of reliability of heat exchanger tubes of the steam genarator is restricted to the fact that the number of heat exchanger tubes which can be plugged for every steam generator annually cannot exceed the permissible number. Consequently, the condition that ensures the implementation of the second principle of reliability of steam generator tubes can be written as a sum of the probabilities of a leak in the heat exchanger tubes, exceeding the permissible number:

$$\sum_{i=k_0}^{5136} P_{Tn}^i < 10^{-7},$$
[9.18]

where n is the number of working heat exchanger tubes, $k0$ is the number of heat exchanger tubes with unacceptable leaks.

The probability level of 10^{-7} per steam generator per year, corresponding to the probability of 10^{-5} per reactor for the operating life of 15 years is conservative, 'rigid' and ensures SG operation at 100%. Equation [9.18] gives the condition which the probability of a leak in heat exchanger tubes must satisfy and which should be used to determine the admissible probability values for the leakage of each steam generator (Table 9.8).

Consideration of the third principle of reliability comes down to determining the conditions under which the availability factor of equipment is equal to 0.972^{135} at the recovery time of the steam generator in case of leaks in heat exchanger tubes over a period of 10 days. Then, the permissible probability of formation of a leak is $4 \cdot 10^{-4}$ per year for a normal operating period resulting in the availability factor of 0.972.

In conducting performance tests of heat exchanger tubes of the steam generator on unit No. 1 of the Kola NPP, approximately 500 calculation variants were produced and processed on the basis of the conservative and realistic approaches. This provided a large amount of information. The main dependence of the probabilities of occurrence of leaks and large-scale destruction on the maximum defect depth was determined for every steam generator and then used to fulfil three principles of reliability

Table 9.8 Permitted levels of failure probability of heat exchanger tubes for steam generator on unit 1 of the Kola NPP

Parameter	Steam generator number					
	1	2	3	4	5	6
Failure probability for 10% margin	$2.31 \cdot 10^{-3}$	$2.30 \cdot 10^{-3}$	$2.13 \cdot 10^{-3}$	$2.4 \cdot 10^{-3}$	$2.34 \cdot 10^{-3}$	$2.36 \cdot 10^{-3}$
Failure probability for 5% margin	$6.18 \cdot 10^{-4}$	$6.13 \cdot 10^{-4}$	$4.89 \cdot 10^{-4}$	$6.86 \cdot 10^{-4}$	$6.38 \cdot 10^{-4}$	$6.52 \cdot 10^{-4}$

and determine the maximum allowable depth of the defect for efficient performance of heat exchanger tbes.

When calculating the probability of formation of open cracks and large-scale destruction of six steam generators on the unit No. 1 of the Kola NPP special attention was given to the inner (shortest) tube of the heat transfer surface as the most loaded. The results of calculations were then conservatively applied to all the heat exchanger pipes of a steam generator. The heat exchanger tube was represented by a structural model which includes: straight sections of tubes, the contact zone with spacing elements, zone of bends, the sealing area of the tube ends into the primary collector. The probability of failure of tubes is generally defined as the sum of probabilities in respect of all criteria (mechanism of destruction), including subcritical corrosion-fatigue crack growth and all structural elements. In this case, it is assumed that heat exchanger tubes are subjected to 100% inspection once in every four years and the dynamics of defects corresponds to linear extrapolation of the data available at the given moment.

The results of comparison of the calculated values (Fig. 9,24, 9.25) with the permissible levels of probability of failure were used to determine the maximum allowable depth of defects ensuring the implementation of each of the three principles of reliability using the conservative and realistic approach, and the reliability criterion in general. Table 9.9 shows the results of calculations of the permissible defect depth with regards to inaccuracy of determining the defect size (equal to 12%) for the heat exchanger surface of steam generator No. 4.

The results of analysis of heat exchanger tubes of the steam generators of the units 3 and 4 of the Novovoronezh NPP and units 1 and 22 of the Kola NPP are presented in Table 9.10.

Table 9.9 Allowable maximum defect depth for the heat exchanger tubes of steam generator No. 4

Reliability principle	Approach to determination of the maximum permissible depth	Maximum depth of defect, % (taking into account 12% inaccuracy)		
		Acceptable for approach?	Average	Minimum
Safe operation of reactor core	Realistic	70	69	
	Conservative	68		
Fulfilment of thermal engineering functions	Realistic	69	67	73
	Conservative	66		
Maintaining constant availability factor of equipment	Realistic	66	65	
	Conservative	64		

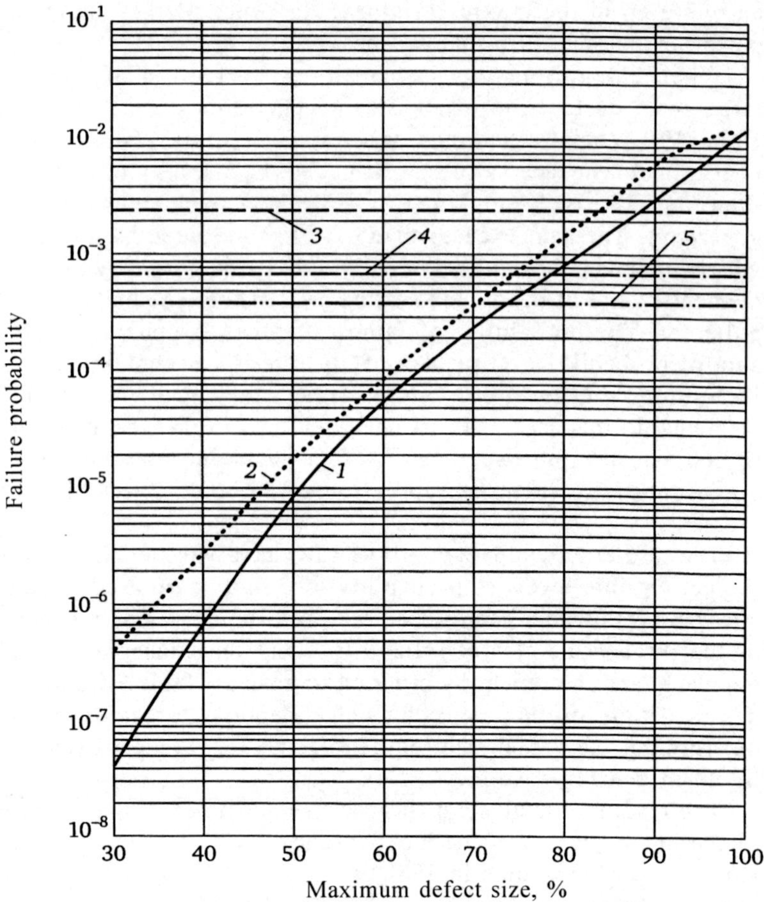

9.24   Probability of failure of heat exchanger tubes, depending on the maximum size of the defect for steam generator No. 4: 1) realistic estimate; 2) conservative estimate; 3) permissible level for 10% margin; 4) permissible level for 5% margin; 5) permissible level in respect of the availability factor.

Thus, based on the reliability test the permissible depth of the defects of tubes is defined. The presented approach allows us to estimate the volume and frequency of eddy current testing to ensure the necessary reliability of tubes SG heat exchanger tubes. However, it should be noted that to assess the feasibility of eddy current testing and preventive plugging of tubes, the effectiveness of this kind of metal inspection should be studied, as well as the influence of the medium of the second circuit on the origin and development of defects in metal heat exchanger tubes.

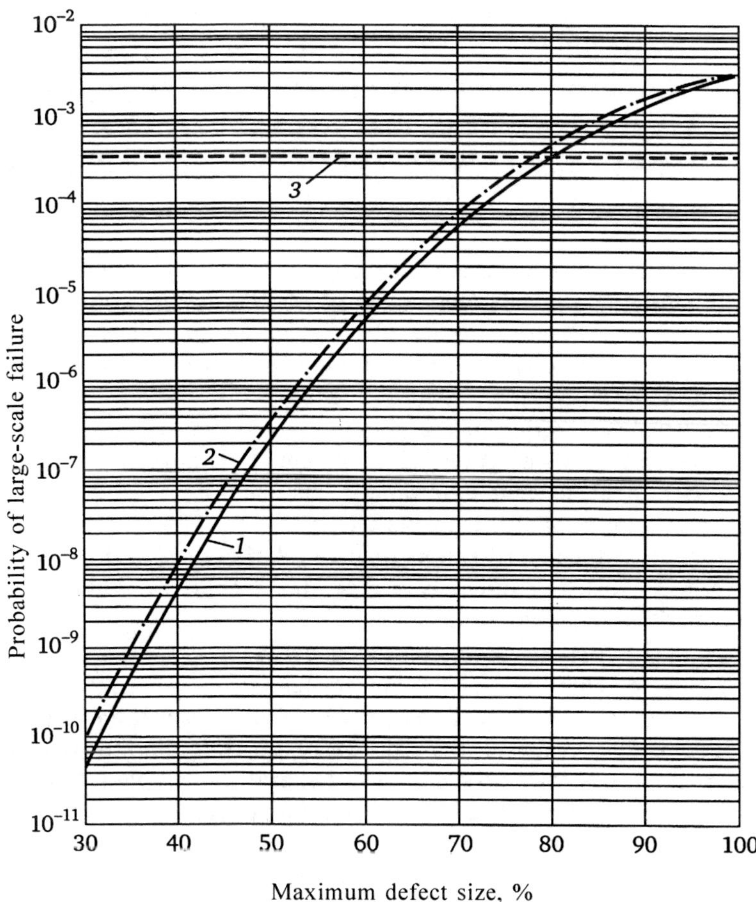

9.25 Probabilities of large-scale failure of heat exchanger tubes in relation to the maximum defect size for steam generator No. 4: 1) realistic appoach; 2) conservative estimate; 3) permissible level.

Table 9.10 Justification of serviceability of steam generator heat exchanger tubes of VVER-440 reactors

Steam generator No.	Minimum permissible defect depth, %			
	Novovoronezh NPP		Kola NPP	
	Unit No. 3	Unit No. 4	Unit No. 1	Unit No. 2
1	60	79	67	67
2	58	71	68	67
3	61	67	69	66
4	55	70	65	68
5	60	74	76	65
6	63	58	75	65

9.4   Application of the generalised method of probabilistic analysis and systematic methodology for analyzing and ensuring integrity of steam generator heat exchanger tubes in nuclear power plants with VVER-1000 and VVER-440 reactors

This section presents the results of application of the generalised method of probabilistic analysis (see section 5.3) and system concept of strength safety[138,139, etc] to ensure integrity of the steam generator heat exchanger tubes in nuclear power plants with VVER-440 and VVER-1000 reactors. The section includes:

– a brief description of the system concept for strength safety;

– description of the guidance document (GD) developed on the basis of the system concept, and conditions of its use;

– results of application of the GD[137];

– associated technical results obtained during the development of system technology ensuring integrity of the steam generator tubes, namely the impact of the technology on:

  – kinetics of defects in the tubes;

  – detectability of eddy current testing;

  – detectability of the pneumohydraulic inspection method.

As already mentioned, the problem of ensuring the integrity of the heat exchanger tubes was solved by the generalised method of probabilistic analysis and system concept of strength safety which was then used to formulate the document[137].

The system concept to ensure the integrity of HETs has been implemented in a number of power units in Russia and Ukraine. In all cases, after applying the recommendations prepared on the basis of the system concept the tube bundles of steam generators worked in these units with high efficiency without unplanned shutdowns cused by unacceptable leakage from the primary and secondary circuits.

Below are the results of applying this approach to ensure the integrity of the steam generator heat exchanger tubes for reactors VVER-1000 and VVER-440 reactors of nuclear power plants in operation in accordance with the document in Ref. 137. These results represent a special case of practical implementation of the system concept to ensure strength, operating life, reliability, survivability and safety. The concept is described in greater detail in Ref. 138.

The system concept is based on a systems approach which allows for detailed consideration of the problem to find fast the optimum solution. The central concept of the system approach is the concept of the system.

The following issues were resolved in considering the problem of ensuring integrity of heat exchanger tubes using the system concept:

1. causes for beynd-design states of tube bundles arising during operation were defined;

2. the target function of the system for ensuring of integrity of the heat exchanger tubes during operation was defined, taking into account the capabilities of modern technical means and technologies;

3. analysis of existing nuclear power technologies and practices ensuring integrity of heat exchanger tubes envisaged primarily in SG operating manuals was performed; effectiveness of these measures was evaluated and based on the results of the work performed were used to design the structure of the system ensuring integrity of the heat exchanger tubes in operation was not affected;

4. the individual elements of the system were upgraded to ensure the integrity of the heat exchanger tubes up to the state in which the system can fulfil its function (purpose).

Causes of damage in the heat exchanger tubes of SG1, SG3, SG44 of unit No. 2 of the Balakovo NPP were analyzed by retrospective network modeling[5,149,etc]. Figure 9.26 shows the course of analysis in the systematised form, using the data described above, as well as the results of additional calculations.

In the first stage the main characteristics of damage to the heat exchanger tubes of SG1, SG3, and SG4 in unit No. 2 of the Balakovo NPP, as well as the information from other blocks (if it was important), were processed. One of the main criteria for correct analysis is the condition that these (this) cause (causes) explain all the features of damage and the damage process.

Hypotheses about the causes (mechanisms) of damage were defined in the second stage. All hypotheses formulated at meetings of working groups and other meetings to discuss the steam generator of unit No. 2 of the Balakovo NPP (8 hypotheses) were considered.

The third stage included analysis of the facts, the survey data, calculations, the results of experimental analysis, and other circumstances that could prove or disprove a particular hypothesis.

As a result of tests carried out in stages 3 and 4, it was found that the root cause of damage such as pits and pits with cracks is the transfer of corrosion products from the condensate feed circuit and their accumulation in the form of deposits and sludge.

Accumulation of corrosion products from the condensate feed circuit in combination with products which arrived as a result of suction, as well as short-term violations of water chemistry were the cause of damage such as cracks (Fig. 9.26).

The development of damage over time is described schematically in Fig. 9.27. The time when the first of damage to the heat exchanger tubes of SG1, SG3 and SG4 (1992) was caused is approximate and may be specified by additional studies.

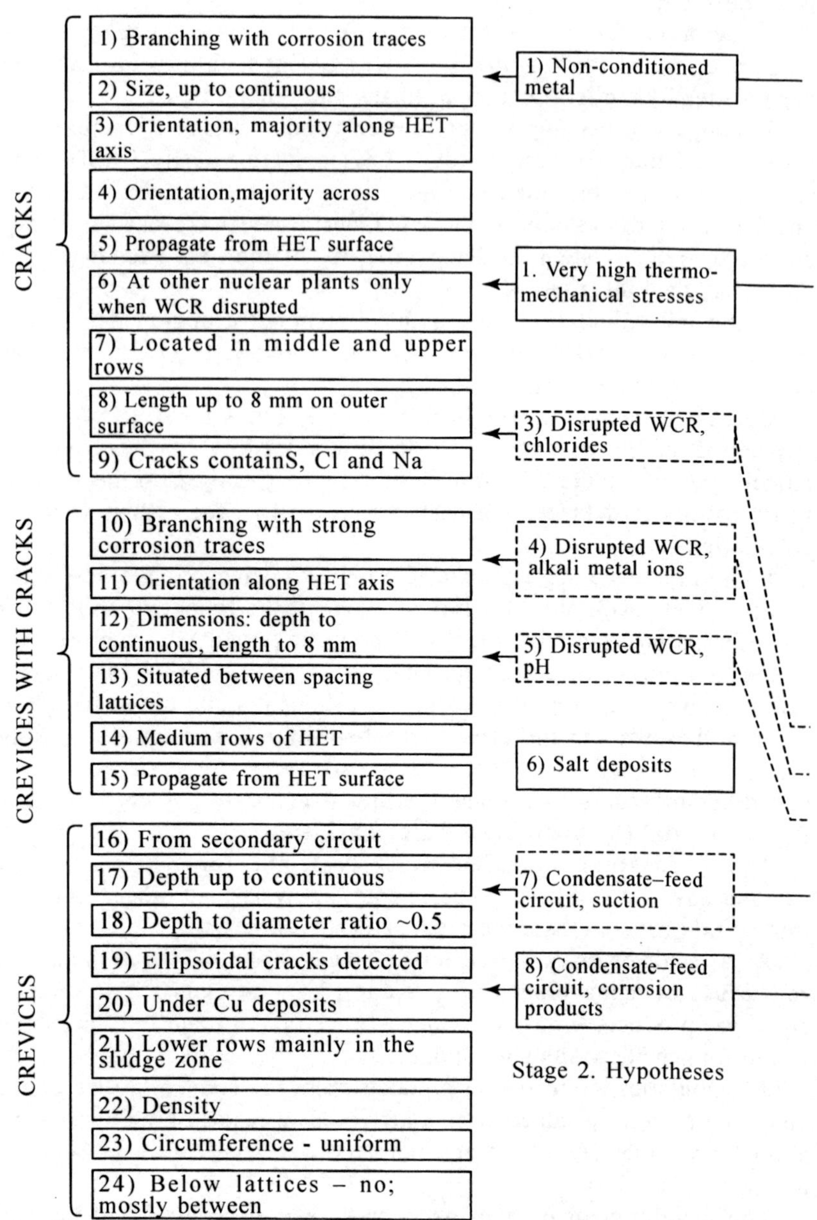

Stage 1. Main features of damage

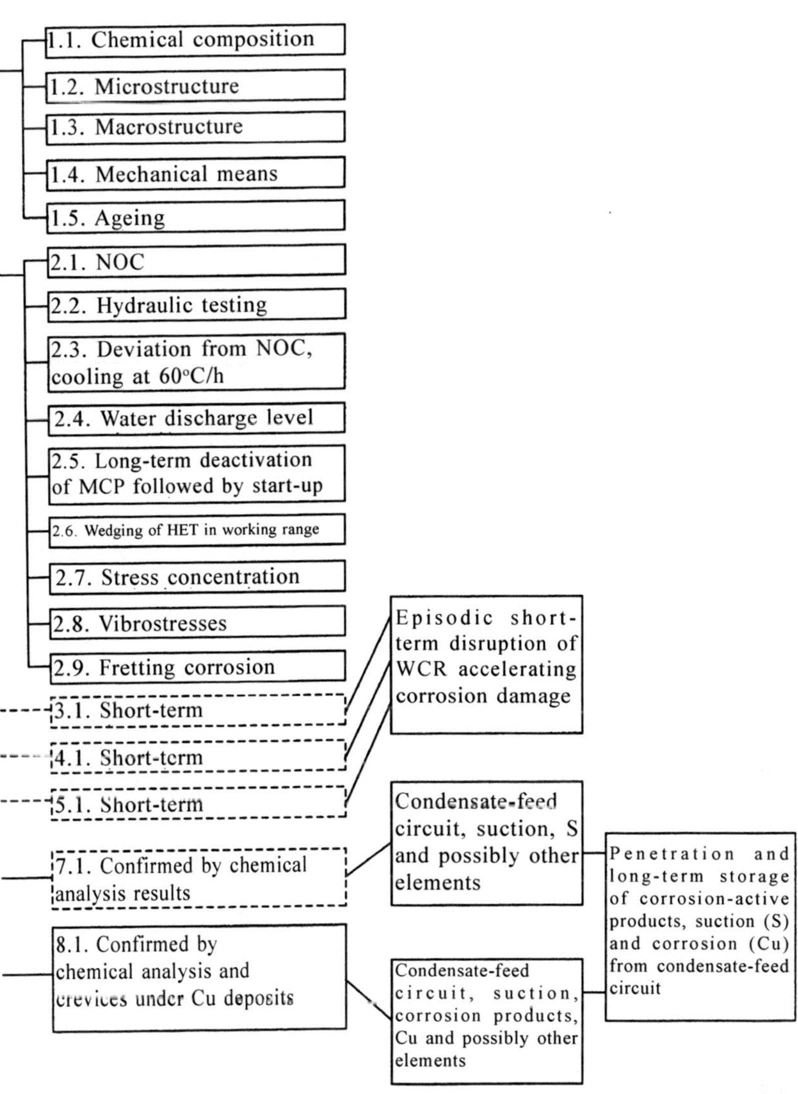

9.28 Analysis of causes of damage by the method of retrospective network modelling.

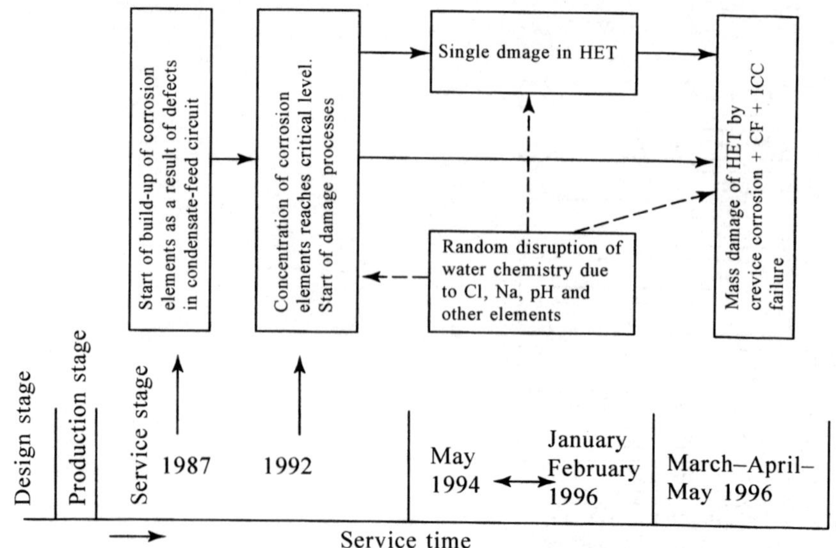

9.27 Development of the process of damage of heat exchanger tubes with time.

After establishing the root cause and the cause of damage, it is clear that to prevent injuries of this type it is necessary:

• to prevent the accumulation of corrosive elements in boiler water of SG from the condensate feed circuit (primarily copper and other elements);

• to perform mechanical and chemical cleaning of the secondary circuit of SG, precluding the accumulation of products falling from the condensate feed circuit.

Next, computational and analytical studies were performed to determine the critical size of the defects in the HETs, the residual life of the tube bundle, as well as the validation of flaw inspection detectability, the influence of inspection detectability and hydraulic tests on the life and reliability of tube bundles, the size of a coolant leakage through through-wall cracks and, finally, a comprehensive assessment of the operational safety of tube bundles taking into account their damage, technical measures adopted in SPM-96 and upcoming operating conditions.

The fundamentally important final results of these studies are presented in Figs. 9.28–9.33.

Figure 9.28 shows the results of calculation of the critical size of defects in pipes during NOC, deviations from NOC and HT conditions. The points on the same graph show the experimental results of the evaluation of the critical size of through-wall defects obtained by Czech experts on full-size tubes[137]. It can be concluded that the calculated and experimental results correspond to each other. The analysis shows that HETs may

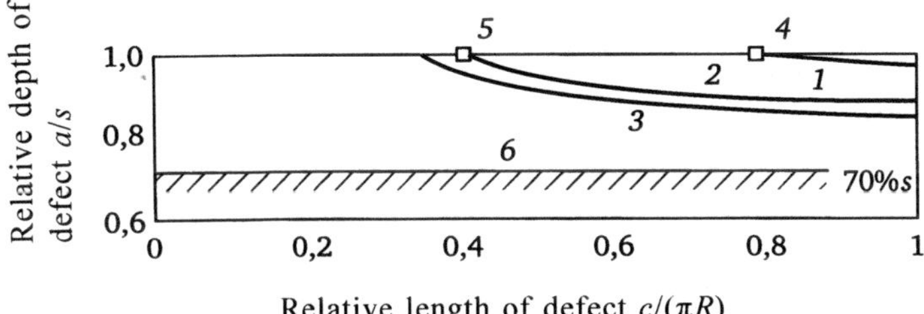

Relative length of defect $c/(\pi R)$

9.28   Longitudinal orientation of the crack. Improved accuracy of the results:1) operating mode; 2) emergency mode; 3) hydraulic tests; 4) operating mode, experimental data[89]; 5) emergency mode, experimental data[89]; 6) rejection level.

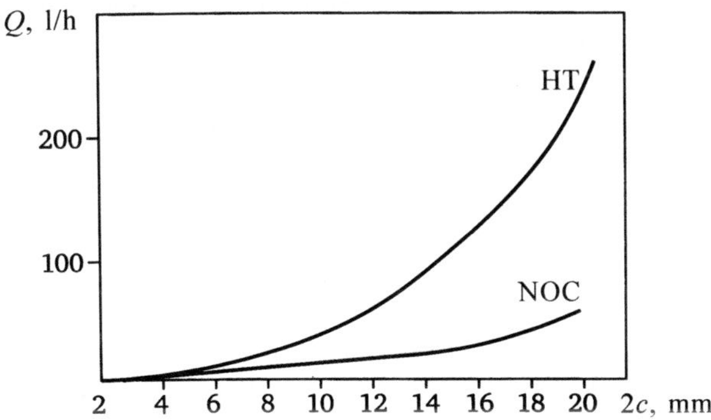

9.29   Dependence of the flow rate of water Q on the length of a continuous crack $2c$.

contain relatively large through-wall cracks due to the relatively low thermomechanical loading of the HETs.

The results of the calculation of leakage through the through-wall cracks are shown in Fig. 9.29.

A continuous longitudinal crack with length $2c$ in the internal wall of a pipe was investigated. The calculation procedure is described in Ref. 138. The calculations took into account the form factor of the crack and the effect of roughness of the crack edges on jet suppression. The calculations were performed for two modes: for normal operating conditions (NOC) and hydraulic tests with the pressure of $p_{HT1} = 24.5$ MPa and $p_{HT2} = 0$.

It is common knowledge that control of SG boiler water radioactivity allows to seal leakages (about 1 l/h) from the primary to secondary circuit. When a leak of about 3 l/h is detected, a shutdown is considered. At the same time, the limiting size of a leak from the primary to secondary circuit shall not exceed 5 l/h.

The results of calculation of the coolant flow through a continuous longitudinal crack in HETs showed that in the NOC mode the coolant flow rate of 5 l/h corresponds to a crack length of $2c = 8.5$ mm. Given that the critical crack size in the NOE mode is greater than 25 mm, it can be stated that the recommended margins for the crack size and the sensitivity of the leak control system are met:

$$2c_{cr}/2c_Q = 5 \text{ l/h} > 2;$$

$$Q(2c_{cr}) / 5 \text{ l/h} \gg 10.$$

In the HT mode, the critical crack size is $2c_{cr} = 16.2$ mm, which corresponds to leakage $Q(2c_{cr}) = 128$ l/h.

A leak of 5 l/h corresponds to a crack with the size $2c = 5.3$ mm.

Thus, the leakage control system at the existing SG of the VVER-1000 with the maximum allowable for operation leakage rate of 5 l/h makes it possible to detect through-wall cracks long before they reach the critical size at which complete rupture of the tube takes place. At the same time, the safety factors satisfy the regulatory guidelines adopted overseas (USA, Germany) and in Russia in the framework of the 'leak before rupture' concept[97].

Large leaks through continuous cracks in the NOC mode (up to 128 l/h) set the conditions for detection and plugging of the tubes on the basis of the HT results.

At the same time, the size of the leak through a crack of about 6 mm in the NOC mode and a 4 mm crack in the HT mode is very small (less than 2 l/h) and may in fact be even smaller because of the influence of the crack shape and clogging of the crack with corrosion products.

An additional uncertainty in the assessment of operating life, reliability and safety is caused by the lack of data on the reliability of flaw inspection by both eddy current method and pneumo-hydraulic (aquarium) inspection.

The detectability of these inspections was assessed by the method described in Sec. 3.1 and Ref. 139. It is based on mathematical analysis of test results (Fig. 9.30) and comparison of the inspection results with monitoring metallographic fractography data (Table 9.11).

The detectability of eddy current testing before chemical cleaning of tubes even for through-wall defects ($a = 100\%s$) is less than 50%. This means that at least half of through-wall defects such as cracks or corrosion pits are not detected in eddy current testing. After chemical cleaning of HETs the probability of detecting defects significantly increased (almost doubled). The estimate of the detectability of eddy current testing in general correlate with the data obtained in the framework of PISC-III programme in the laboratory conditions involving about 15 groups of NDT inspectors with eddy current devices[140].

If carried out immediately after reactor shutdown, the PHI method does not offer 100% detection of through-wall defects. 100% detection of through-wall defects was achieved by PHI conducted only after chemical

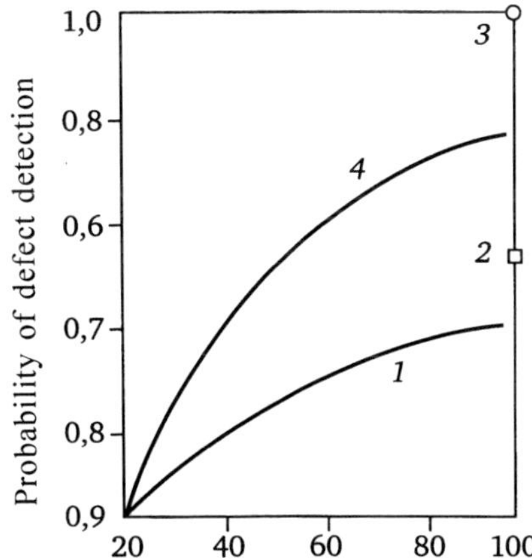

Depth of defect, % to wall thickness

9.30   Results of evaluation of the detectability of ECT (eddy current testing) and pneumo-hydraulic inspection (PHI) of PGV-1000 heat exchanger tubes using the SPM-96 results: 1 – ECT in SPM-96; 2 – PHI; 3 – HT+PHI; 4 – ECT in SPM-98 (specified on the basis of data from SPM-98).

Table 9.11 Comparison of the sizes of defects identified during ECT and MI of HET of steam generator No. 2 at the Balakovo NPP

Steam generator No.	Number of HET (according to AER)		Location of defect	Depth of defect a, % of pipe thickness s		Inaccuracy $(a_{ECT}-a_{MI})$, % of s
	Row	Column		ECT	MI	
1	2	−60	NR3; 000	94	53	+41
	92	60	NR3; 000	80	≈ 1	+79
	92	60	NR6; 000	–	100	−100
	1	61	NR2; 000	–	100	−100
	1	61	Between HP4-ПР5	–	88	−88
3	102	60	NR3; −320	80	2	78
	102	60	NR3; 000	–	100	−100
	96	−60	NR3; −290	100	100	0
	96	−60	NR3; 000	–	100	−100

Comment:
1. (+) and (−) in the last column indicate over-rejection and under-rejection, respectively according to the result of eddy current testing (ECT).
2. NR3 (6;2) is the spacing lattice No. 3 (6;2), 320 (290) is the distance from the lattice to the defect, to collector (−) and from collector (+)
3. AER is the organisation contracted to carry out ECI
4. MI – metallographic investigatons

cleaning of HETs and their HT at excess pressure of 24.5 MPa in the primary circuit coupled with lack of counter pressure in the secondary circuit.

The low detectability of the results of eddy current testing is also indicated by direct comparison of eddy current testing data with the results of metallographic fractographic analysis of HETs cut out on the basis of the eddy current testing results (Table 9.11). The data in Table 9.11 are not sufficient for making statistically convincing conclusions, but preliminary conclusions on the low detectability of quality inspection and a large number of over-rejected and under-rejected cases can be made.

These data on the actual detectability of eddy current testing and PHI are crucial for the understanding of the state of the structure, evaluation of the causes of damage and justification of the forecast of its behaviour.

Soon after unit 2 of the Balakovo NPP was commissioned after scheduled–preventive maintenance (SPM 96), repeated eddy current testing on the HET was performed in SPM 98. The results of inspection again showed a large number of damaged HETs. At meetings of the working group the participants unanimously expressed the view that the rate of corrosion cracking in HETs is very high and the steam generators should be immediately replaced..

However, our analysis of the actual detectability of ECT and of the residual defects in the tube bundles after the first and second ECTs, supplemented by estimates of the kinetics of cracks, convinced experts on the possibility of continued SG operation. Thus, graphs were presented of the initial defectiveness of a tube bundle prior to ECT in SPM 96, residual defects after ECT during SPM 96 and plugging of the HET with defects to more than 70% of tube thickness, and also for the residual defects after SPM in 1996 during which ECT was repeated and defects were plugged in more than 75% of tube thickness (Fig. 9.31).

Before the final technical measures in SPM 96 were specified and a decision was made to continue steam generator operation, the residual life and safe operation of the tube bundles of the steam generators were estimated, taking into account the damage detected in SPM 96 and the detectability of inspection.

The residual life was estimated for the tubes without damage, for the tubes with damage in the form of cavities, and for the tubes with cracks. Figure 9.32 shows the kinetics of propagation of a surface crack with the depth equal to 70% of wall thickness and for a through-wall crack with the initial length of 11 mm. Thus, in 20 years of operation in the nominal conditions (including those with the water chemistry regime) the growth of cracks did not exceed 1%. Analysis of the residual life showed that when the tubes with defects in more than 70% of wall thickness are plugged, the steam generator can be operate up to the end of the design operating life (i.e. 30 years). Curves of the integral probability of the loss of integrity of tubes are shown in Fig. 9.33.

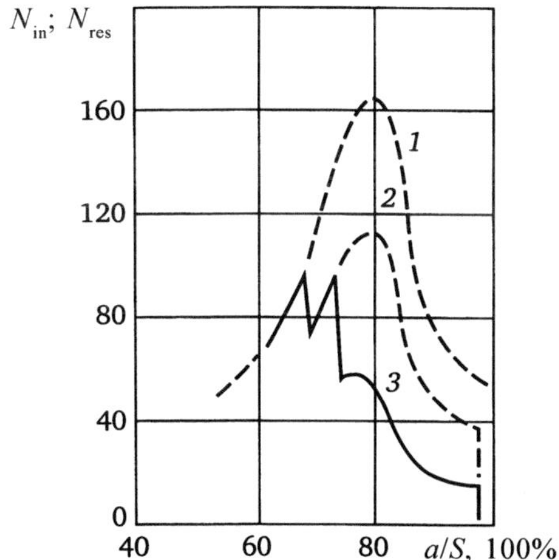

9.31   Number of defects in HETs of the 2PG-1 steam generator: 1) prior to SPM-96; 2) after plugging during SPM-96, 97; 3) after plugging during SPM-98.

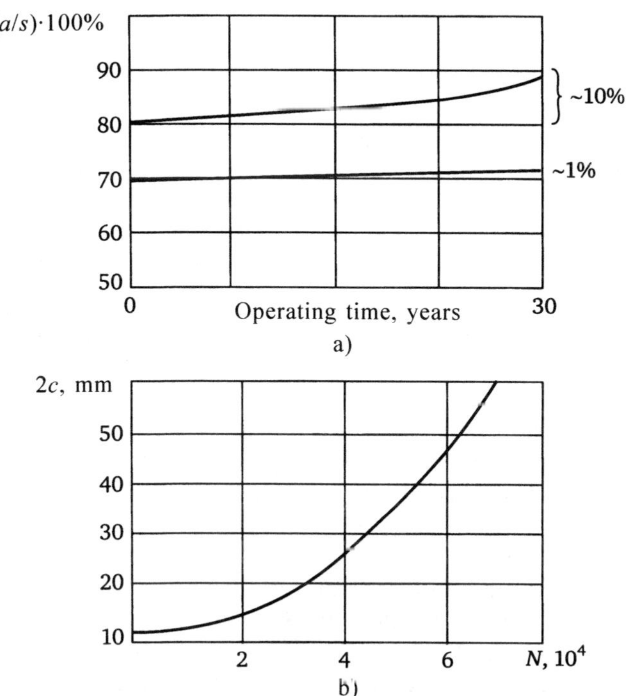

9.32   Kinetics of propagation of surface cracks in relation to the initial size at a safety factor of 10 (a) and dependence of through-wall crack growth on the number of cycles N (b).

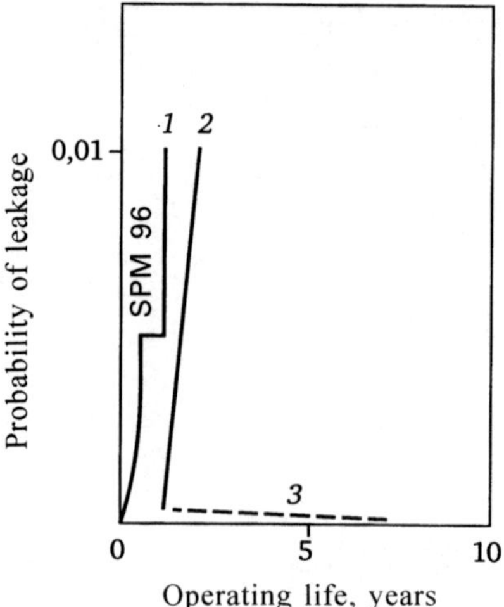

9.33   Curves of integral probability of finding tubes with leaks: 1) prior to SPM 96; 2) after plugging defective tubes during SPM 96, without washing the tube bundle; 3) after plugging defective tubes, washing the tube bundle and operation in boiler water with nominal parameters.

Safe operation of tube bundles 2SG1, 2SG3 and 2SG4 was evaluated taking into account the limited detectability of defects by ECT. Safe operation means that the integrity of the HETs with sudden fracture of one or several tubes with leakage from the primary to secondary circuit greater than 5 l/h cannot take place. The analysis results described above show that sudden appearance of such a leak can be the result of a combination of the following events:

Normal operation at nominal conditions + propagation of a large crack in one or more HETs with the leak of less than 5 l/h due to bl21ckage of the tube by its corrosion products and (or) because of the special form of the crack + sudden break of the steam line with the steam pressure drop in the second circuit to 0 at a pressure in the primary circuit p_{HT1} = 17.64 MPa.

As shown by the calculations of critical crack dimensions, the regime of hydraulic testing with the pressure p_{HT1} = 24 MPa, p_{HT2} = 0 is a more stringent regime for the identification of tubes with large cracks compared with the above scenario.

Due to HT, detection of pipes with leaks revealed by HT, and plugging of one or several pipes, sudden rupture of one or several tubes becomes impossible. At the same time, the growth of critical cracks in the HT mode to critical cracks in the NSC mode + steam line fracture is nine heating–cooling cycles. The influence of hydraulic pressure on the time period of absolutely safe operation is shown in Fig. 9.34.

This shows that the control of leakage with reactor shutdown at a leak of 5 l/h also allows timely crack detection before cracks reach the critical

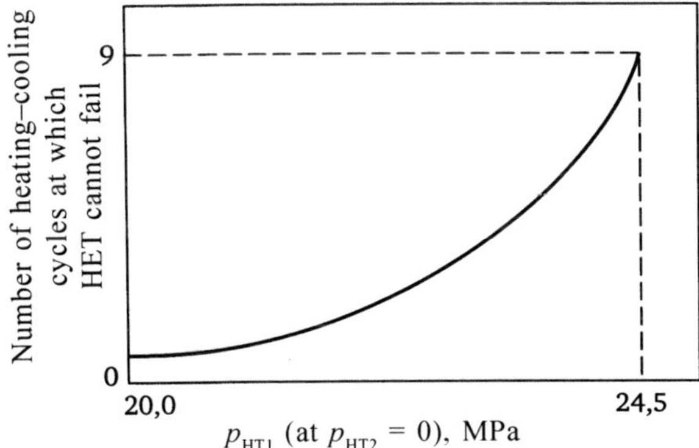

p_{HT1} (at $p_{\mathrm{HT2}} = 0$), MPa

9.34   Dependence of the time of absolutely safe HET operation (expressed in the number of loading cycles in the heating–cooling mode) on the pressure in hydraulic testing.

values of the safety margin adopted in international practice under the 'leak before break' concept.

The results of the analysis of safe operation of tube bundles 2SG1, 2SG3 and 2SG4 are summarised in Fig. 9.35. The probability of the following events is described here:

- emergence of a leak < 5 l/h (P_l);
- emergence of a leak > 5 l/h ($P_{g.l.}$);
- sudden appearance of a large leak due to tube rupture in the normal operating conditions (NOC) (P_f^{NOC});
- sudden appearance of a large leak in the emergency mode characterised by steam line fracture (P_f^{SLF})

Probabilities of events $P_{g.l.}$, P_f^{NOC} and P_f^{SLF} (variants No. 10–12, 14–16 and 18) are estimated approximately because of the lack of statistical data on crack length. In the estimates of the events $P_{g.l.}$ and P_f^{SLF} the detectability of leakage control was assumed to be 0.9 (due to possible clogging of cracks with corrosion products). The reliability of inspection of thermal–hydraulic cycles was also 0.9 (due to the lack of an instrumental system for controlling and taking into account the thermal–hydraulic modes (such as SAKOR) [5].

Event P_f^{SLF} in the variant number 20 was assessed as impossible under the condition that there is no more than nine heating–cooling cycles between two hydraulic tests ($p_{\mathrm{HT1}} - 24.5$ MPa and $p_{\mathrm{HT2}} - 0$).

Thus, the aggregate results of the calculations described above suggest that the life and safe operation of tube bundles in SG1, SG3 and SG4 of unit 2 of the Balakovo NPP are ensured under the following conditions:

- Conducting HT with pressure $p_{\mathrm{HT1}} - 24.5$ MPa, $p_{\mathrm{HT2}} - 0$ and then plugging all leaky pipes;
- reactor shutdown when a leak of 5 l/h from the primary to secondary circuit is detected and leaky pipes are plugged;

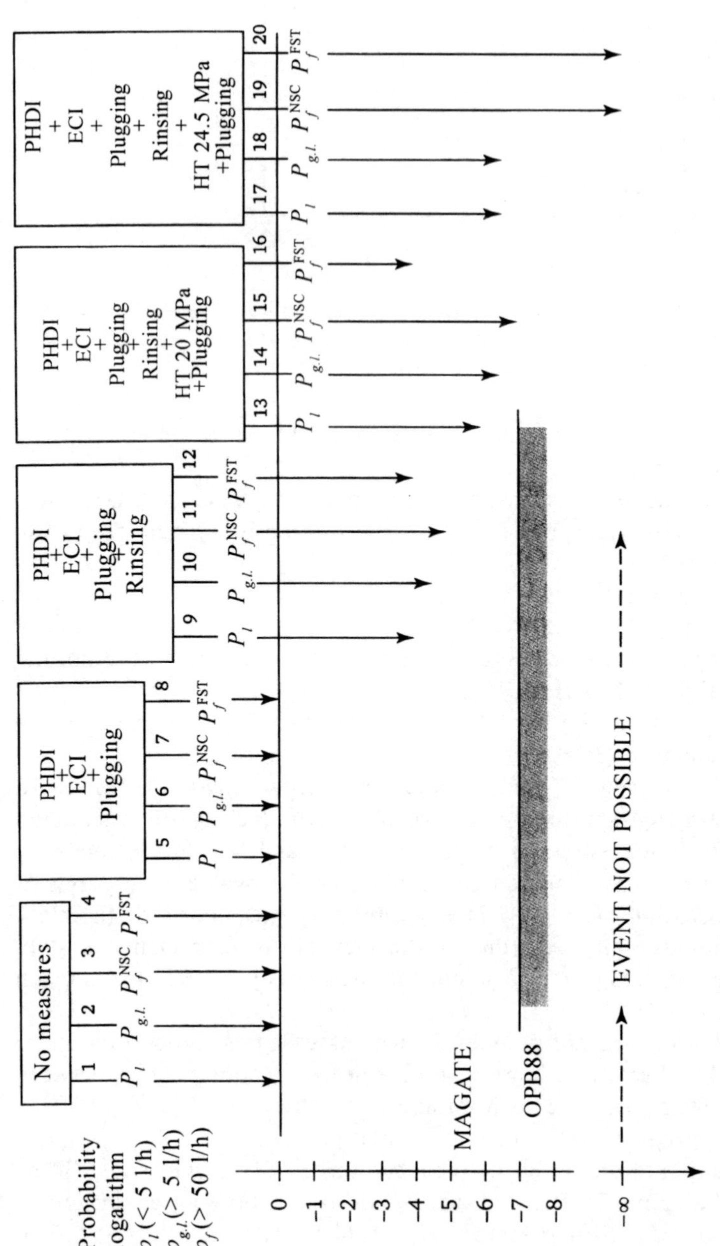

9.35 Result of safe service of tube bundles of steam generators.

• plugging all leaky pipes detected by other methods (ECT, PHI, etc.) in SPM;

• control of operating modes of tube bundles on SG1, SG3, SG4 of unit 2 of the Balakovo NPP after SPM 96-97 until the next SPM.

From these data it also follows that the safe operation of SG1, SG3, and SG4 can be ensured also up to the end of the design life taking into account the reference inspections carried out in subsequent SPM and implementation of appropriate measures arising from the results of these inspections.

The safety conditions must be determined more accurately on the basis of the operation results and SPM. HT pressure p_{HT1} = 24.6 MPa can be reduced (to 18–20 MPa), if the conditions precluding sudden steam pipe rupture are accepted.

The above-described technology used to define the residual life, reliability and safe operation and development on its basis of appropriate technical measures proved to be highly efficient when applied to the HETs of the steam generator on unit No. 2 of the Balakovo NPP.

However, given the large number of similar structural elements – each steam generator has 11 000 HETs, it is possible to check the accuracy of probability forecasts. So according to the prediction [150] made in SPM 96, 2–3 HETs should become depressurised in the section of the steam generator with the highest degree of damage during a single campaign. In fact, inspection of integrity, carried out in SPM 98, showed that only two tubes were depressurised.

The steam generators of unit No. 2 of the Balakovo NPP after SPM 98 worked for another life cycle in 1999 and were then replaced by new ones because of the large number of plugged tubes and not due to tube reliability reasons.

In conclusion, it should be noted that the positive results obtained on unit No. 2 of the Balakovo NPP using a system technology for assessment and ensuring the required operating life, reliability and safety have been also used in other nuclear power plants since 2003, including the RD EO-0552-2004.

The experience obtained in the framework of these studies is summarised in Ref. 137.

9.5 Guidance Document RD EO-0552-2004 'Guidelines on the application of system methodology to ensure the integrity of steam generator heat exchanger tubes of NPP with VVER-440 and VVER-1000 reactors'

The technology to ensure the integrity of the steam generator tubes on the basis of the system concept of strength safety (hereinafter – system technology) was developed in 1996–1997 to ensure reliable operation of unit 2 of the Balakovo NPP. Due to the positive experience with using this technology at the Balakovo NPP and increasing relevance of the problem of ensuring the integrity of the steam generator tubes in other plants, it became necessary to generalise the experience obtained at the Balakovo NPP and apply it to all nuclear power units with VVER-440 and VVER-1000 reactors with damaged steam generator tubes.

This work was done in 2003 and resulted in a guidance document RD EO-0552-2004 'Guidelines on the application of system methodology to ensure the integrity of steam generator heat exchanger tubes of NPP with VVER-440 and VVER-1000 reactors (hereafter guidelines). In 2004, the guidelines were adopted by all nuclear power plants with VVER reactors for experimental use until 2007 inclusive.

In fact, the guidelines were adopted only for those units of nuclear power plants which contained a steam generator with damaged tubes. There were six such units in Russia. In one unit, the guidelines were not used for organisational reasons.

Four inventions were made in developing the system technology to ensure the integrity of the pipes and are protected by patents [145–148]. These patents form the basis of RD EO-0552-2004.

Results of application of the guidelines

The most severe consequences of unreliable work of steam generator tubes are associated with the need for an unscheduled shutdown of the unit to eliminate leaks between circuits. Thus, one of the main objectives of the guidelines is to ensure the reliability of the steam generator tubes on the basis of the criterion of resistance to the formation of leaks in the circuits (primary–secondary leakage) during operation (with the unit at power) or, in other words, avoiding unplanned shutdowns of the units due to leaks un the circuits. The above problem can be solved with high efficiency through by the application of the system technology prescribed by the guidelines.

In all cases, the operation of the steam generator tube bundles of these units after the adoption of the guidelines was successful, with no unplanned shutdowns due to unacceptable leakage from the primary to secondary circuits.

Figure 9.36 shows the change in the number of unplanned shutdowns of the units before the guidelines were adopted and after that. The Figure shows that all the units where the guidelines were adopted did not have any unplanned shutdowns (curves 1, 2, 4 and 5). The exception is one unit on which the guidelines were not adopted for a number of organisational reasons (curve 3). On this unit there were two unplanned shutdown (curve 3) during this period.

An unscheduled shutdown occurred on unit 3 of the Novovoronezh nuclear power plant in 2008 and was associated with the termination of the use of the guidelines for organisational reasons during the planned

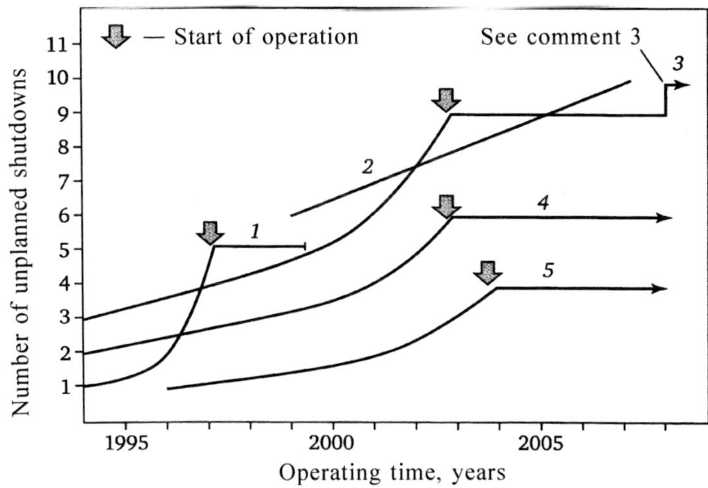

Comment. 1. Total application time of the guidelines – more than 100 SGs/year; 2. Curve 2 – the guidelines were not used; 3. Failure because use of the guidelines was interrupted.

9.36 Dynamics of unscheduled shutdowns prior to and after applying RD EO-0552-2004: 1) Balakovo nuclear power plant unit 2; 2) Kalinin nuclear power plant, unit 1 (the guidelines were not used); 3) Novovoronezh nuclear power plant, unit 3: 4) Novovoronezh nuclear power plant, unit 4: 5) Kola nuclear power plant, units 1 and 2.

shutdown in 2007 for routine maintenance. This case illustrates that all the guidelines should be used with high care, otherwise the process of unplanned degradation of steam generator tubes would resume.

Other results obtained during the development and application of system technology to ensure the integrity of steam generator tubes
Termination of unexpected process of stress corrosion cracking of steam generator tubes

Figure 9.37 shows the dynamics of growth in the number of leaky pipes in a steam generator with damaged pipes. It is seen that before applying the guidelines the number of leaky pipes was increasing with acceleration. After applying the guidelines, the rate of growth in the number of leaky pipes fell almost to zero.

Guaranteed elimination of the possibility of sudden pipe rupture with a large leak of coolant during operation

Excluding the possibility of sudden pipe rupture with a large leak of coolant during operation is of great importance for the safety of NPP as this may be accompanied by radiation leaks and creation of conditions for the initiation of an accident with nuclear consequences (in the case of fracture of two or more pipes).

Adoption of the full volume of the guidelines excludes the possibility of pipe rupture during operation. The probability of pipe rupture becomes zero after complete implementation of the guidelines. Some results of evaluating the reliability of steam generator pipes in relation to measures taken are shown in Fig. 9.35. It can be seen that the set of events of the last variant (chemical washing (CW), hydraulic tests (HT) at a pressure of 240 atm, the pneumo-hydraulic method for inspecting density (PHI), and plugging of leaking HETs ensures that the steam generator tubes cannot fail even in the most severe conditions – in the case of a pressure drop in the SG secondary circuit (e.g. as a result of a steam generator rupture).

Increasing the reliability of eddy current testing. Following the recommendations of the guidelines full, the reliability of ECT doubles (curves in Fig. 9.38).

Increasing the detectability of the pneumo-hydraulic method for monitoring the density and integrity of the steam generator pipe. The detectability of the pneumo-hydraulic method for monitoring the density and integrity of the steam generator pipes after implementation of the guidelines in full increased from 50% to 100% (points in Fig. 9.38).

Reduced cost of operation of steam generator with damaged tubes. As a result of application of the guidelines, other results were also achieved that have both technical and economic importance to the plant, namely:

1) increase in power generation by eliminating unplanned shutdowns of units because of leaks between in SG circuits, as well as by reducing the timing of planned maintenance work in which the eddy current testing of steam generator pipes was in the 'critical path' of these works;

2) reduction in maintenance costs of steam generators due to the fact that the volume of eddy current testing and plugging of leaky pipes is reduced.

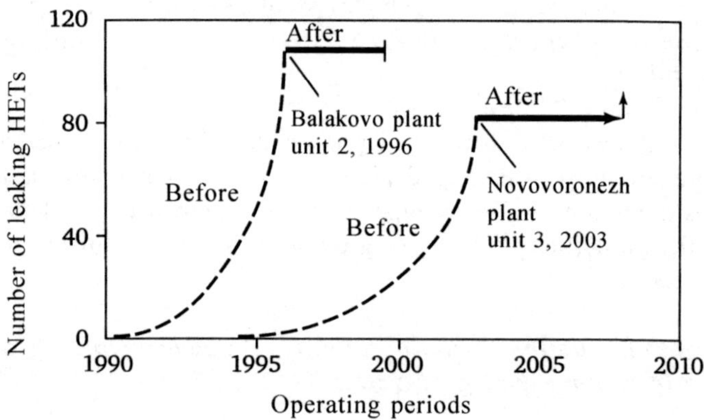

9.37   Increase the number of leaky pipes in a steam generator prior to applying the system concept and after.

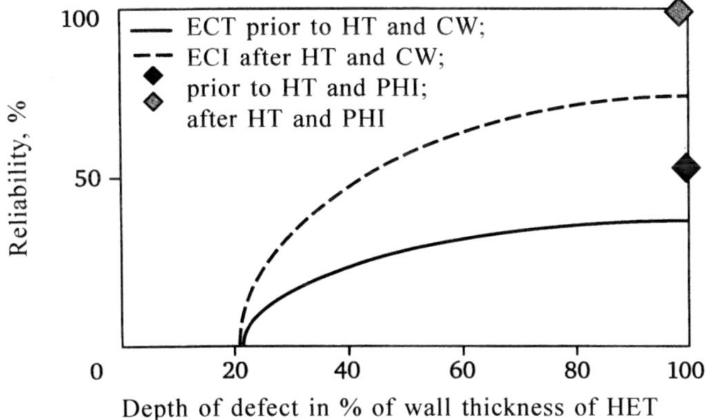

9.38   Detectability of ECT and PHI after the inspection of a nuclear power plant.

3) there is no need to replace the steam generators because steam generators can operate reliably not only during the design life of 30 years, but also for significantly longer periods of time.

9.6   Conclusions

Application of probabilistic methods (generalised method for determining the probability of fracture) and the system concept of strength safety at nuclear power plants with VVER reactors to ensure the integrity of the steam generator heat exchanger tubes with extensively damaged tube bundles results in technical and economic benefits, namely:

- improvement of the reliability of the steam generator tube bundles;
- no unplanned shutdowns of power units due to leaks in HETs;
- increased safety of SG operation;
- sudden appearance of large leaks, significantly greater than 5 l/h, is no longer possible;
- the level of leakage in the steam generators after adoption of the whole complex guidelines recommendations is either not measured or is on the lower acceptable level;
- state of tube bundles is stabilised in the event of a minor leakage and the leak intensity does not increase during operation, even if cycles are associated with shutdowns of units;
- eliminating the need for large numbers of ECT;
- increased lifetime of steam generators based on the criterion 'number of plugged HETs' and reduced amount of plugging the leaky pipes;
- shortage of electric power because of leaks falls to zero, and SG maintenance costs are reduced.

Table A1.1 Chemical composition of main types of structural steel for reactors produced in Russia

No.	Steel grade	Chemical composition, %									
		Carbon	Chromium	Nickel	Titanium	Molybdenum	Silicon	Manganese	Vanadium	Sulphur	Phosphorus
1	22K	0.16–0.24	–	–	–	–	0.15–0.30	0.35–0.65	–	0.04	0.04
2	10GN2MFA	0.08–0.12	–	2.0	–	1.0	–	–	–	0.01	0.01
3	15Kh2MFA	0.14–0.16	2.0	–	–	1.0	–	–	1.0	0.01	0.01
4	15Kh2NMFA	0.14–0.16	2.0	1.0	–	1.0	–	–	1.0	0.01	0.01
5	08Kh18N10T	0.06–0.12	17–19	9–11	0.6–0.8	–	0.8	2.0	–	0.01	0.01
6	Kh18N10T	0.06–0.12	17–19	9–11	0.6–0.8	–	0.8	2.0	–	0.01	0.01

Application of the Markov model for forecasting reliability of PWR pipelines

For NPP pipelines, including for both PWRs and BWRs, the most widely used model for forecasting the probability of breakage during operation is the Markov model[1,2]. The opportunities for applying this method to PWR main pipelines (such as for 600 mm diameter pipelines made of carbon steel, with austenitic steel cladding) were analysed.

Both the OPDE database and the pipe failure database of a PWR NPP (the referent plant (RP)) were used for the application of this method, .

A general Markov model is given in Ref. 1 and a set of differential equations was built based on this model. The model involves four different states: Success (S), Detectable Flaw (F), Detectable Leak (L) and Rupture (R). The solutions to the differential equations represent the time-dependent probabilities of a pipe system occupying each of these states, and these solutions can be determined either numerically or analytically. Further, to determine the system failure rate, or hazard rate, the system reliability function for the generic model was first determined using the reference data, and the hazard rate as a function of the reliability function was then determined, according to the definition of the hazard rate explained below. Since we are primarily concerned with pipe ruptures and seek to estimate pipe rupture frequencies, any state except R is considered to be a success state.

Using this concept, the reliability function of the Markov model, $r\{t\}$, is given by:

$$r\{t\} = 1-R\{t\} = S\{t\} + F\{t\} + L\{t\} \qquad [A2.1]$$

The hazard rate for pipe ruptures, $h\{t\}$, is given by:

$$h\{t\} = -\{1/r(t)\}*\{dr(t)/dt\}* = \{1/[1-R]\}*\{dR(t)/dt\} \qquad [A2.2]$$

A thorough evaluation of types of events included in both databases (for the category of registrations referring to welds on an ASME Class 1 system) led to the conclusion that five generic categories of events can be said to exist:

1. Success state.

2. Minor flaws inside the welding seam, such as gas intrusions, lack of fusion, pores, weld root failures, which are found using volumetric tests (ultrasonic and X-ray tests), and which can evolve into a more severe surface degradation. These types of welding failures are not found in the OPDE database.

3. Signs of visible deterioration (using surface tests, including penetration liquid and magnetic particle tests) such as part cracks and full cracks, representing indications of degraded conditions but without active leaks. After a check of registered Reportable Events, it was concluded that this type of event can also be included in this second category. These types of part cracks, full cracks and Reportable Events can also be found in the OPDE and RP pipe failure databases.

4. An increasing spectrum of leaks, called P/H leaks, small leaks, leaks and large leaks.

5. Rupture and severance events. A rupture is a major structural failure resulting in a significant through-wall flow rate. A severance event implies a 360° circumferential, through-wall crack. In the OPDE database it was found that these events take place in pipe diameters < 60 mm, as a result of fabrication errors, water hammer events and stress corrosion cracking.

Table A2.1 Summary of all the categories of events in the RP pipe failure and OPDE databases, for events related to welds on ASME Class 1.

Category	RP events	OPDE events
1*	-	-
2	18	-
3	3	46
4	-	115
5	-	115

*Category 1 is the success state, **Only for pipes with a diameter less than 60 mm

To estimate the Markov model parameters for Category 2 flaws, only the RP data were considered. Given the particularities of the ISI program at the RP, the Markov model in Figure A2.1 was used, based on the following assumptions taken from the observed data collection process at the RP:

- the leak before break model applies, which states that the transition from success to break state cannot happen directly, at least in the inspected pipe components.

- some form of indication such as weld root failure, weld layer stratification or wall thinning, or part-crack, has to exist before a leak can happen, at least in the inspected pipe components.

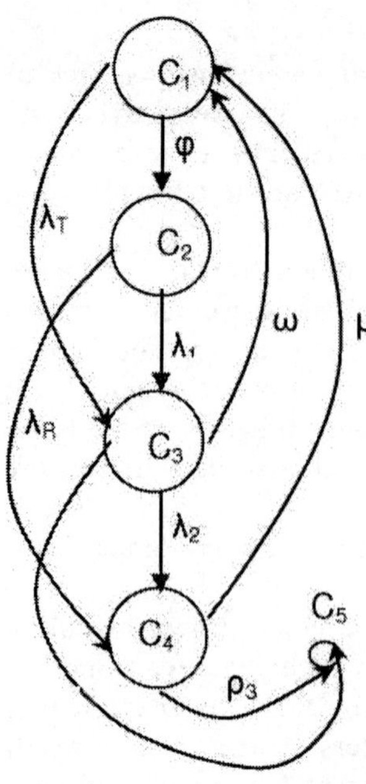

Pipe Element States
C1 –Success, no detectable damage
C2 – Category 2 events, welding failures
C3 – Category 3 events, part cracks, full cracks, Reportable Events
C4 – Category 4 events, through-wall leaks
C5 – Rupture or severance events

State Transition Rates
φ – Category C2 event occurrence rate
λ_1 – Part crack failure rate, given welding failures
λ_2 – Leak failure rate, given part crack failures
λ_R – Leak failure rate given welding failures
λ_T – Part crack failure rate given success state
ρ_3 – Rupture failure rate given leaks
ρ_2 – Rupture failure rate given part crack failures
w – Repair rate of part crack failures
μ – Repair rate of leaking failures

A2.1 **The generalised schema of the pipeline condition within the framework of Markov model.**

- repair actions are performed only on visible pipe component deterioration and on leak seepages, and the repair action fully restores the component to its original successful state.

– an existing degradation within the limits of Category 2 described here is detected with 0.9 probability of detection, given that the pipe component is inspected.

– an existing degradation within the limits of Category 3 described here is detected with 1.0 probability of detection.

– the major contributor to a pipe failure is considered to be the weld zones. This is in accordance with Ref. 2 and is confirmed by statistics in the OPDE database for the relevant pipe category: from a total of 327 eventst in OPDE for PWR plants and for ASME Class 1 systems, 167 events concerned welds, with the rest concerning other pipe components such as elbows, straight pipe elements, fittings, nozzles, etc. Consequently, the proportions of welds to other pipe elements contributing to failure are approximately equal.

– the causes of failure considered in this study are: stress corrosion cracking, design/manufacturing/construction errors, and fatigue caused by vibrations or temperature.

Since the five states are mutually exclusive, at any given time moment t,

$$C_1 + C_2 + C_3 + C_4 + C_5 = 1$$

[A2.3]

A number of factors are used to define events C_i. The technique used in their definition is complex and in each case differ. As an example, we consider below a definition technique using only two factors: λ_1 and λ_2.

The failure rate λ_j for Category 3 type crack part events, taken from volumetric and surface inspections of the RP pipe failure database, is determined as[1,4]:

$$\lambda_1 = n_2 / N_{1*}T_2 = 3/250{\cdot}22 = 5.4E{-}4 \qquad [A2.4]$$

where n_2 is the number of occurrences of Category 3 (part-cracks) depending on the chosen ISI quality and normative documents used; $n_2 = 3$; N_1 is the number of pipe components providing the n_2 occurrences (taken from all surface and volumetric tests on different locations for ASME class 1); $N_1 = 250$; T_2 is the time during which n_2 occurrences were collected; $T_2 = 22$ years.

For Leak events failure rate from Category 3 events λ_2, another formula is used:

$$\lambda_2 = (n_{\text{leaks}} \cdot \%P_{1,\text{fc}}) / (n_{\text{ev}} \cdot \%P_{\text{lc,fc}} \cdot N_R \cdot T), \qquad [A2.5]$$

where: n_{leaks} is the number of Category 4 events (leaks) with specified failure causes from OPDE database in Table A2.1; $\%P_{1,\text{fc}}$ is the percentage of leaks occurred due to specified causes, from the total leak events OPDE database in table 1: n_{ev} is the total number of failure events for welds in PWR plants, ASME Class 1, with specified failure causes OPDE database in Table A2.1; $\%P_{\text{lc,fc}}$ is the percentage of cracks and leaks occurred due to specified causes, from the total events OPDE database in Table A2.1; N_R is the number of reactors which were observed within the framework of the OPDE database in Table A2.1; T is the time window within the framework of the OPDE database.

A similar approach is also used to define the other factors.

In summary, we note the following drawbacks to the approach described above:

The process of damage (ageing, degradation) of a pipeline during operation is continuous. It can be described by a known mathematical equation. Therefore, it is not rational to divide a uniform process of damage into five discrete conditions.

The accepted discrete conditions (C_1, C_2 .... C_5) in the Markov model do not have an absolutely clear physical sense. Thus, if we admit that the definitions of conditions C_i are correct, the basic equation of the technique, ($C_1 + C_2 + C_3 + C_4 + C_5 = 1$), is incorrect.

In reality, there will always be defects in equipment and pipelines. And whether they are detected or not depends on the NDE techniques applied, inspection quality, the volume of inspection, the normative documents establishing the threshold of sensitivity and on the rejection size of defects (i.e. the size at which repair is deemed necessary). Therefore, conditions C_i are not homogeneous and the summation of their probabilities is not possible. In particular, it is impossible to prove that their sum is equal to 1.

The destruction or non-destruction of a component (or its element) is studied by fracture mechanics. Unfortunately, the Markov model does not allow to apply methods of the fracture mechanics and this also reduces the value of the approach described above.

The model has a large number of factors and parameters (see Figure A2.1), varying which it is possible to receive any beforehand given result. The randomness of definition of these factors and parameters (see, for example the equations A2.3, A2.4 and A2.5) reduces even more the value of received results according to reliability of pipelines.

References

1.   Fleming K.N., Markov models for evaluating risk-informed in-service inspection strategies for nuclear power plant piping systems, *Reliability Engineering and System Safety* **83** (2004) 27–45.
2.   Thomas H.M., Pipe and vessel failure probability, *Reliability Engineering* **2** (1981) 83–124.
3.   OPDE 2007:2 *Coding Guidelines* [OPDE-CG] & OPDE 2007:2 User's Guide, December 2007.
4.   Transactions of seminar within the framework of the international program 'Ageing and probabilistic safety analysis' (December, 10–11, 2010, GKKW, Switzerland).

Russian Scientific Research Institute of Operation of Nuclear Power Plants (VNIIAES)

The name of the software package:
Software package PN-1.1

Strength reliability. Determination of the probability of failure, leaks and defects in equipment and piping of nuclear power plants, optimisation of non-destructive testing and technical maintenance during operation

Computer: IBM PC, 1.5 GHz processor or faster, 512 MB RAM or greater.

Operating system: Windows 2000, 2003, XP, Vista.

The name of the authors: G.V. Arkadov, A.F. Getman, A.Yu. Kuz'michevskii.

Developing organisation: VNIIAES

1   Brief description

The composition of the software package:
- 1_CRIT - software module for calculating the critical and permissible sizes of defects (cracks); (there is also a module 1_CRIT_add for additional graphical confirmation of the accuracy of the results);
- 2_ROST-ZIKLCRIT – software module for calculating the kinetics of crack growth during operation under cyclic loading;
- 3_OST_DEF – software module for calculating the residual defectiveness and its kinetics in service;
- 4_NAD_OD – software module for calculating the reliability characteristics: safety, the number of failed elements of the same type (on the basis of the criterion of failure, leak or defects), the function of the intensity of failure, the function of the density of distribution of operating time to failure in service;
- 5_NAD_OBSH_HR – software module for calculating the reliability characteristics taking into account the statistical functions of residual

defectiveness, strength properties and stresses in the brittle state of the components;

- 6_NAD_OBSH_VJAS – software module for calculating the reliability characteristics taking into account the statistical functions of residual defectiveness, strength properties and stresses in the ductile state of the components;
- 7_OPTIM_ISI – software module for calculating the optimum frequency of non-destructive inspection;
- 8_OPTIM_GH – software module for optimisation of hydraulic tests (frequency and pressure);

Purpose and areas of application

Calculations of the probability of failure, leaks and defects of equipment and piping of nuclear power plant, optimisation of non-destructive inspection and technical maintenance.

The results of calculations are used to form the following classes of output data:
- the values of the critical defect sizes, graphical representation;
- the values of the permissible dimensions of discontinuities, graphical representation;
- the kinetics of crack growth in service in relation to the initial state of the crack and further conditions of propagation under cyclic loading; graphical comparison of various crack growth curves under different conditions;
- the characteristics of residual defectiveness, including the reliability and probability parts of residual defectiveness;
- characteristics of the variation of residual defectiveness during operation;
- the reliability characteristics, including the characteristics of safe operation, the number of failed elements of the same type (on the basis of the criterion of failure, leak or defect), the function of the rate of failure, the function of the distribution density of operating time to failure during operation (graphical representation);
- optimum frequency of non-destructive inspection taking into account the quality of inspection, reliability of equipment and consequences of its failure.

Limitations of application

The calculations are quite resourse-intensive and, therefore, the rate of calculation and construction of graphical representations depend on the capacity of the computer used for this purpose. The minimum requirements: IBM PC with a Pentium 1.5 GHz processor or greater, 515 MB RAM or greater.

The accuracy of calculations in practice is determined mainly by the accuracy of definition of the physical–mechanical properties of materials and operating conditions.

Information about the constants

All the constants used in the calculations are defined by the user explicitly directly in the calculation model.

1. GOST 27.002-89, Reliability in Engineering. Terms and definitions.
2. Federal law: On industrial safety of hazardous production objects, 21.07.97, No. 116-FZ.
3. The main provisions for the safety of NPPs. OPB88/97.
4. Standards of strength calculation of equipment and pipelines of nuclear power plants, Moscow, Metallurgiya, 1989.
5. Getman, A.F., Operating life of nuclear vessels and piping, Moscow, Energoatomizdat, 2000.
6. Rules for design and safe operation of equipment and pipelines of nuclear power plants PN AEG-7-008-89, Moscow, Energoatomizdat, 1998.
7. Code ASME, US Nuclear Regulatory, USA.
8. Code RSEM, EDF, France.
9. Getman, A.F., The safety concept 'leak before break' for pressure vessels and piping of nuclear power plant, Moscow, Energoatomizdat, 1999.
10. Rodionov, A., Elaboration of reliability data for ageing PSA, Proceedings of the Eighth International Conference on Probabilistic Safety Assessment and Management, May 14–18, 2006, ASME Press, USA.
11. Trifanov, A., Status of research project 'Incorporating Ageing Effects into PSA', Proceedings of EC Workshop on Use of Probabilistic Safety Assessment (PSA) for Evaluation of Impact of Ageing Effects on the Safety of Nuclear Power Plants, 15-16 November 2007, Budapest, Hungary, EUR 23078 EN, EC DG JRC, Petten, Netherlands, 2008.
12. Getman, A.F. and Kozin, N., Non-destructive testing and safe operation of pressure vessels and piping, Moscow, Energoatomizdat, 1998.
13. Bolotin, V.V., Prediction of life of machines and structures, Moscow, Mashinostroenie, 1984.
14. Safety of nuclear power stations, Rosenergoatom (Russia) – EDF (France), 1994.
15. Arkadov, G.V., et al., in: Proceedings of the International Seminar on European programme 'Ageing and PSA', Moscow, October, 2008.
16. Arkadov, G.V., Getman, A.F. and Usanov, A.I., Technologies of maintenance of a resource and safety of equipment in pipelines of atomic power stations during operation, 16th International conference on nuclear engineering, Orlando, Florida, USA, May 11–15, 2008.
17. Vora, J.P., Nuclear Plant Aging Research (NPAR) Program Plan, NUREG-1144.

Rev. 2, US NRC, June 1991.

18. Methodology for the Management of Ageing of Nuclear Power Plant Components Important to Safety, Technical Report Series No. 338, 1992.

19. Data Collection and Record Keeping for the Management of Nuclear Power Plant Ageing, Safety Series No. 50-P-3, 1991.

20. Safety Aspects of Nuclear Power Plant Ageing, TECDOC 540, 1990.

21. A Review of Equipment Aging, Theory and Technology, 1980.

22. European Commission, Nuclear Safety and Environment. Safe Management of NPP Ageing in the European Union, Final Report, EUR 19843 EN, 2001, 363 p.

23. Simola, K., Probabilistic Methods in Nuclear Power Plant Component Ageing Analysis, VTT Publication 94, Espoo, 1992.

24. Pulkkinen, U. and Uryas'ev, S., Optional Operational Strategies for an Inspected Component", in: European Safety and Reliability Conference '92, Copenhagen, 10-12 June 1992, p. 13, Work Report VTT / SAH 1/92.

25. Vesely, W.E., Risk Evaluations of Aging Phenomena: the Linear Aging Reliability and its Extensions, NUREG/CR-4769, 1987.

26. Subudhi, M., Gunter, W., Shier, W., Fullwood, R., Lofaro, R. and Taylor, J., Aging study of Boiling Water Reactor Residual Heat Removal System, USNRC, June 1989.

27. Ermakov, S.M., Monte Carlo Methods and Related Issues, Moscow, Nauka, 1971.

28. Mukhin, O., Simulation Systems Manual, Perm, RCI PSTU, 1994.

29. Bong Lee, J. and Park, J.H., in: 18th International Conference on Structural Mechanics in Reactor Technology (SMiRT 18) Beijing, China, August 7–12, 2005, SMiRT18-M02-6.

30. Makhutov, N., Getman, A.F., et al., Risk Theory and its Applications in engineering, Moscow, Znanie, 2007.

31. Rodionov, A.N., Influence of Aging Effects of Systems, Structures and Components of Nuclear Power Plants in the Probabilistic Safety Analysis, Thesis for the Degree of Candidate of technical sciences, Paris – Moscow – Obninsk, 2009.

32. Rodionov, A.N. and Patrik, M., EC JRC Network on the Use of PSA for the Evaluation of Ageing Effects on the Safety of Energy Facilities. Activities and Results, Second International Symposium on Nuclear Power Plant Life Management, Shanghai, China, 15–18 October 2007, Vienna, Austria: IAEA, 2008, 17 p.

33. Lannoy, A., Evaluation et Maitrise du Vieillissement Industriel, Lavoisier, Paris, France, 2005.

34. Atwood, C., LaChance, J., Martz, H., et al., Handbook of Parameter Estimation for Probabilistic Risk Assessment,NUREG/CR-6823 US NRC, USA, 2003.

35. Rodionov A., Models and Data Used for Assessing the Ageing of Systems, Structures and Components (European Network on Use of Probabilistic Safety Assessment (PSA) for Evaluation of Ageing Effects to the Safety of Energy Facilities), EUR 22483 EN, EC DG JRC, Petten, Netherlands, 2007.

36. Klugel, J.-U., Investigation of Time-Dependent Trends in Plant Specific Data for Active Components, Proceedings of EC Workshop on Use of Probabilistic Safety Assessment (PSA) for Evaluation of Impact of Ageing Effects on the Safety of Nuclear Power Plants, 15–16 November 2007, Budapest, Hungary, EUR 23078 EN, EC DG JRC, Petten, Netherlands, 2008.

37. Procaccia, H., Clarotti, C. and Piepszownik, L., Fiabilite des Equipements et Theorie de la Desision Statistique Frequentielle et Bayesienne, Eyrolles, France, 1992.

38. Kelly, D., Repair as New or Same as Old: Practical Issues in Choosing the Appropriate Stochastic Process to Model Failure, Proceedings of EC Workshop on Use of

Probabilistic Safety Assessment (PSA) for Evaluation of Impact of Ageing Effects on the Safety of Nuclear Power Plants, 15–16 November 2007, Budapest, Hungary, EUR 23078 EN, EC DG JRC, Petten, Netherlands, 2008.

39. Rodionov A., Age-dependent Reliability Models. Statistical Data Analysis and Parameters Estimation, Proceedings of EC Workshop on Use of Probabilistic Safety Assessment (PSA) for Evaluation of Impact of Ageing Effects on the Safety of Nuclear Power Plants, 2–5 October 2006, Bucharest, Romania, EUR 22514 EN, EC DG JRC, Petten, Netherlands, 2006.

40. Vesely, W.E., et al, Component Unavailability versus Inservice Test Interval (ITI): Evaluation of Component Aging Effect with Applications to Check Valves, NUREG/CR-6508, US NRC, USA, June 1997.

41. Lafaro, L., et al., NUREG/CR-5268, US NRC, USA, 1989.

42. Radulovich, R., et al., Aging Effects on Time Dependent NPP Component Unavailability: An Investigation of Variations from Static Calculation, Nuclear Technology, Vol. 112, October 1995, 21–40.

43. Higgins J., et al., Operating Experience and Aging assessment of Component Cooling Water Systems in PWR, NUREG/CR-5052, US NRC, USA, June 1988.

44. Bacha M., et al., Estimation de Modèles de Durées de Vie Fortement Censurée, Eyrolles, Paris, France, 1998.

45. Rodionov A., Feasibility of Ageing PSA. The Key Tasks and Major Difficulties of Development. JRC Network on Incorporating Ageing Effects into Probabilistic Safety Assessment Applications, Proceedings of the Kick off meeting, EUR 21382 EN, EC DG JRC, Petten, Netherlands, October 2004.

46. Lyonnet, P., La Maintenance: Mathematiques et Methods, Lavoisier, Paris, France, 1992.

47. Risk Spectrum PSA Professional – User Manual, Relcon AB, Sweden, 2005.

48. Wels, H., Generic Ageing Characteristics of Conventional Power Plants. Lessons Learned from Data Analysis, A Model for Life Extension Planning, NRG Report 911569/07.81244/C, Arnhem, Netherlands, February 2007.

49. Priiiotization of TIRGALEX – Recommended Components for Further Aging Research. NUREG/CR-5248. US NRS, USA, 1988.

50. Blombach, J., et al., Does Ageing of NPPs Require the Incorporation of Time Dependent Failure Rates in PSA Models, Proceedings of 31st ESREDA Seminar on Ageing, Smolenice Castle, Slovakia, 2006.

51. Antonov, A.V., et al., Method of Accounting for Apriori Information in Determining the Reliability of the Equipment of Nuclear Power Plants, Obninsk, IPPE, 1982.

52. Antonov, A.V., Assessing the Reliability Characteristics of Components and Systems with Combined Methods NPS, Moscow, Energoatomizdat, 1993.

53. Rzhanitsyn, A.R., Determination of Safety of Structures, Stroit. Promyshlennost', 1947.

54. Rzhanitsyn, A.R., Analysis of Structures, Taking into Account the Plastic Properties of Metals, Moscow, Stroiizdat, 1979.

55. Kogaev, V.P., Strength Calculations for Stresses Variable with Time, Moscow, Mashinostroenie, 1977.

56. Sharyi, N.V., et al., The strength of Basic Equipment and Pipelines of VVER reactors, Moscow, IzdAT, 2004.

57. Tutnov, A.A. and Tkachev, V.V., Calculation of the Probability of Early Brittle Fracture of Pressure Vessels, Atomnaya Energiya, 1988, vol. 64, No. 3, 188–194.

58. The Program for the PC. The Calculation of the Probability of Leaks and Breaks in Cylindrical Pressure Vessels of Cyclic and Static Loadings, MAVR-1.1, Kurchatov

Institute, Moscow, 1991.

59. Grigoriev, V.A., Applied Value of the Criterion in Assessing the Reliability of the Operating Life of Elements of Reactor Facilities, Proceedings of the Seventh International Conference Material Science Issues in the Design, Manufacture and Operation of NPP equipment, 17–21 June 2002, St. Petersburg, 2002

60. Grigoriev, V.A., Failure Characteristics in the VVER type Reactor. Regional Traning Course on NPP maintenance, 26 August–13 September, Obninsk, 1996.

61. Bolotin, V.V. Operating Life of Machines and Structures, Moscow, Mashinostroenie, 1990.

62. Gorbatykh, V.P., The Influence of a Combination of Ammonia and Copper Alloys in Corrosion Cracking of Steam Generator Tube Bundles Made From Austenitic Steel, IAEA Regional Workshop Materials, Udomlya, 27–30 November 2000, Moscow, VNIIAES EREC, 2001, 123–130.

63. Sobol', I.M., Numerical Monte Carlo Methods, Moscow, Nauka, 1973.

64. Ermakov, S.M. and Mikhailov, G.A., The Course of Statistical Modelling, Moscow, Nauka, 1976.

65. Rules of Drawing up Design Schemes and the Determination of the Parameters of Loading of Structural Elements with Known Defects. Guidelines MR 125-02-95. NGOs TSNIITMASH. Moscow, 1995.

66. Cherepanov, G.P., Mechanics of Brittle Fracture, Moscow, Nauka, 1974.

67. Parton, V.Z. and Morozov E.M., The Mechanics of Elastic-Plastic Fracture, Springer-Verlag, 1993.

68. Shiratori, M., et al., Computational Fracture Mechanics, Moscow, Nauka, 1986.

69. Plyuvinazh G., Mechanics of Elastic-Plastic Fracture, Springer-Verlag, 1993.

70. Malinin, N.N., Applied Theory of Plasticity and Creep, Moscow, Mashinostroenie, 1968.

71. Gnedenko, B.V., et al., Mathematical Methods in Reliability Theory, Moscow, Nauka, 1965.

72. Kapoor, K. and Lamberson, L., Reliability and System Design, Springer-Verlag, 1980.

73. Rules for Inspection of Welded Joints and Deposits on Sections and Structures of Nuclear Power Plants, Experimental and Research Nuclear Reactors and Installations, PC 1514-72, Moscow, Metallurgiya, 1974.

74. Zherebenkov, A.S. and Timofeev, B.T., Analysis of Defects in Welded Joints of Steel 22K with Respect to the Conditions of Manufacture of Steam Generators, Vopr. Sudostroeniya, Ser. Svarka, No. 29, 1980.

75. Volchenko, V.N., Probability and Reliability Assessment of the Quality of Metal, Moscow, Metallurgiya, 1987.

76. Crutzen, S., PISC I, Evaluation of NDT Techniques for High Pressure Components, Nucl. Eng. and Design, 1985, Vol. 86.

77. Nichols, R., et al., NDE in Relation to Structural Integrity. Intern. J. of Pressure Vessels, 1989, Vol. 35, No. 1–4.

78. Nichols, R. and Crutzen, S., Ultrasonic Inspection of Heavy Section Steel Components. The PLSC II Final Report, Elsevier Applied Science Publ. Ltd, 1988.

79. Nichols, R., Crutzen S., Miller A. PISC III: status report / / Proceed. Third International Conference on Material Science Problems in NPP Equipment Production and Operation. 17-22 June 1994, Moscow and St Petersburg, Prometey, IAEA, 1994. Vol. 1–8.

80. Kyssmaul, K. and Mletzko, U., Action 2 of PISC III, with Emphassis on Phase I

(Advanced Sizing Techniques), ibid.

81.   Doktor, S.A., Lemaitre, P. and Grutzen, S., Status Report: PISC III Action 4 on Austenitic Steel Testing, ibid.

82.   Reale, S. and Tognarelli, L., Structural Integrity Approach for PISC Results Evaluation and Comparison, ibid.

83.   Crutzen, S., et al., The European Networks: NTSC, AMES, ENIO, ibid, 76–86.

84.   Shcherbinsky, V.T. and Aleshin, N.P., Ultrasonic Testing of Welded Joints. Moscow, Stroiizdat, 1989.

85.   Gurvich, A.K., Reliability of Flaw Control as the reliability of Complex 'Flaw – Operator – Medium', Defektoskopiya, 1992, No. 13.

86.   Corby, D.M., Packman, P.P. and Pearson, H.S., Accuracy and Precision of Ultrasonic Shear Wave Flow Measurements as a Function of Stress on the Flow, Materials Evaluation, 1970, vol. 28. No. 15.

87.   Denisov, L.S., Improving the Quality of Welding in the Construction Industry, Moscow, Stroiizdat, 1982.

88.   Shastin, A.G., et al., Reflection of Ultrasound from Natural Defects on the Surface, Defektoskopiya, 1986.

89.   Getman, A.F., Ensuring Durability and Service Life of Nuclear Reactors in Operation, Based on a System Approach, in: Improving the Operating Efficiency of Nuclear Power Plants, Moscow, Energoatomizdat, 1989.

90.   Getman, A.F. and Zubov, V.Yu., Methodology and Some Results on the Evaluation of Failure Probability of pipelines DN500 of nuclear power plants with VVER-440 reactors, in: Reliability of Pipeline and Pressure Vessels of NPP, Obninsk, 1989, 14–21.

91.   PNAEG-7-010-89, NPP Equipment and Pipelines. Control Rules.

92.   Guidelines MR 108.7-86. M. TsNIITMash.

93.   Daniel, K., Application of Statistics in Industrial Experiment, Mir, Moscow, 1979.

94.   Patent number 2245385. Method for Determining the Reliability of NDT Results (Authors: Makhutov, N.A., et al), 2006.

95.   Trukhanov, V.M., Methods of Ensuring Reliability of Engineering Products, Moscow, Mashinostroenie, 1995.

96.   Arkadov, G., et al., The Software Package PN1.1, Determination of Probability of Failure, Leaks and Defects in Equipment and Pipelines of Nuclear Power Plants, Optimization of Non-destructive Testing and Maintenance During Operation.

97.   Makhutov, N.A., et al., Patent No. 2243585, Method for Determining the Probability of Defect Detection, Initial and Residual Defects Using the Results of Non-destructive Testing.

98.   Gaponov, A.A., Condition of Pipelines DN500 of the Main Circulation Loop of VVER reactors after 100 000 h of Operation, Teploenergetika, 1989, No. 1, 43–44.

99.   Report of the Kola NPP, Study of changes of Structure and Properties of Metal DN500 Pipes Made of Steel 08Kh18N10T in Kola NPP after 100 000 operating hours, Inv. No. 14 542, 1990.

100. PPI NAS Ukraine report, Investigation of the Cyclic Strength and Deformability of Structural Materials of NPP Equipment, Taking into Account Structural, Technological and Operational factors, Kiev, 1987.

101. Beznosikov, S.A., et al., Evaluation of the Properties of the Base Metal and Welded Joints of NPP Pipelines after Long Operation, Proceedings of Workshop Ageing of NPP Component materials, St Petersburg, 28 February–2 March 1995, 221–232.

102. Karzov, G.P., et al., The Influence of Prolonged Exposure to Operating Temperatures and Strain Aging on the Material Properties of Nuclear Power Plant Reactor Vessels

and Piping, International Journal of Pressure Vessel and Piping, 1993, vol. 53, No. 2, 195–216.

103.   Ehrnsten, U., et al., Properties of Aged Ti-stabilized Stainless Steels, Proceedings of the Seventh International Conference on Material Issue in Design, Manufacturing and Operation of Nuclear Power Plants Equipment, volume 2, St Petersburg, 17–21 June 2002, 122–127.

104.   Methods of Determining Allowable Sizes of Discontinuities of Metal Equipment and Pipeline during Plant Operation, M-02-91, VNIIAES, NIKIET.

105.   Tkachev, V.V., Methods of Evaluating the Effectiveness of Control and Optimization of Metal of Pressure Vessels and Piping of NPP, Atomnaya Energiya, 1993, vol. 74, No. 5.

106.   Tkachev, V.V., Methods of Pptimizing the Control of Metal and Hydraulic Test-based Probabilistic Analysis and Analysis of Structural Strength, in: Pipeline Safety. Reports of the International Conference 17-21.09.1995, Institute of Atomic Energy, Moscow, 1995, 81–91.

107.   Tutnov, A.A. and Loskutov, O.D. The approach to the Optimization of Non-destructive Testing Regulation of Components of Reactor Plants, ibid, 215–225.

108.   IAEA-TECDOC-1400, Improvement of In-service Inspection in Nuclear Power Plant, July 2004.

109.   ENIQ report No. 23, European Framework Document for Risk-informed In-service Inspection, EUR 21581 EN, Luxembourg, Office for Official Publications of the European Communities, 2005.

110.   IAEA Regional Workshop on Qualification of In-Service Inspection Systems and Risk-Informed In-Service Inspection. Zagreb, Croatia, 18–22 October 2004, Chapman, V., Structural Reliability Models (SRM). Link between RI-ISI & Inspection Qualification, Inspection Interval.

111.   Alzbutas, R., Development and Optimization of RI-ISI program, Lithua-ND, Petten, 2008.

112.   Abagyan, A.A. Getman, A.F., Questions of Strength of NPP Equipment during Operation, Metallurg. Prod., 1986, No. 3, 47–53.

113.   News of Nuclear Energy, 1/1989, International Newsletter of the Association Interatomenergo, Rivkin, E., et al., Methods for Determination of Allowable Defects in Metal of Equipment and Pipes during Operation of nuclear power plants.

114.   Lee, T.H. and Park, J.Y., Intergranular Crack Formation on Alloy 600 SG Tubing, Argonne National Laboratory, SMIRT-18, 2005.

115.   Tutnov, A.A., et al., Assessing the Impact of Periodic Technical Inspection on the Strength Reliability of the Circuit Elements of VVER-1000 reactor, Atomnaya Energiya, 2005, No. 4.

116.   Majumdar, S., Assessment of Current Understanding of Mechanism of Initiation, Arrest and Reinitiation of Stress Corrosion Cracks in PWR Steam Generator Tubing, NUREG/CR-5752, ANL, Argonne, 2000.

117.   Karwoski, K., et al., Regulatory Perspective on Steam Generator Tube Operating Experience, Nuclear Pressure Equipment Expertise and Regulation Symposium 2005, US NRC, Rockville, 2005.

118.   Berge, Ph., et al., Lessons Learned from TubesPpulled from French Steam Generators, Third International Steam Generator and Heat Exchanger Conference, Toronto, 1998.

119.   Myong-Ho Song, et al., Safety Review on Recent Steam Generator Tube Failure in Korea and Lessons Learned, IAEA Technical Meeting on Steam Generator Problems, Repair and Replacing, Prague, 2003

120. Meeting to Discuss Results of Spring 2005 Steam Generator Inspection at Oconee Nuclear Station, Unit 1, ml051940468, US NRC, Washington, 2005.

121. Staehle, R.W. and Gorman, J.A., Predicting the Occurrence of Corrosion Failures in Nuclear Power Components with Emphasis on Application of Alloy 690, Proceedings of the 2003 Nuclear Safety Conference, NUREG/CP-0185, US NRC, Washington, 20–22 October, 2003.

122. Malinovski, D.D., et al., Operating Experience with Model F Westinghouse Steam Generator, NEA / CSNI Unipede Specialist Meeting on Operating Experience with Steam Generators, Brussels, 16–20 September, 1991

123. Environmental Assessment of Diablo Canyon SG Replacement Projects, 2003.

124. Staehle, R.W. and Gorman, J.A., Quantitative Assessment of Submodes of Stress Corrosion Cracking on the Secondary Side of Steam Generator Tubing in Pressurized Water Reactors, Corrosion, part 1 (vol. 59, No. 11, 2003), part 2 (vol. 60, No.1, 2004), part 3 (vol. 60, No. 12, 2004), NACE, Houston.

125. Fruzetti, K., Materials Reliability Program. Effect of Zinc Addition on Mitigation of PWSCC of Alloy 600 (MPR78), Final Report 1003522, EPRI, Palo Alto, 2002.

126. Maeda, N., et al., Optimization of Operation and Maintenance of Nuclear Power Plant by Probabilistic Fracture Mechanics, Nuclear Engineering and Design, 2002, Vol. 214, 1–12.

127. ERPI NP-7493 (1991), Statistical Analysis of Stream Generator Tube Degradation.

128. Chung, H.S., et al., A Study on the Integrity Assessment of Detected SG Tube, KERPI, 2000.

129. Wu, W.F. and Syau, J.J., A Study of Risk-Based Non-Destructive In-Service Inspection, Nuclear Engineering and Design, 1995, Vol. 158, 409–415.

130. Mendenhall, W., et al., Probability and Statistics, 11th ed., Thomson, 2003.

131. Berens, A.P., NDE Reliability Data Analysis, Metals Handbook, 19889, 9th ed., Vol. 17, 689–701.

132. Davis, J., ANL / CANTIA: A Computer Code for Steam Generator Integrity Assessments, Argonne National Laboratory, NUREG/CR-6786, 2001.

133. Grigoriev, V.A., Strength of Basic Equipment and Pipelines of VVER Reactors, Moscow, IzdAT, 2004.

134. Defining Initiating Events for Purposes of Probabilistic Safety Assessment, IAEA-TECDOC-719, IAEA, September, 1993.

135. GOST 24722-81, Pressure Water Nuclear Reactors (VVER), General Technical Requirements.

136. Romaniv, O.N. and Nikiforovich, G.N., Mechanics of Corrosion Failure of Structural Alloys, Moscow, Metallurgiya, 1986.

137. RD EO-0552-2004, Procedure Recommendations for the Application of the System Methodology of Ensuring Integrity of Heat Exchanger Pipes of Steam Generators of Operating Nuclear Power Stations with VVER-1000 and VVER-440 Reactors, 2004.

138. RD SK-01.04, System Concept of Ensuring Strength, Operating Life, Reliability and Safety of Equipment and Piping of Nuclear Power Stations, KTsNBRAS, 2003.

139. Getman, A.F., System Method of Ensuring Strength of Equipment and Piping of Nuclear Power Plants in Service, in: Reliability and Endurance of Machines and Installations, 1986, No. 10.P9, 4 [89]

140. Analysis of Integrity of the Heat Exchanger Pipes of VVER Nuclear Reactors, Bulletin of Technical Proposals, Moscow, ISKO AES MKhO Interatomenergo, 1996, No. 14

141. M-02-91, Procedure Recommendations for the Determination of the Dimensions of

Permissible defects in Metal of Equipment and Piping in Service, Moscow, VNI-IAES, NIKIET, 1991.

142. RD TPR, Procedure Recommendations For the Application of the Concept TPR In NPP, VNIIAES, 1995.

143. Bueth, M., et al., Final Results of PISC III, Round Robin Test on Steam Generator Pipe Inspection, in: the Transactions of the First International Conference on NDE in Relation to Structural Integrity for Nuclear and Pressurised Components, October 20-22, 1998, Amsterdam.

144. Arkadov, G.V., et al., The Results of the System Concept of Strength Security to Ensure the Integrity of the Heat Exchanger Tubes of Steam Generators of NPP with VVER-440 and VVER-1000, Proceedings of the International Conference Resurse-2009 (Kiev, May 20–22, 2009).

145. Makhutov, N.A., et al., Patent 2243565 RF, MKP[7] G01N 35/00, Method for Determining the Reliability of Non-destructive Testing of Defects Determining the Quality of Manufacturing, Reliability and Safe Operation of Products, Otkrytiya i Izobreteniya, 2003.

146. Makhutov, N.A., et al., Patent 2243585 RF, MPK[7] G05B 23/02, Method for Determining the Probability of Defect Detection, Initial and Residual Defects using the Results of Non-destructive Testing, Otkrytiya i Izobreteniya, 2004.

147. Makhutov, N.A., et al., Patent 2263296 RF, MPK[7] G01N 3/00, Method for Determination of Residual Defects in the Product after Two or More Non-destructive Tests, Otkrytiya i Izobreteniya, 2003.

148. Makhutov, N.A., et al., Patent 2243523 RF, MPK[7], G01M 3/12, Method of Hydraulic (pneumatic-hydraulic) Testing of Pressure Vessels and Pipelines to Ensure Total Reliability and Safety of their Operation, Otkrytiya i Izobreteniya, 2003.

149. Getman, A.F., et al., Ensuring Integrity of Heat Exchanger Pipes of Steam Generators of Nuclear Power Bands with VVER Reactors on the basis of System Methodology, in: International Scientific Conference Safety, Economy and Efficiency of Nuclear Power Plants, 2010.

150. Technical Report: Justification of the Possibilities and Conditions of Safe Service of Heat Exchange Tubes of Steam Generators of Unit 2 of the Balakovo NPP, VNI-IAES, 1996.

151. Guidelines for Analysis of Data Related to Ageing of Nuclear Power Plant Components and Systems, EC DG JRC Intitute for Energy, Petten, 2008.

152. Antonov, A., et al., Application of Generalised Linear Model for Time-Dependent Trend Assessment - a Case Study for the Ageing PSA Network, Reliability Engineering and System Safety, Elsevier Oxford, 2009, vol. 94/6, 1021–1029.

153 Guideline Document, Procedure Recommendations for Optimisation of In-service Non-destructive Inspection, Rosenergoatom, 2007.

154. Bergunker, V.D., Integrity of Heat Exchanger Pipes in Vertical and Horizontal Steam Generators, in: Proceedings of the Seventh Conference on Horizontal Steam Generators, October 3–5, 2006.